Lecture Notes in Artificial Intelligence   4520

Edited by J. G. Carbonell and J. Siekmann

Subseries of Lecture Notes in Computer Science

T0241816

Martin V. Butz   Olivier Sigaud
Giovanni Pezzulo   Gianluca Baldassarre (Eds.)

Anticipatory Behavior in Adaptive Learning Systems

From Brains to Individual and Social Behavior

 Springer

Series Editors

Jaime G. Carbonell, Carnegie Mellon University, Pittsburgh, PA, USA
Jörg Siekmann, University of Saarland, Saarbrücken, Germany

Volume Editors

Martin V. Butz
Department of Cognitive Psychology III
University of Würzburg
Röntgenring 11, 97070 Würzburg, Germany
E-mail: butz@psychologie.uni-wuerzburg.de

Olivier Sigaud
AnimatLab, Laboratoire d'Informatique de Paris 6 (LIP6)
Univ. Pierre et Marie Curie
104 avenue de Président Kennedy, 75 015 Paris, France
E-mail: Olivier.Sigaud@lip6.fr

Giovanni Pezzulo
Gianluca Baldassarre
Institute of Cognitive Science and Technology (ISTC-CNR)
Via S. Martino della Battaglia, 44, 00185 Roma, Italy
E-mail: {giovanni.pezzulo,gianluca.baldassarre}@istc.cnr.it

Library of Congress Control Number: 2007933376

CR Subject Classification (1998): I.2.11, I.2, F.1, F.2.2, J.4

LNCS Sublibrary: SL 7 – Artificial Intelligence

ISSN   0302-9743

ISBN-13   978-3-540-74261-6 Springer Berlin Heidelberg New York

Springer is a part of Springer Science+Business Media

springer.com

© Springer-Verlag Berlin Heidelberg 2007

Typesetting: Camera-ready by author, data conversion by Scientific Publishing Services, Chennai, India
Printed on acid-free paper   SPIN: 12108282   06/3180   5 4 3 2 1 0

Preface

Anticipatory behavior in adaptive learning systems is steadily gaining the interest of scientists, although many researchers still do not explicitly consider the actual anticipatory capabilities of their systems. Similarly to the previous two workshops, the third workshop on anticipatory behavior in adaptive learning systems (ABiALS 2006) has shown yet again that the similarities between different anticipatory mechanisms in diverse cognitive systems are striking. The discussions and presentations on the workshop day of September 30th, 2006, during the Simulation of Adaptive Behavior Conference (SAB 2006), confirmed that the investigations into anticipatory cognitive mechanisms for behavior and learning strongly overlap among researchers from various disciplines, including the whole interdisciplinary cognitive science area.

Thus, further conceptualizations of anticipatory mechanisms seem mandatory. The introductory chapter of this volume therefore does not only provide an overview of the contributions included in this volume but also proposes a taxonomy of how anticipatory mechanisms can improve adaptive behavior and learning in cognitive systems. During the workshop it became clear that anticipations are involved in various cognitive processes that range from individual anticipatory mechanisms to social anticipatory behavior. This book reflects this structure by first providing neuroscientific as well as psychological evidence for anticipatory mechanisms involved in behavior, learning, language, and cognition. Next, individual predictive capabilities and anticipatory behavior capabilities are investigated. Finally, anticipation relevant in social interaction is studied.

Anticipatory behavior research on cognitive, adaptive systems aims at exploiting the insights gained from neuroscience, linguistics, and psychology for the improvement of behavior and learning in artificial cognitive systems. However, this knowledge exchange is expected to become increasingly bidirectional. That is, the insights gained during the design and evaluation of different anticipatory cognitive mechanisms and architectures may also provide insights into how anticipatory mechanisms can actually shape, guide, and control natural brain activity. This book reveals many interesting and thought-provoking connections between distinct cognitive science areas. We strongly hope that these connections do not only lead to a deeper understanding of the functioning of anticipatory processes but also enable a more effective, bidirectional knowledge exchange and consequently more effective scientific progress in the natural and artificial cognitive systems research disciplines.

April 2007

Martin V. Butz
Olivier Sigaud
Giovanni Pezzulo
Gianluca Baldassarre

Organization

The third workshop on Anticipatory Behavior in Adaptive Learning Systems (ABiALS 2006) was held during the Ninth International Conference on Simulation of Adaptive Behavior (SAB 2006) on September 30th, 2006 in Rome, Italy. We are grateful to the organizers of SAB 2006 for giving us the possibility to hold the third workshop during their conference. This volume is an enhanced compilation of the workshop contributions and discussions. The organizers of the conference were Stefano Nolfi (chair), Gianluca Baldassarre, Raffaele Calabretta, John Hallam, Davide Marocco, Orazio Miglino, Jean-Arcady Meyer, and Domenico Parisi.

We are also more than grateful to our program committee members for providing us with careful reviews of the diverse contributions and often additional helpful comments and suggestions. Due to their hard work, we were able to organize two reviewing processes, one to evaluate and improve the original workshop contributions and a second one to improve the contributions (including modified resubmissions of the workshop's contributions) for this enhanced post-workshop proceedings volume. We are convinced that all accepted contributions provide new, highly stimulating insights into the realm of anticipatory mechanisms for adaptive cognitive systems. We would also like to thank Marjorie Kinney for proofreading all the contributions in this volume. Finally, we would like to thank all our colleagues, friends, and families that helped us during the production of this book with discussions, comments, suggestions, and general support.

This work was supported by the EU funded projects *MindRACES – from Reactive to Anticipatory Cognitive Embodied Systems*, FP6-STREP-511931, and *ICEA – Integrating Cognition Emotion and Autonomy*, FP6-IP-027819.

Executive Committee

Martin V. Butz

Department of Cognitive Psychology III, University of Würzburg, Germany

Olivier Sigaud

AnimatLab, Laboratoire d'Informatique de Paris 6 (Lip6), France

Giovanni Pezzulo

Istituto di Scienze e Tecnologie della Cognizione, Consiglio Nazionale delle Ricerche (ISTC-CNR), Roma, Italy

Gianluca Baldassarre

Istituto di Scienze e Tecnologie della Cognizione, Consiglio Nazionale delle Ricerche (ISTC-CNR), Roma, Italy

Program Committee

Christian Balkenius	Lund University Cognitive Science, Lund University, Sweden
Edoardo Datteri	Department of Physics, University "Federico II" of Naples, Italy and Department of Philosophy, University of Pisa, Italy
Oliver Herbort	Department of Cognitive Psychology III, University of Würzburg, Germany
Pier Luca Lanzi	Dipartimento di Elettronica e Informazione, Politecnico di Milano, Italy
Ralf Moeller	Computer Engineering Group, Faculty of Technology, Bielefeld University, Germany
Tony Prescott	Department of Psychology, The University of Sheffield, UK
Jesse Reichler	Department of Computer Science, University of Illinois at Urbana-Champaign, IL, USA
Alexander Riegler	Center Leo Apostel (CLEA), Vrije Universiteit Brussel, Belgium
Deb Roy	Cognitive Machines Group, MIT Media Laboratory, Cambridge, MA, USA
Samarth Swarup	Department of Computer Science, University of Illinois at Urbana-Champaign, IL, USA
Stewart W. Wilson	Prediction Dynamics, Concord, MA, USA

Table of Contents

Anticipatory Individual Behavior

Anticipatory Social Behavior

Anticipations, Brains, Individual and Social Behavior: An Introduction to Anticipatory Systems

Martin V. Butz[1], Olivier Sigaud[2], Giovanni Pezzulo[3],
and Gianluca Baldassarre[3]

[1] University of Würzburg, Röntgenring 11, 97070 Würzburg, Germany
butz@psychologie.uni-wuerzburg.de
[2] Animat Lab, University Paris VI,
104 Av du Président Kennedy, 75016 Paris, France
olivier.sigaud@lip6.fr
[3] ISTC-CNR, Via S. Martino della Battaglia, 44 - 00185 Rome, Italy
{gianluca.baldassarre,giovanni.pezzulo}@istc.cnr.it

Abstract. Research on anticipatory behavior in adaptive learning systems continues to gain more recognition and appreciation in various research disciplines. This book provides an overarching view on anticipatory mechanisms in cognition, learning, and behavior. It connects the knowledge from cognitive psychology, neuroscience, and linguistics with that of artificial intelligence, machine learning, cognitive robotics, and others. This introduction offers an overview over the contributions in this volume highlighting their interconnections and interrelations from an anticipatory behavior perspective. We first clarify the main foci of anticipatory behavior research. Next, we present a taxonomy of how anticipatory mechanisms may be beneficially applied in cognitive systems. With relation to the taxonomy, we then give an overview over the book contributions. The first chapters provide surveys on currently known anticipatory brain mechanisms, anticipatory mechanisms in increasingly complex natural languages, and an intriguing challenge for artificial cognitive systems. Next, conceptualizations of anticipatory processes inspired by cognitive mechanisms are provided. The conceptualizations lead to individual, predictive challenges in vision and processing of event correlations over time. Next, anticipatory mechanisms in individual decision making and behavioral execution are studied. Finally, the book offers systems and conceptualizations of anticipatory processes related to social interaction.

1 Introduction

The presence of anticipatory mechanisms and representations in animal and human behavior is becoming more and more articulated in the general, interdisciplinary research realm of cognitive systems. Hereby, anticipatory processes receive different names or are not mentioned explicitly at all. Commonalities between these processes are often overlooked. The workshop series "Anticipatory

M.V. Butz et al. (Eds.): ABiALS 2006, LNAI 4520, pp. 1–18, 2007.

Behavior in Adaptive Learning Systems" (ABiALS) is meant to uncover these commonalities, offering useful conceptualizations and thought-provoking inter-connections between the research disciplines involved in cognitive systems research.

After the publication of the first enhanced post-workshop proceedings volume in 2003 [13], research has progressed in all involved areas. Somewhat unsur-prisingly, neuroscience and cognitive psychology are continuously revealing new influences of anticipations in cognition and consequent behavior and learning. Individual and, even more strongly, social behavior seem to be guided by antici-patory mechanisms, in which predictions of the future serve as reference signals for efficient perceptual processing, behavioral control, goal-directed behavior, and social interaction.

In the previous volume we offered an encompassing definition of anticipatory behavior: "A process, or behavior, that does not only depend on the past and present but also on predictions, expectations, or beliefs about the future." [14, page 3]. While this definition might clarify anticipatory behavior, anticipatory mechanisms can clearly come in a variety of forms, influencing a variety of be-havioral and cognitive mechanisms.

This introduction first provides an overview over the possible beneficial in-fluences of anticipatory mechanisms and how these influences might be real-ized most efficiently. It then surveys the contributions included in this volume. First, known cognitive mechanisms involved in anticipatory processes in the brain and in language evolution are surveyed. Moreover, a fundamental challenge for artificial cognitive systems is identified. Next, individual anticipatory behav-ioral processing mechanisms are addressed, including several conceptualizations, frameworks, the effective generation of predictions, and effective behavior execu-tion. Finally, the book moves on to interactive, social systems and investigates the utility of anticipatory processes within.

2   Potential Benefits of Anticipatory Behavior Mechanisms

During the discussion sessions at the workshop day in Rome in September 2006, it became clear that there are multiple facets and benefits of anticipatory mecha-nisms. These can be conceptualized by their nature of representation and general influence on cognitive processes, as proposed previously [15]. Additionally, rep-resentations of time-dependent information and consequent knowledge gain can be distinguished based on their respective benefits for behavior and learning. These aspects are re-considered in the following sections.

2.1   The General Nature of Anticipatory Mechanisms

In many cases, it has become clear that anticipation itself is often slightly mis-understood, particularly due to the non-rigorous usage in habitual language. Therefore, we have offered an explicit distinction of different processing aspects of anticipations and have focused the workshop effort more on explicitly anticipatory mechanisms in cognitive systems.

First of all, anticipations can very generally be divided into *implicit* and *explicit* anticipatory systems. In implicit anticipatory systems, very sophisticated but reactive control programs are evolved or designed—potentially leading to intelligent, implicitly anticipatory system behavior. That is, albeit these systems do not have any explicit knowledge about future consequences, their (reactive) control mechanisms are well-designed so that the systems appears to behave cleverly, that is, in implicit anticipation of behavioral consequences and the future in general. This workshop, however, focuses more on explicitly anticipatory systems, in which current system behavior depends on actual explicit representations of the future. Cognitive psychology and neuroscience have shown that explicit anticipatory representations exist in various forms in animals and humans [44,26]. Thus, we are interested in anticipatory programs that generate predictions and utilize knowledge about the future to control, guide, and trigger maximally suitable and efficient behavior and learning.

Explicit anticipatory systems may be divided further into systems that use:

- Payoff Anticipations;
- Sensory Anticipations;
- State Anticipations.

Payoff anticipations characterize systems that have knowledge of behaviorally-dependent payoff and can base action selection on that representation. That is, different payoff may be predicted for alternative actions, which allows the selection of the current best action, as done in model-free reinforcement learning [78]. Sensory anticipations can be characterized as anticipatory mechanisms that support perceptual processing. State anticipatory processes enhance behavior decision making and execution exploiting anticipatory representations [15].

2.2 How Anticipations Can Help

To conceptualize and distinguish different sensory and state anticipatory mechanisms further, it is worthwhile to consider the question of *how* anticipations may affect cognitive processes (cf. also [26]). Thus, we now discuss how anticipatory mechanisms may influence adaptive behavior and, particularly, how such mechanisms may be beneficial for adaptive behavior. From a computationally oriented perspective the question arises how predictions, predictive representations, or knowledge about the future can influence sensory processing, learning, decision making, and motor control. Several different "*how* aspects" may be distinguished, which are first listed and then discussed:

- Useful information can be made available sooner, stabilizing and speeding-up behavior.
- Predictions can be compared with actual consequences, improving sensory processing, enabling predictive attention, and focusing model learning.

- The possibility to execute internal simulations can improve learning and decision making.
- Goal-oriented behavior can be triggered by currently desirable and achievable future states, yielding more flexible decision making and control.
- Anticipatory representations of information over time can be behaviorally useful.
- Models and predictions of the behavior of other agents may be exploited to improve social interaction.

Information Availability. Cognitive systems often face a serious timing and time delay issue. Sensory information is simply too slow to be processed and to arrive in time at the relevant behavioral control centers of the brain to ensure system stability. Behavioral experiments and simulations confirm that humans must use forward model information to stabilize behavioral control [21,61]. In psychology, the *reafference principle* [83] conceptualizes the existence of a forward model, proposing that efferent motor activity also generates a reafference, which specifies the expected action-dependent sensory consequences. Advanced motor control uses predictive control approaches that can yield maximally effective control processes [16].

Thus, cognitive systems should use re-afferent predictions that depend on activated efferences. These predictions can be used to avoid system instabilities due to delayed or missing sensory feedback. Interestingly, such stabilization effects come into play even with stabilizing recursive mathematical equations, making them "incursive" [22]. In sum, since future information can be predicted and thus be made available before actual sensory information arrives, system control and stability can be optimized by incorporating predicted feedback information.

Predictions Compared with Actual Consequences. Once subsequent sensory information is available, though, the predicted information can be compared with the real information to determine information novelty and thus information significance. Hoffmann [43,44] provides various pieces of evidence from psychological research that suggest that many cognitive processes, and especially learning, rely on comparisons between predictions and actual observations. One fundamental premise of his anticipatory behavior control framework is the comparison of anticipated with actual sensory consequences. These comparisons may be based on Bayesian models [53,20], which suggest that information integration in the brain is dependent on certainty measures for each source of information, and thus also most likely for forms of predicted information.

The first benefit of such a comparison is the consequent, continuous adaption of behavior based on the difference between predicted and actual behavioral consequences, as was also proposed in the reafference principle [83]. Hereby, the difference measure gives immediate adaptive control information, in addition to the current sensory state information. Also control theory relies on such comparisons to improve system measurements and system control, most explicitly realized in the Kalman filtering principle [51,36].

The filtering principle can also be applied to detect unexpected changes in the environment and consequently trigger surprise mechanisms. For example, based on a novelty measure that depends on the reliability of current predictions and actual perceived sensory information [59], surprise may be triggered if the current observation significantly differers from the predicted information. Surprise-based behavioral mechanisms can then improve system behavior, enabling a faster and more appropriate reaction to surprising events.

Surprise-dependent processes can also be used to improve predictive model learning itself. For example, surprise-like mechanisms were shown to be useful to detect important substructures in the environment [9], which furthermore is useful to partition the environment into partially independent subspaces. This capability was used, for example, to efficiently solve hierarchical reinforcement learning problems [6,75]. Other mechanisms train hierarchical neural networks based on failed predictions or based on activity mismatch between predicted and perceived information [74,67].

Internal Simulations. Both aspects considered so far are mainly of the nature of sensory anticipations, that is, sensory processing is improved, enhanced, compared with, or substituted by anticipatory information. On the other hand, anticipatory information can also be used beyond the immediate prediction of sensory consequences to improve behavior and learning. Interactions with the experienced environment are often re-played or projected into the future by means of an internal predictive environmental model [18,32,40]. Two types of internal simulations can be distinguished: online and offline simulations. Online simulations depend on the current environmental circumstances and can improve immediate decision making. Offline simulations resemble reflective processes that re-play experienced environmental interactions to improve learning, memory, and future behavior.

Current decision making can be influenced by simulating the consequences of currently available alternatives. In its simplest but least computationally costly form, *preventive state anticipations* [19] may be employed, which simulate the usually occurring future events based on habitual behavior. The mechanism only triggers preventive actions if the habitual behavior is expected to lead to an undesirable event. In doing so, undesirable states can often be avoided with only linear additional computational effort—linearly predicting the future of what "normally" happens. Advanced stages of such anticipatory decision making leads to planning approaches that consider many possible future alternatives before making an actual decision [5,15,77].

In contrast to such online, situation-dependent simulation approaches for action decision making, offline simulation, that is, the simulation of events that are not necessarily related to the current situation, have been shown to be useful for memory consolidation as well as for behavioral improvement. An example for memory consolidation is the wake-sleep algorithm [41], which switches between online learning phases, in which data inputs are stored in internal activation patterns, and offline learning phases, in which internally generated memory traces lead to memory generalization and consolidation. A similar structure is exhibited in bidirectional neural networks, originally applied to visual structuring tasks [67] where the emergent activity patterns resembled neuronal receptive fields in the visual cortex.

However, there are also behaviorally-relevant types of simulation, as exemplified in the DYNA-Q system in model-based reinforcement learning [77,78] and related sub-symbolic generalizing implementations of the same principle [5,10,76]. Hereby, an internal environmental model is exploited to execute internal "as if" actions and to update internal reinforcement estimates. Interestingly, from the behavior observation alone, it is often hard to determine if behavior is anticipatory due to previous offline simulations and resulting memory consolidation or due to online, situation-dependent planning simulations [12].

In summary, internal environmental simulations can help to make better immediate decisions, improve action decision making in general, and to learn and generalize the predictive environmental model itself.

Goal-initiated Behavior. Internal simulations, however, do not appear to be the whole story in the realization of efficient, flexible, adaptive behavior. Rather, behavior appears to be generally goal-directed, or rather goal-initiated [43,44,82]. That is, the activation of a desired goal state precedes and triggers actual behavioral initiation and execution. Cognitive psychological research confirms that an image of a goal, which is currently achievable, such as some immediate action consequences, is present before actual action execution is initiated [56]. Moreover, concurrently executed actions interfere mainly due to goal representation interferences, as shown in various bimanual behavioral tasks [60,55].

Thus, goal representations appear to trigger behavior, which is thus never reactive but always anticipatory. This is essentially the tenet of the *ideomotor principle*, proposed over 150 years ago [37,81,48]. This principle is now most directly used in inverse modeling for control, in which a goal state and the current state trigger suitable motor commands as output [50,57,62,80]. To further tune the inverse model capabilities, coupled forward-inverse modules can enable the choice of the currently most suitable inverse models amongst alternatives [84,34].

Additionally, it has been shown that goal-initiated behavior can efficiently resolve and exploit redundancies in the activated goal representation(s). For example, concrete goal states may be chosen based on redundant alternatives [72]. Also motor paths may be chosen based on current alternatives dependent on anticipated movement effort [8]. In this architecture, additional task constraints can be easily accounted for, for example, realizing efficient obstacle avoidance or compensating for inhibited joints [8,38]. A recent combination with reinforcement learning mechanisms enables the motivation-dependent goal activation, effectively unifying payoff with state anticipations [39].

Predictive Representations. Besides immediate influences on sensory processing and behavior, predictive representations need to be considered in more detail, which are often neglected in current adaptive behavior research. Representations need to be generated that identify dependencies in time rather than in space or between current input dimensions.

Recurrent neural networks have been applied in this respect, beginning with the famous Elman networks [23]. Recently, successful motor control patterns were published not only for hierarchical, self-organizing forward-inverse control structures [35] but also for the generation of believable behavioral patterns in real robot applications [46,45]. Additionally, the LSTM network approach [42,30] proved to be able to efficiently relate regular recurring patterns over time. Echo-state networks [47], on the other hand, are able to efficiently detect dynamic patterns over time.

Applications of predictive representations in artificial cognitive systems appear imminent. Hierarchical clusters of captured dynamics to, for example, cluster linguistic structures into recurring phonemes, syllables, words, and sentences appear demanding. In this respect, a hierarchical sequence learning architecture was shown to exhibit interesting, dynamically growing characteristics [11]. Current performance of various recurrent neural network approaches and hierarchical approaches can be found elsewhere [31,24].

Social Anticipations. The last aspect of beneficial influences of anticipatory mechanisms lies in social interaction. Recently, there has been increasing evidence that social beings show strong capabilities to represent the behavior of other animals by means of mirror neurons [71]. Hereby, neural activity is shown to represent not only one's own behavioral patterns, such as a grasping action, but also similar behavioral patterns executed by another animal.

Studies show that the animals hereby not only mirror the actual action but also the purpose (that is, the goal) of the action [29]. Gallese strongly suggests that mirror neurons are the key component to develop mutually beneficial interpersonal relations and empathy mechanisms [28,27]. Arbib relates the mirror system and consequent imitative capabilities to language evolution [1].

Regardless of the representation used, it seems obvious that, in order to effectively interact with conspecifics, avoid betrayal, but exploit mutual possible benefit, it is necessary both to be able to individuate the conspecifics with which interaction will take place and to be able to predict the behavior and current goals of the other individual. Only then does trust and mutually beneficial behavior seems possible beyond evolutionary determined self-less behavior [69].

3   Overview of the Book

The taxonomy presented in the last section is reflected in the workshop contributions. Additionally, as the title suggests, the book moves from brain and cognitive evidence for anticipatory mechanisms to individual and social anticipatory behavior systems. This general train of thought, however, is not only reflected by the paper distribution in this volume, but it is also reflected in various contributions themselves.

3.1 Anticipations in Brains, Language, and Cognition

In the next chapter, Jason Fleischer [26] surveys neural correlates of anticipatory processes in the brain, linking neural activity patterns identified in neuroscience research to anticipatory processes and research in adaptive behavior. First, he gives an overview of neuroscientific research paradigms and points out the difficulty in the different methodologies. He then focuses on three important brain areas: (1) the cerebellum, which is mainly involved in motor learning and control, (2) the basal ganglia, which is involved in reward-based learning, sequential action selection, and timing issues, and (3) the hippocampus, which is involved in sequential representations and memory formation. All three areas are known to also represent anticipatory aspects of behavior and learning. Fleischer concludes that the insights gained with respect to the distinct structures of the three regions as well as their involvement in anticipatory processes should provide helpful guidelines to design future anticipatory, brain-inspired artificial cognitive systems.

Samarth Swarup and Les Gasser [79] survey anticipatory aspects in language. They suggest that the more complex the language, the more anticipatory and social components appear to be involved in it. They take an evolutionary approach and first identify the minimal conditions for the emergence of a proto-language. Then, they analyze various languages in animals and identify the complexity of the structure of a language and the symbolic character of a language as the two main criteria for overall language complexity. Finally, they propose that overall language complexity increases along an anticipation axis from implicit over payoff and sensory, to state, and to social anticipations. Theories of natural and artificial language evolution are surveyed from this perspective. In conclusion, the paper proposes that the study of the minimal conditions for the emergence of language and the anticipatory component within may lead towards the design of artificial social agents that are able to learn to interact by a form of communication that emerges within the agent society itself.

Alexander Riegler [70] then provides a slightly controversial but thought provoking essay on the potential problem of superstitious machines. He points out that an artificial system that attempts to process all information available is destined to start believing in non-existing correlations. Such false beliefs about interdependencies in the world may then lead to superstition and potentially mental illness in the machine. The solution is not to follow an information processing paradigm for the design of artificial cognitive agents, but rather an anticipatory constructivist approach, which focuses on the validation of internally generated, relevant anticipatory representations. Thus, instead of constructing artificial cognitive systems as datamining machines, we should focus on machines that construct an internal reality that represents only relevant interactions and dependencies of the environment.

3.2 Individual Anticipatory Frameworks

The subsequent contributions focus on anticipatory mechanisms and artificial cognitive system frameworks that include anticipatory components.

Giovanni Pezzulo et al. [65] compare the ideomotor principle from the field of psychology with the test operate test exit (TOTE) system from cybernetics. Both principles have a goal-directed nature with an emphasis on behavior and learning. Studies of a visual search system, a developmental arm control system, and a motivational model-based reinforcement learning system show that the ideomotor principle and the TOTE specify very similar behavioral principles. Moreover, the comparisons point out that both principles are rather underspecified and highlight additional mechanisms necessary to realize actual implementations.

Vladimir Red'ko et al. [68] then propose the "animat brain" framework for the design of artificial cognitive control systems. The framework is based on functional systems that contain a coupled system of a forward model predictor and an inverse model actor. Comparisons with other approaches highlight the potentially high flexibility of the "animat brain" approach due to the combination of reinforcement learning with hierarchically linked functional systems.

Aregahegn Negatu et al. [64] introduce an autonomous agent architecture termed the "learning intelligent distribution agent (LIDA) system", which is also inspired by cognitive processes. Their system incorporates payoff, sensory, and state anticipatory mechanisms. It it able to build associative and procedural memory structures based on schema mechanisms, it realizes selective attention based on global workshop theory [3,4], and it is able to select actions based on its current internal drives and reinforcement learning principles. Simulations of the system show competent behavioral and adaptive capabilities illustrating automation and deautomation due to an anticipatory measure of prediction failure and consequent allocation of attentional resources.

Giovanni Pezzulo and Gianguglielmo Calvi [66] introduce a framework that can be used to simulate and evaluate schema-based anticipatory behavior mechanisms. Schema-based design, which is inspired by cognitive psychology research, is theoretically analyzed emphasizing goal-orientedness, flexibility of application, selectivity of information, and excitability, which depends on current drives and contextual input. Moreover, cooperative competition between schemas as well as pragmatic and epistemic (that is, information seeking) aspects of schema activity are investigated. Pezzulo and Calvi then introduce the computational platform "AKIRA Schema Language (AKSL)", which allows the implementation of concurrent resource-competitive schema systems. Exemplars show that the system masters action selection, attentional mechanisms, category formation, the simulation of future behavior, grounding schema activity in behavioral patterns, and hierarchical action control. The paper concludes with a proposal to use AKSL to shed further light on the question *when* anticipatory mechanisms are really beneficial for the improvement of cognitive process and behavior.

3.3 Learning Predictions and Anticipations

The next section of the book introduces several approaches to learning predictions and correlations in time. Often, it is proposed that sensorimotor contingencies are learned, that is, action-dependent sensory changes.

Wolfram Schenck and Ralf Möller [73] teach a moving camera head to predict sensory changes dependent on self-induced camera movements. They distinguish between two learning tasks: learning to predict future visual input and learning to predict the predictable visual areas in the input. To do so, their algorithms learn an action-dependent mapping of visual input rather than to predict the visual input directly. The task is successfully accomplished with a real camera head plus simulated fovea image (a retinal mapping), showing impressive learning and consequent action-dependent image mapping capabilities. The anticipatory component comes in handy here both for learning the mapping as well as for identifying predictable sensory input, working on the direct comparison of anticipated and consequently perceived actual input.

Jérémy Fix et al. [25] move higher up in the visual processing realm and tackle the task of memorizing the location of stimuli, which were previously focused upon. The task to maintain a coherent internal memory of stimulus locations despite the drastic perceptual changes due to saccadic eye movements is certainly non-trivial. To solve the problem, the authors introduce an interactive model of working memory, which maintains currently perceived inputs dependent on focus and predictions, and long-term memory, which predicts perceived inputs and is updated by working memory activity. Hereby, simulations show that anticipations are mandatory to be able to maintain a coherent memory of stimuli locations in the environment, independent of current eye focus. A complete and coherent memory can only be maintained when anticipatory mechanisms are applied.

Stefano Zappacosta et al. [85] propose a testbed for recurrent neural networks and related systems to integrate information in time. The task is to scan an object or a wall while moving around it or along it, respectively. The recurrent network is trained to classify the object scanned, investigating prediction robustness, noise-robustness, and different aspects of generalization capabilities of the network in question. Elman networks, leaky integrator neural networks, and echo state networks are exemplary introduced as suitable network candidates. An Elman network is then evaluated on two testbed instances: a wall task in which two different wall patterns need to be distinguished, and an object task in which three different objects are perceived. The testbed, possibly with additional action-information of movement type and speed in the future, seems to be a valuable tool to test and compare the capabilities of different time-series classification algorithms on somewhat real-world robotic classification tasks.

Philippe Capdepuy et al. [17] investigate the more symbolic challenge of event anticipation. The information-theoretic measures based on constant and consistent time delays as well as on contingency, that is, proximity in time, are used to automatically detect interesting event dependencies. Although only the predictive capabilities are investigated, the authors discuss the importance of such capabilities for anticipatory action decision making and propose also the involvement of epistemic verification actions that could be triggered for the verification of hypothesized event dependencies. Despite currently unresolved scalability as well as subsymbolic issues, the paper shows that the employed information-theoretic measures are highly capable of detecting consistent event contingencies and time-delay relationships.

3.4 Anticipatory Processes in Behavioral Control

Predictive capabilities alone are not sufficient for anticipatory behavior, though. The following papers address different aspects of goal representations and predictions that directly influence actual behavior.

Kiril Kiryazov et al. [52] present an integrated behavioral architecture that uses symbolic analogical reasoning to make action decisions. The system is mounted onto the Aibo real-robot platform and solves the task of finding interesting objects in a house-like environment. Besides the anticipatory decision making capabilities based on analogy, the system applies selective attention mechanisms as well as top-down anticipatory perception mechanisms to filter out relevant information in the environment. Although it is hard to compare the current capabilities of the platform with other architectures due to the many hardware and setup dependent factors, the resulting anticipatory behavior aspects realized on an integrated real-robot platform are highly promising.

Toshiyuki Kondo and Koji Ito [54] present a recurrent neural network architecture with neuromodulatory biases that shows to be able to reach targets under various force fields. The network weights and connectivity evolve by means of a genetic algorithm. It is shown that the anticipatory biases are beneficial to achieve more robust reaching behavior under differing force fields. The results suggest that recurrent self-stabilization mechanisms can be highly beneficial for adaptation in gradually changing environmental circumstances. Future evaluations appear necessary to further shed light on the emergent representations and control components in such evolved recurrent neural network structures.

Arnaud Blanchard and Lola Cañamero [7] study how positive and negative goal states can be efficiently remembered in order to enable optimal behavioral control. They use a developmental approach that learns to classify goals based on a reinforcement learning derived scheme. Their aim is to use a minimal amount of memory by remembering only maximally suitable and unsuitable states in the environment—leaving the task to reach these states to a goal-directed control architecture. Their real robot implementation of the system is able to identify suitable goals as well as undesirable goals efficiently with a very low memory requirement. Future work intends to enhance the goal identification mechanism to be able to identify multiple and more distinct goals. Moreover, the goal generation mechanism will be interfaced with a motivational component, which will generate drives and correspondingly desired goal states as well as goal-directed motor control mechanism, which will be able to reach currently desirable goal states.

Arshia Cont et al. [2] use predictive system capabilities for the generation and improvisation of music. The paper provides a thorough overview of anticipatory cognition identified in music theory, suggesting that musical processing is highly anticipatory based on veridical expectations, schematic expectations, dynamic adaptive expectations, and conscious expectations. All four types interact concurrently and competitively. The remainder of the paper then focuses on the integration of payoff and state anticipations into a music generating and improvisation architecture, working either in self listening mode or in interaction mode, respectively. The provided results of the imitation of a Bach piece are impressive and promise fruitful future integrations of anticipatory mechanisms for automatized music generation and improvisation.

3.5 Anticipatory Social Behavior

After the study of different aspects of individual anticipatory behavior, the last chapters of this book address the importance of anticipatory mechanisms for efficient social interaction.

Mario Gómez et al. [33] introduce an anticipatory trust model in open distributed systems. A theoretical taxonomy of trust distinguishes between direct trust, which is about previously experienced service quality of another agent, and advertisement- and recommendation-based forms of trust, which are about the suggested service quality of another agent by yet other agents. The different measures are combined into a global trust measure—essentially the weighted average of the individual measures. Experiments are carried out in a simulated market environment with trading agents. The results stress the importance of stability and the capability to identify properties of other individuals, in order to be able to develop effective notions of trust. Moreover, they show that if the system is able to predict the behavior of other agents, the agent is able to adapt to changes in the environment more effectively.

Gerben Meyer and Nick Szirbik [63] study anticipatory alignment mechanisms in multi agent systems with petri nets. Conceptualizations are carried out within belief propagating networks, studying three types of alignment policies: on-the-fly alignment, pre-interaction alignment, and alignment induced by a third party. The mechanisms are illustrated within a business information system, sketching out constraint transactions of goods and money between multiple agents. It is shown that the state anticipatory mechanism is able to yield more efficient agent interaction executions. The integration of trust mechanisms for more efficient agent communication appears imminent. Moreover, the proposition of actual simulations in real-world game-like scenarios with other artificial agents, but also with expert players, promises to be highly revealing for future applications.

Emilian Lalev and Maurice Grinberg [58] study two recurrent neural network architectures playing the iterated prisoner's dilemma. While the first model used backward-oriented reinforcement learning methods, the second network basis its move decisions on generated predictions about future games. Thus, the latter network anticipates the behavior of the opponent player. The results suggest that human players use anticipatory capabilities to guide their decision process within the game. As with actual human participants, the cooperation rate of the latter network depended on a so-called cooperation index, which quantifies the likelihood that the opponent player cooperates. Thus, the results suggest that anticipatory connections are mandatory for efficient human-like network interaction within the iterated prisoner's dilemma game.

The final paper in this series studies the benefits of anticipating the behavior of another robot agent. Birger Johansson and Christian Balkenius [49] placed two real robots in differently complex arenas with the task of switching places with each other. The results show that in very simple environments without obstacles, a goal-directed behavioral strategy without any consideration of the opponent player, except for a reactive hard-coded obstacle avoidance mechanism, yielded the most efficient behavior. However, in more complex environments, in which robot interference is inevitable and harder to resolve, anticipatory mechanisms yielded the fastest behavior. In this case, the anticipatory mechanism predicted the behavior of the opponent robot and resolved possible trajectory conflicts online. Thus, it is shown that higher complex environments can make more complex, cooperative, anticipatory mechanisms beneficial. In very simple interactive environments, on the other hand, ignorance of the opponent or cooperative player can also be more effective, since no expensive contemplations and communicative interactions are necessary.

4   Conclusions

Research on anticipatory behavior mechanisms can be found in a variety of research areas. Indications for anticipatory mechanisms in the brain, and their influences on cognition and resulting individual and social behavior, continue to accumulate. It is hoped that anticipatory research in general, and this enhanced and re-reviewed post-workshop proceedings volume in particular, will contribute to a general understanding of anticipatory mechanisms in cognitive systems.

This introduction conceptualized different anticipatory mechanisms providing a taxonomy of how anticipatory mechanisms may improve adaptive behavior and learning. The overview of the contributions of this volume exposes important correlations of anticipatory behavior mechanisms between different research disciplines. These include neuroscience, cognitive psychology, linguistics, individual and social adaptive behavior research, music theory, business research with trading agents, and research in cognitive modeling.

The book can certainly only provide a glimpse at the different aspects of anticipations in these various disciplines. However, we believe that the contributions reveal and develop many highly correlated recurring anticipatory mechanisms and they identify many anticipatory principles that are highly beneficial to improve individual and social adaptive behavior. Thus, we hope that the articles in this volume will be inspiring for researchers in the cognitive systems area and lead to the offspring of many fruitful future research projects and interdisciplinary collaborations amongst scientists interested both in a deeper understanding of natural cognitive systems and in the further development, design, and application of adaptive, flexible, and efficient artificial cognitive systems.

Acknowledgments. This work was supported by the EU project **MindRACES**, FP6-511931.

References

1. Arbib, M.: The mirror system, imitation, and the evolution of language. In: Dautenhahn, K., Nehaniv, C.L. (eds.) Imitation in animals and artifacts, pp. 229–280. MIT Press, Cambridge, MA (2002)
2. Cont, A., Shlomo Dubnov, G.A.: Anticipatory model of musical style imitation using collaborative and competitive reinforcement learning. In: Butz, M.V., Sigaud, O., Pezzulo, G., Baldassarre, G. (eds.) Anticipatory Behavior in Adaptive Learning Systems: From Brains to Individual and Social Behavior, Springer, Heidelberg (2007)
3. Baars, B.: A Cognitive Theory of Consciousness. Cambridge University Press, Cambridge, MA (1988)
4. Baars, B.J.: The conscious access hypothesis: Origins and recent evidence. Trends Cogn Sci. 6, 47–52 (2002)
5. Baldassarre, G.: Forward and bidirectional planning based on reinforcement learning and neural networks in a simulated robot. In: Butz, M.V., Sigaud, O., Gérard, P. (eds.) Anticipatory Behavior in Adaptive Learning Systems. LNCS (LNAI), vol. 2684, pp. 179–200. Springer, Heidelberg (2003)
6. Barto, A.G., Mahadevan, S.: Recent advances in hierarchical reinforcement learning. Discrete Event Dynamic Systems 13, 341–379 (2003)
7. Blancharnd, A.J., Cañamero, L.: Anticipating rewards in continuous time and space: A case study in developmental robotics. In: Butz, M.V., Sigaud, O., Pezzulo, G., Baldassarre, G. (eds.) Anticipatory Behavior in Adaptive Learning Systems: From Brains to Individual and Social Behavior, Springer, Heidelberg (2007)
8. Butz, M.V., Herbort, O., Hoffmann, J.: Exploiting redundancy for flexible behavior: Unsupervised learning in a modular sensorimotor control architecture. Psychological Review (in press)
9. Butz, M.V., Swarup, S., Goldberg, D.E.: Effective online detection of task-independent landmarks. In: Online Proceedings for the ICML'04 Workshop on Predictive Representations of World Knowledge, p. 10 (2004)
10. Butz, M.V.: Anticipatory learning classifier systems. Kluwer Academic Publishers, Boston, MA (2002)
11. Butz, M.V.: COSEL: A cognitive sequence learning architecture. IlliGAL report 2004021, Illinois Genetic Algorithms Laboratory, University of Illinois at Urbana-Champaign (2004)
12. Butz, M.V., Hoffmann, J.: Anticipations control behavior: Animal behavior in an anticipatory learning classifier system. Adaptive Behavior 10, 75–96 (2002)
13. Butz, M.V., Sigaud, O., Gérard, P. (eds.): Anticipatory Behavior in Adaptive Learning Systems: Foundations, Theories, and Systems, LNCS (LNAI), vol. 2684. Springer, Heidelberg (2003)
14. Butz, M.V., Sigaud, O., Gérard, P.: Anticipatory behavior: Exploiting knowledge about the future to improve current behavior. In: Butz, M.V., Sigaud, O., Gérard, P. (eds.) Anticipatory Behavior in Adaptive Learning Systems. LNCS (LNAI), vol. 2684, pp. 1–10. Springer, Heidelberg (2003)
15. Butz, M.V., Sigaud, O., Gérard, P.: Internal models and anticipations in adaptive learning systems. In: Butz, M.V., Sigaud, O., Gérard, P. (eds.) Anticipatory Behavior in Adaptive Learning Systems. LNCS (LNAI), vol. 2684, pp. 86–109. Springer, Heidelberg (2003)
16. Camacho, E.F., Bordons, C. (eds.): Model predictive control. Springer, Heidelberg (1999)

17. Capdepuy, P., Polani, D., Nehaniv, C.L.: Construction of an internal predictive model by event anticipation. In: Butz, M.V., Sigaud, O., Pezzulo, G., Baldassarre, G. (eds.) Anticipatory Behavior in Adaptive Learning Systems: From Brains to Individual and Social Behavior, Springer, Heidelberg (2007)

18. Damasio, A.R.: Descartes' Error: Emotion, Reason and the Human Brain. Grosset/Putnam (1994)

19. Davidsson, P.: A framework for preventive state anticipation. In: Butz, M.V., Sigaud, O., Gérard, P. (eds.) Anticipatory Behavior in Adaptive Learning Systems. LNCS (LNAI), vol. 2684, pp. 151–166. Springer, Heidelberg (2003)

20. Deneve, S., Pouget, A.: Bayesian multisensory integration and cross-modal spatial links. Journal of Physiology - Paris 98, 249–258 (2004)

21. Desmurget, M., Grafton, S.: Forward modeling allows feedback control for fast reaching movements. Trends in Cognitive Sciences 4, 423–431 (2000)

22. Dubois, D.M.: Mathematical foundations of discrete and functional systems with strong and weak anticipations. In: Butz, M.V., Sigaud, O., Gérard, P. (eds.) Anticipatory Behavior in Adaptive Learning Systems. LNCS (LNAI), vol. 2684, pp. 110–132. Springer, Heidelberg (2003)

23. Elman, J.L.: Finding structure in time. Cognitive Science 14, 179–211 (1990)

24. Fernández, S., Graves, A., Schmidhuber, J.: Sequence labelling in structured domains with hierarchical recurrent neural network. In: Proceedings of the 20th International Joint Conference on Artificial Intelligence, IJCAI 2007, vol. 20 (2007)

25. Fix, J., Vitay, J., Rougier, N.P.: A distributed computational model of spatial memory anticipation during a visual search task. In: Butz, M.V., Sigaud, O., Pezzulo, G., Baldassarre, G. (eds.) Anticipatory Behavior in Adaptive Learning Systems: From Brains to Individual and Social Behavior, Springer, Heidelberg (2007)

26. Fleischer, J.G.: Neural correlates of anticipation in cerebellum, basal ganglia, and hippocampus. In: Butz, M.V., Sigaud, O., Pezzulo, G., Baldassarre, G. (eds.) Anticipatory Behavior in Adaptive Learning Systems: From Brains to Individual and Social Behavior, Springer, Heidelberg (2007)

27. Gallese, V.: The manifold nature of interpersonal relations: The quest for a common mechanism. Philosophical transactions of the Royal Society of London. Series B, Biological sciences 358, 517–528 (2003)

28. Gallese, V.: The 'shared manifold' hypothesis: From mirror neurons to empathy. Journal of Consciousness Studies: Between Ourselves - Second-Person Issues in the Study of Consciousness 8, 33–50 (2001)

29. Gallese, V., Goldman, A.: Mirror neurons and the simulation theory of mindreading. Trends in Cognitive Sciences 2, 493–501 (1998)

30. Gers, F.A., Schraudolph, N., Schmidhuber, J.: Learning precise timing with LSTM recurrent networks. Journal of Machine Learning Research 3, 115–143 (2002)

31. Graves, A., Schmidhuber, J.: Framewise phoneme classification with bidirectional LSTM and other neural network architectures. Neural Networks 18, 602–610 (2005)

32. Grush, R.: The emulation theor of representation: Motor control, imagery, and perception. Behavioral and Brain Sciences 27, 377–396 (2004)

33. Gomez, M., Carbo, J., Benac-Earle, C.: An anticipatory trust model for open distributed systems. In: Butz, M.V., Sigaud, O., Pezzulo, G., Baldassarre, G. (eds.) Anticipatory Behavior in Adaptive Learning Systems: From Brains to Individual and Social Behavior, Springer, Heidelberg (2007)

34. Haruno, M., Wolpert, D.M., Kawato, M.: Mosaic model for sensorimotor learning and control. Neural Computation 13, 2201–2220 (2001)

35. Haruno, M., Wolpert, D.M., Kawato, M.: Hierarchical mosaic for movement generation. In: Ono, T., Matsumoto, G., Llinas, R., Berthoz, A., Norgren, R., Nishijo, H., Tamura, R. (eds.) Excepta Medica International Coungress Series, vol. 1250, pp. 575–590. Elsevier Science B.V, Amsterdam, The Netherlands (2003)
36. Haykin, S.: Adaptive filter theory. 4th edn. Prentice Hall, Upper Saddle River, NJ (2002)
37. Herbart, J.F.: Psychologie als Wissenschaft neu gegründet auf Erfahrung, Metaphysik und Mathematik. Zweiter, analytischer Teil. August Wilhem Unzer, Königsberg, Germany (1825)
38. Herbort, O., Butz, M.V.: Encoding complete body models enables task dependent optimal behavior. International Joint Conference on Neural Networks (in press)
39. Herbort, O., Ognibene, D., Butz, M.V., Baldassarre, G.: Learning to select targets within targets in reaching tasks.In: 6th IEEE International Conference on Development and Learning (submitted)
40. Hesslow, G.: Conscious thought as simulation of behaviour and perception. Trends in cognitive sciences 6, 242–247 (2002)
41. Hinton, G.E., Dayan, P., Frey, B.J., Neal, R.M.: The wake-sleep algorithm for unsupervised neural networks. Science 268, 1158–1161 (1995)
42. Hochreiter, S., Schmidhuber, J.: Long short-term memory. Neural Computation 9, 1735–1780 (1997)
43. Hoffmann, J.: Vorhersage und Erkenntnis: Die Funktion von Antizipationen in der menschlichen Verhaltenssteuerung und Wahrnehmung. [Anticipation and cognition: The function of anticipations in human behavioral control and perception.]. Hogrefe, Göttingen, Germany (1993)
44. Hoffmann, J.: Anticipatory behavioral control. In: Butz, M.V., Sigaud, O., Gérard, P. (eds.) Anticipatory Behavior in Adaptive Learning Systems. LNCS (LNAI), vol. 2684, pp. 44–65. Springer, Heidelberg (2003)
45. Ito, M., Noda, K., Hoshino, Y., Tani, J.: Dynamic and interactive generation of object handling behaviors by a small humanoid robot using a dynamic neural network model. Neural Networks (in press)
46. Ito, M., Tani, J.: On-line imitative interaction with a humanoid robot using a dynamic neural network model of a mirror system. Adaptive Behavior 12, 93–115 (2004)
47. Jaeger, H., Haas, H.: Harnessing nonlinearity: Predicting chaotic systems and saving energy in wireless communication. Science 304, 78–80 (2004)
48. James, W.: The principles of psychology. vol. 1, Holt, New York (1890)
49. Johansson, B., Balkenius, C.: An experimental study of anticipation in simple robot navigation. In: Butz, M.V., Sigaud, O., Pezzulo, G., Baldassarre, G. (eds.) Anticipatory Behavior in Adaptive Learning Systems: From Brains to Individual and Social Behavior, Springer, Heidelberg (2007)
50. Jordan, M.I., Rumelhart, D.E.: Forward models: Supervised learning with a distal teacher. Cognitive Science 16, 307–354 (1992)
51. Kalman, R.E.: A new approach to linear filtering and prediction problems. Transactions of the ASME-Journal of Basic Engineering 82, 35–45 (1960)
52. Kiryazov, K., Petkov, G., Grinberg, M., Kokinov, B., Balkenius, C.: The interplay of analogy-making with active vision and motor control in anticipatory robots. In: Butz, M.V., Sigaud, O., Pezzulo, G., Baldassarre, G. (eds.) Anticipatory Behavior in Adaptive Learning Systems: From Brains to Individual and Social Behavior, Springer, Heidelberg (2007)
53. Knill, D.C., Pouget, A.: The bayesian brain: The role of uncertainty in neural coding and computation. Trends in Neurosciences 27, 712–719 (2004)

54. Kondo, T., Ito, K.: An intrinsic neuromodulation model for realizing anticipatory behavior in reaching movement underunexperienced force fields. In: Butz, M.V., Sigaud, O., Pezzulo, G., Baldassarre, G. (eds.) Anticipatory Behavior in Adaptive Learning Systems: From Brains to Individual and Social Behavior, Springer, Heidelberg (2007)

55. Kunde, W., Weigelt, M.: Goal congruency in bimanual object manipulation. Journal of Experimental Psychology: Human Perception and Performance 31, 145–156 (2005)

56. Kunde, W., Koch, I., Hoffmann, J.: Anticipated action effects affect the selection, initiation, and execution of actions. The Quarterly Journal of Experimental Psychology. Section A: Human Experimental Psychology 57, 87–106 (2004)

57. Kuperstein, M.: Infant neural controller for adaptive sensory-motor coordination. Neural Netw. 4, 131–145 (1991)

58. Lalev, E., Grinberg, M.: Backward vs. forward-oriented decision making in the iterated prisoners dilemma: A comparison between two connectionist models. In: Butz, M.V., Sigaud, O., Pezzulo, G., Baldassarre, G. (eds.) Anticipatory Behavior in Adaptive Learning Systems: From Brains to Individual and Social Behavior, Springer, Heidelberg (2007)

59. Lorini, E., Castelfranchi, C.: The unexpected aspects of surprise. International Journal of Pattern Recognition and Artificial Intelligence 20, 817–835 (2006)

60. Mechsner, F., Kerzel, D., Knoblich, G., Prinz, W.: Perceptual basis of bimanual coordination. Nature 414, 69–73 (2001)

61. Mehta, B., Schaal, S.: Forward models in visuomotor control. Journal of Neurophysiology 88, 942–953 (2002)

62. Mel, B.W.: A connectionist model may shed light on neural mechanisms for visually guided reaching. Journal of cognitive neuroscience 3, 273–292 (1991)

63. Meyer, G.G., Szirbik, N.B.: Anticipatory alignment mechanisms for behavioural learning in multi agent systems. In: Butz, M.V., Sigaud, O., Pezzulo, G., Baldassarre, G. (eds.) Anticipatory Behavior in Adaptive Learning Systems: From Brains to Individual and Social Behavior, Springer, Heidelberg (2007)

64. Negatu, A., DMello, S., Franklin, S.: Cognitively inspired anticipatory adaptation and associated learning mechanisms for autonomous agents. In: Butz, M.V., Sigaud, O., Pezzulo, G., Baldassarre, G. (eds.) Anticipatory Behavior in Adaptive Learning Systems: From Brains to Individual and Social Behavior, Springer, Heidelberg (2007)

65. Pezzulo, G., Baldassarre, G., Butz, M.V., Castelfranchi, C., Hoffmann, J.: From actions to goals and vice-versa: Theoretical analysis and models of the ideomotor principle and tote. In: Butz, M.V., Sigaud, O., Pezzulo, G., Baldassarre, G. (eds.) Anticipatory Behavior in Adaptive Learning Systems: From Brains to Individual and Social Behavior, Springer, Heidelberg (2007)

66. Pezzulo, G., Calvi, G.: Schema-based design and the akira schema language: An overview. In: Butz, M.V., Sigaud, O., Pezzulo, G., Baldassarre, G. (eds.) Anticipatory Behavior in Adaptive Learning Systems: From Brains to Individual and Social Behavior, Springer, Heidelberg (2007)

67. Rao, R.P.N., Ballard, D.H.: Predictive coding in the visual cortex: A functional interpretation of some extra-classical receptive-field effects. Nature Neuroscience 2, 79–87 (1999)

68. Redko, V.G., Mikhail, S., Burtsev, K.V.A., Manolov, A.I., Mosalov, O.P., Nepomnyashchikh, V.A., Prokhorov, D.V.: Project animat brain: Designing the animat control system on the basis of the functional systems theory. In: Butz, M.V., Sigaud, O., Pezzulo, G., Baldassarre, G. (eds.) Anticipatory Behavior in Adaptive Learning Systems: From Brains to Individual and Social Behavior, Springer, Heidelberg (2007)

69. Ridley, M.: The Origins of Virtue: Human Instincts and The Evolution of Cooperation. Penguin Books (1996)
70. Riegler, A.: Superstition in the machine. In: Butz, M.V., Sigaud, O., Pezzulo, G., Baldassarre, G. (eds.) Anticipatory Behavior in Adaptive Learning Systems: From Brains to Individual and Social Behavior, Springer, Heidelberg (2007)
71. Rizzolatti, G., Fadiga, L., Gallese, V., Fogassi, L.: Premotor cortex and the recognition of motor actions. Cognitive Brain Research 3, 131–141 (1996)
72. Rosenbaum, D.A., Loukopoulos, L.D., Meulenbroek, R.G.J., Vaughan, J., Engelbrecht, S.E.: Planning reaches by evaluating stored postures. Psychological Review 102, 28–67 (1995)
73. Schenck, W., Möller, R.: Training and application of a visual forward model for a robot camera head. In: Butz, M.V., Sigaud, O., Pezzulo, G., Baldassarre, G. (eds.) Anticipatory Behavior in Adaptive Learning Systems: From Brains to Individual and Social Behavior, Springer, Heidelberg (2007)
74. Schmidhuber, J.: Learning complex extended sequences using the principle of history compression. Neural Computation 4, 234–242 (1992)
75. Simsek, Ö., Barto, A.G.: Using relative novelty to identify useful temporal abstractions in reinforcement learning. In: Proceedings of the Twenty-First International Conference on Machine Learning (ICML- 2004), pp. 751–758 (2004)
76. Stolzmann, W., Butz, M.V., Hoffmann, J., Goldberg, D.E.: First cognitive capabilities in the anticipatory classifier system. From Animals to Animats 6: In: Proceedings of the Sixth International Conference on Simulation of Adaptive Behavior, pp. 287 296 (2000)
77. Sutton, R.S.: Reinforcement learning architectures for animats. From Animals to Animats: In: Proceedings of the First International Conference on Simulation of Adaptive Behavior, pp. 288–296 (1991)
78. Sutton, R.S., Barto, A.G.: Reinforcement learning: An introduction. MIT Press, Cambridge, MA (1998)
79. Swarup, S., Gasser, L.: The role of anticipation in the emergence of language. In: Butz, M.V., Sigaud, O., Pezzulo, G., Baldassarre, G. (eds.) Anticipatory Behavior in Adaptive Learning Systems: From Brains to Individual and Social Behavior, Springer, Heidelberg (2007)
80. Todorov, E.: Optimality principles in sensorimotor control. Nature Review Neuroscience 7, 907–915 (2004)
81. von Helmholtz, H.: Handbuch der physiologischen Optik. vol. III. Leopold Voss, Leipzig translated by The Optical Society of America in 1924 from the third germand edn. 1910, Treatise on physiological optics, vol. III (1867)
82. von Hofsten, C.: An action perspective on motor development. Trends in Cognitive Science 8, 266–272 (2004)
83. von Holst, E., Mittelstaedt, H.: Das Reafferenzprinzip. Naturwissenschaften 37, 464–476 (1950)
84. Wolpert, D.M., Kawato, M.: Multiple paired forward and inverse models for motor control. Neural Networks 11, 1317–1329 (1998)
85. Zappacosta, S., Nolfi, S., Baldassarre, G.: A testbed for neural-network models capable of integrating information in time. In: Butz, M.V., Sigaud, O., Pezzulo, G., Baldassarre, G. (eds.) Anticipatory Behavior in Adaptive Learning Systems: From Brains to Individual and Social Behavior, Springer, Heidelberg (2007)

Neural Correlates of Anticipation in Cerebellum, Basal Ganglia, and Hippocampus

Jason G. Fleischer

The Neurosciences Institute
10640 John Jay Hopkins Drive
San Diego, CA
fleischer@nsi.edu

Abstract. Animals anticipate the future in a variety of ways. For instance: (a) they make motor actions that are timed to a reference stimulus and motor actions that anticipate future movement dynamics; (b) they learn to make choices that will maximize reward they receive in the future; and (c) they form memories of behavioral episodes such that the animal's future actions can be predicted by current neural activity associated with those memories. Although these effects are clearly observable at the behavioral level, research into the mechanisms of such anticipatory learning are still largely in the early stages. This review, intended for those who have a computational background and are less familiar with neuroscience, addresses neural mechanisms found in the mammalian cerebellum, basal ganglia, and the hippocampus that give rise to such adaptive anticipatory behavior.

1 Introduction

Anticipatory behavior occurs when actions depend not only on past and present but also on predictions, expectations, or beliefs about the future [4]. While the volume in your hands is largely concerned with the computational or theoretical aspects of anticipatory behavior, the researchers who participate in the Anticipatory Behavior in Adaptive Learning Systems meeting are also interested in the processes that allow animals to generate such behavior. This paper will provide an overview of some types of neural activity in the mammalian brain that are highly correlated with particular forms of anticipatory behavior. Behaviorally correlated neural activity is generally interpreted as evidence that the brain area where the activity appears is involved in producing the behavior. Therefore, the study of activity correlated with anticipatory behavior can potentially reveal the neural mechanisms underlying behavioral production.

1.1 The Neural Correlates of Behavior

Behavioral neuroscience[1] is the study of how the mechanisms of the nervous system give rise to the behaviors observed at the level of the whole organism.

---

[1] Some people prefer terms such as biological psychology, systems neuroscience, or cognitive neuroscience to describe roughly the same idea.

M.V. Butz et al. (Eds.): ABiALS 2006, LNAI 4520, pp. 19–34, 2007.

Questions of mechanism are always difficult to address, but the nature of the nervous system, with its nearly-innumerable interacting components (humans have on the order of 10^{15} synapses) and its largely unknown molecular machinery, is particularly difficult. Creating a theory about the mechanism of a behavior, or the contribution a brain region makes to a behavior, is often only achieved by piecing together several indirect lines of evidence. The most common forms of evidence to look at are:

1. Anatomy: It is fairly clear where sensory and motor information arrives at or departs from the central nervous system. Neurons that are only a few synapses away from a sensory or a motor neuron are most likely to process that kind of information. Likewise, once a theory of function has been well-established for a particular brain region, it naturally may suggest functions of other areas that produce input for that region or receive output from it.
2. Lesion studies: Brain-damaged human patients or targeted lesions created in experimental animals can define which brain areas are at least necessary for a behavior. However, the problem arises that there are often multiple, parallel systems performing similar functions that can be difficult to disassociate from each other. In addition, knowing that a lesion to a particular region disrupts a behavior does not mean the region is responsible for the behavior. For example, the signals responsible for producing the behavior could merely transit through that region rather than originating or being processed there. Also the lack of disruption when a lesion occurs does not indicate the region is not involved in the behavior under some circumstances — parallel pathways could be compensating for the damaged region.
3. Imaging studies: Using magnetic resonance imaging, positron emission tomography, or similar techniques the neural activity of an entire region may be studied at once. This allows one to look for behaviors that correspond to activity in the region being studied; such correlated activity would at least suggest that the region is involved in that behavior. However, these methods suffer from the problem that it is not always clear what is going on at a mechanistic level. For instance measuring blood flow in an MRI only tells how metabolically active that brain region is, not whether it is affecting other regions and, if so, whether its effect is inhibitory or excitatory.
4. Single-unit recording: Electrodes are inserted into the brain and the action potentials fired by a single neuron are recorded, thus allowing one to study behavioral correlates of single neurons. However, the problem here is the inverse of the one above — it is difficult to record enough neurons simultaneously to understand much about how activity progresses through the nervous system.

Finally, it is important to note that while behavioral neuroscience is striving to understand the functioning of the nervous system in producing behavior, this does not imply that the goal is to assign one function to each and every neuron or brain region. Clear behavioral correlates are usually obtained only once the experiment has been highly simplified and the subject over-trained.

Neural activity is much messier when performing natural tasks, because most parts of the brain are likely to be involved in many behavioral functions. Yet clear progress can be, and has been made in understanding several aspects of behavioral production in the mammalian nervous system. It just requires the steady accumulation of multiple lines of evidence.

1.2 Scope

There is strong evidence that several brain regions are important in anticipating the future over different time-scales. This review focuses only on anticipation over time-scales of seconds or less. Long time-scale (minutes and longer) cognitive planning and circadian timing of behaviors is beyond the scope of this review, although the former are typically seen as involving frontal cortical areas of the brain [39,17], and the latter are believed to be primarily related to genetic transcription/translation auto-regulatory loops and involve a brain area known as the suprachiasmatic nuclei [17].

The brain areas and behavioral functions that will be reviewed in this paper are listed below:

Cerebellum: involved in many aspects of motor learning. This review looks at its involvement in both timing motor movements in relationship to stimuli, and in computing forward kinematic models that predict the results of motor commands.

Basal ganglia: involved in learning, timing, and sequential action selection. Most strikingly, parts of the basal ganglia contain neurons whose activity correlates with predictions of the reward structure of the environment.

Hippocampus: (and surrounding areas) involved in learning and memory. Particularly, the hippocampus is necessary for creating memories that are sequential in nature, such as memories of route navigation. There are strong activity correlates of these kinds of memories in the hippocampus, and these correlates encode information about future events in the behavioral sequence.

2   Cerebellum - Motor Actions and Timing

The cerebellum is located to the rear of the brain underneath the cerebral cortex. Although it contains only about 10% of the brain's volume it has roughly half of all neurons. It has a regular, layered organization that basically consists of repetitions of the same circuit. Yet it contains several distinct regions, each receiving input from different parts of the brain, thus suggesting that the cerebellum is performing the same kind of calculations, but on a variety of data.

A cerebellar circuit (see figure 1) receives its input through two pathways: climbing fibers that excite single Purkinje cells directly, and mossy fibers that excite the parallel fibers joining cerebellar circuits together. The Purkinje cells excite deep cerebellar nuclei which in turn produce the cerebellum's output. The parallel fibers also stimulate inhibitory cells (not shown in the figure for clarity) that prevent Purkinje cells from firing; this produces a center excitation and surrounding inhibition effect. Only Purkinje cells that receive high levels of input from their climbing fibers as well as the parallel fibers will fire. Nearby circuits will be prevented from firing due to the inhibitory cells that take their input from the parallel fibers.

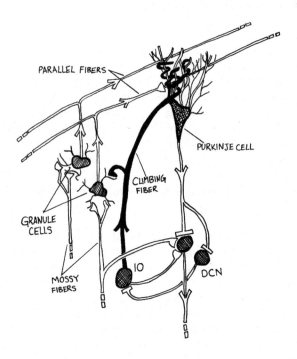

Fig. 1. A schematic depiction of a single cerebellar circuit. Input goes through the inferior olive (IO) via the climbing fibers to the Purkinje cells, which excite the deep cerebellar nuclei (DCN) that produce the cerebellum's output. Another input pathway goes via the mossy fibers to the granule cells and their parallel fibers that lead to other cerebellar circuits. The parallel fibers also excite cells (not shown for clarity) which inhibit Purkinje cells.

The cerebellum has long been implicated in the feedback control of motor activity [22], but it may also be involved in anticipatory forms of motor control. This section contains a review of the cerebellum's possible role in timing motor actions and in predicting the kinematic results of those actions. These are problems of anticipation in the millisecond to second range, resulting from the need of the animal to produce smooth, controlled behavior while interacting with the physical world.

The cerebellum is involved in the timing of motor actions that are repetitive, such as rhythmic tapping. An ability to accurately reproduce a rhythm is a basic form of motor timing which can be used for producing anticipatory motor behavior. One way to measure this ability is to start a test subject tapping in time to a metronome and ask them to continue the beat after the metronome stops. Patients with cerebellar damage are impaired at this task [14,15]. Patients with damage to medial parts of the cerebellum have deficits that could be described as execution (motor) error, and patients with damage to lateral parts of the cerebellum had deficits in timing their tapping but not in executing the motor action. This demonstrates that the cerebellum is involved not only in producing the motor performance, as would be predicted by the standard theory of cerebellar function, but is also involved in timing the rhythm.

The cerebellum is also involved in learning the timing of motor actions that are not repetitive, but must be executed in precise temporal relation to an external stimulus. The classic example of this is eyeblink conditioning [24,48], where a tone cue (conditioned stimulus, CS) begins at a fixed interval (100 ms - 3 s) before a puff of air (noxious unconditioned stimulus, US) is delivered to the eye. Rabbits learn that the CS predicts the appearance of the US, and shut their eyelids in response to the CS after 100-200 trials. After conditioning they shut their eyes with a delay such that the response peaks on average at the time the US is presented. Cerebellar lesions abolish the proper timing of a conditioned eyeblink response.

Another form of prediction that seems to be performed in the cerebellum is the computation of where a limb will be in the near future — this is known as a forward model. Optimal motor control theory requires the presence of a forward computation of what the effect of a motor command will be given the current state and motor commands [51]. A forward model provides the nervous system with a prediction of what the body state will be like in the near future. This prediction can be used to more accurately estimate actual body position.

Alternatively, a forward model can allow the production of movements that are faster than using only feedback control [17]. Animals have muscles arranged in antagonistic pairs so producing a desired motion without overshoot requires applying an early breaking force. While this calculation can in theory be performed using only feedback control, the long delays in the nervous system and the dynamic properties of muscles and proprioception favor the use of forward motion models for choosing when to apply the breaking force.

The cerebellum may coordinate eye and hand movements through the use of forward models [27]. Maximum eye-hand coordination is achieved by having the eyes slightly lead the hand, and the motor system is therefore hypothesized to use forward-model information from the ocular motor system to increase accuracy in producing hand movements. Imaging studies have demonstrated cerebellar involvement in these types of tasks [28]. Several studies reviewed in [2] demonstrate that patients with cerebellar damage have difficulty adapting to predictive (forward) motor control problems, but are equivalent to controls when adapting to reactive (feedback) motor control problems.

Mauk and Buonomano [23] hypothesize that both interval timing and feedforward control deficits in cerebellar patients may be due to the same mechanism. They argue that tasks that involve stopping and starting, such as tapping and blinking, would have particularly noticeable effects if the feed-forward models controlling antagonist muscle timing were disturbed. Thus they offer us the controversial possibility that both types of deficits result from a disturbance of the forward model.

It should be noted that the cerebellum is not the only brain region thought to be involved in creating forward motion models in the brain. The parietal cortex and various motor regions have neural activity that appears to encode limb positions in a variety of coordinate systems. They also have neural activity that encodes limb kinematics and dynamics in the near future; i.e., forward models. A good review of this material can be found in [45]. This does not, however, conflict with the view that the cerebellum is involved with computing forward models. It appears that the cerebellum is particularly involved with adapting these models during behavior — all of the citations offered above regarding cerebellar forward models have that flavor to their results.

3   Basal Ganglia — Dopamine and Reward

The basal ganglia are a collection of midbrain structures, including the regions that are the primary topics of this section: the striatum, the substantia nigra, and the ventral tegmental area. The basal ganglia is involved in learning and action selection, and disorders of this region lead to devastating diseases such as Parkinson's, Huntington's, and Attention Deficit Hyperactivity Disorder.

The basal ganglia are also interesting because of their unique anatomy and physiology, which is illustrated in figure 2. Input converges on the striatum from many different cortical locations. The striatum in turn projects to several other structures inside the basal ganglia. Information flows in several anatomically segregated, parallel loops through the basal ganglia before projecting back out to the cortex by way of the thalamus [26].

This region is the primary source of the neurotransmitter dopamine in the brain. The dopaminergic neurons, named after the neurotransmitter they release when firing, are located in the substantia nigra pars compacta of the basal ganglia and in the nearby ventral tegmental area (VTA). From these small structures the dopaminergic neurons project their axons widely throughout the brain, but primarily to striatum, where most of their input originates, and frontal cortical regions. The disorders mentioned above in relation to the basal ganglia are all disorders of dopamine function.

The basal ganglia's involvement in learning may come from neurons in this region whose activity seems related to the computational theory of reinforcement learning. Reinforcement learning is in turn based on observations of animal behavior at the cognitive level, thus completing a nice circle from observation of behavior, through computational theory, to observations of neural mechanisms that seem to be responsible for the behavior.

Reinforcement learning is of interest from the anticipatory learning viewpoint because the animal is learning to predict rewards that it will receive in the future in order to make decisions about actions in the present to obtain the best possible reward. Thorndike [49] originally formulated the principle of reward learning in

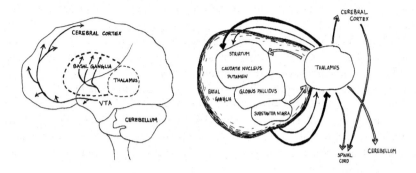

Fig. 2. *Left*: Dopaminergic neurons located in the substantia nigra of the basal ganglia and the adjacent ventral tegmental area project widely (arrows) across the brain. Dopamine release is associated with rewarding or pleasurable stimuli. Right: Information flows through the basal ganglia in segregated loops that pass through the thalamus and onward through sensory and motor cortical regions and also directly to motor neurons in the spinal cord. Connections within basal ganglia (not shown) maintain the segregation of the loops.

the following way, "Any act which in a given situation produces satisfaction becomes associated with that situation so that when the situation recurs the act is more likely than before to recur also." This principle was further elaborated on in the Rescorla-Wagner model of Pavlovian conditioning [38]. This is an error-correction model where the associative strength between conditioned stimuli and the unconditioned stimulus is increased only to the extent to which the US is unpredicted by the CS. Learning slows progressively as the US becomes more predicted by the appearance of the CS. The temporal difference (TD) learning rule [47] is an extension of this principle to the operant conditioning domain[2]. TD learning is also an error-correction model based on the animal's experiences. The TD rule is particularly powerful because learning can still take place even when the reward does not arrive every time the correction action is taken, and even when the reward arrives at an arbitrary delay after the correct action.

Temporal difference learning is algorithmically simple: $\Delta V(t) \propto V(t-1) - R(t)$, where t is time, V is the predicted reward, and R is the actual reward received. The ΔV term is the reward prediction error, and is used in a reinforcement learning algorithm to drive the system to change its action selection mechanism in a way that allows the learner to obtain more reward over time.

Dopaminergic neurons have activity which can be related to the reward prediction error of TD learning [41,50]. Dopamine is released phasically with rewarding stimuli; that is, midbrain neurons release dopamine with short latency after the rewarding stimulus, and this increase in dopamine release persists for only a

---

[2] In Pavlovian conditioning reward always occurs after the CS is presented, whereas in operant conditioning the animal must make a correct action or it will receive no reward.

short time. The release of dopamine does not appear, however, to be just related to reward or pleasure [29]. The phasic dopamine signal has several properties that make it a candidate for carrying a TD-style reward prediction error, as is shown in figure 3. Dopamine is released when the animal receives an unexpected reward, which would correspond to a positive reward prediction error. Dopamine is no longer released for a rewarding stimulus once the animal has learned to expect reward in a particular situation — the reward would be fully predicted in this case and the prediction error should be zero. Finally, if the animal does not receive the reward when it expects to receive it, then there is a depression of dopamine release, which is consistent with the negative prediction error that would occur in that situation.

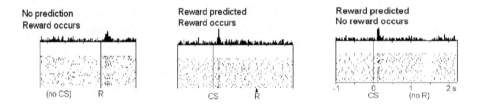

Fig. 3. Rastergrams showing how the activity of dopaminergic neurons in VTA is similar to the reward prediction error of TD learning. VTA neurons respond strongly if the animal receives an unpredicted reward (*left*). When the animal has learned that CS predicts reward then VTA neurons respond strongly at the CS and not at the US (*middle*). When the animal does not receive the reward as expected, the firing rates of VTA neurons depresses at the time of the expected reward (*right*). Each dot is an action potential fired by the neuron being recorded, and each row represents one task trial. Time is on the x-axis. The top of each graph is a histogram showing the number of action potentials that occurred over all trials at each time point. This figure is modified from http://scholarpedia.org/article/Reward_Signals

The dopaminergic system is also an interesting candidate for carrying a reward prediction error because it projects widely across many brain areas in a uniform fashion. Dopamine producing neurons fire synchronously and do not have any systematic differences in the locations to which they project. Therefore the dopamine signal reaches many brain regions with no systematic difference in the signal from brain region to brain region. This is consistent with usefulness of a reward prediction error in any situation in which the animal is trying to maximize the amount of reward it receives.

However, not everyone believes that the short-latency dopamine signal is a reward prediction error. An alternative hypothesis is that the short-latency dopamine response is involved in novelty-detection and attentional mechanisms. It has been suggested that the initial burst of dopaminergic firing could represent an essential component in the process of switching attentional and behavioral selections to unexpected, behaviorally important stimuli [36]. This fits in with the role of dopamine dysfunction in Attention Deficit Hyperactivity Disorder. Note that this theory of function is also of interest from an anticipatory learning standpoint; one possible mechanism that can be used in deciding which stimuli are novel and worth giving attentional resources is to pick ones which cannot be predicted [8,40,18].

If we accept the hypothesis that the short-latency dopamine response is a reward prediction error, it still raises the question of how this signal is being used to produce behavior. One mechanism that is suggested by the analogy of dopamine and TD error is the use of an actor-critic architecture [47]. In this form of model the critic learns to predict the reward structure of the environment and produces a learning signal used by a separate actor to learn what actions will produce the most reward over time. The exact form and location of the actor and critic components has been the subject of several theories [13,16,35,31,12]. But most of these models point at the probable involvement of the striatum because it is both the largest input to and largest output from the dopaminergic neurons.

There are also neural activity correlates of reward terms in several areas outside the basal ganglia [42,43] including orbitofrontal cortex, prefrontal cortex, cingulate cortex, perirhinal cortex, parietal cortex, premotor cortex, frontal and supplementary eye fields, and the superior colliculus. This list includes regions involved in executive function, sensory processing, and even motor control. Many of these regions receive substantial dopaminergic projections. There is a broad range of reward-related activity in these regions, including stronger activity when rewarded, stronger activity when the reward is more certain, and a ramp-up of activity during the waiting period culminating in highest activity when rewarded. It is likely that all of these signals are useful in different ways in producing adaptive behavior that maximizes reward. It is also likely that there are multiple, parallel systems that use these signals just as there are multiple, parallel locations that carry this reward information.

The basal ganglia are also hypothesized to be involved in the problem of action sequencing. Neurons in basal ganglia show activity that encodes serial movement order, for instance when a rat is engaged in grooming actions that have a natural sequence [1]. Actor-critic models of the basal ganglia, such as those discussed above, possess the useful computational ability to learn action sequences that lead to reward at a time much later than the action. Interestingly, the interval timing task discussed in the cerebellum section can also be seen as an action sequencing problem. There is evidence that the basal ganglia are involved in interval timing performance as well [3]: patients with dopamine disorders have problems with interval timing tasks, and the basal ganglia are hypothesized to produce interval timing through coincidence detection of the relative phases of many oscillators located in different brain regions. This close relationship between the functions of basal ganglia and cerebellum in learning, timing, and motor control has been noted by others [6,15].

4   Hippocampus - Episodic Memory

Episodic memory can be defined as the integration of sensory events ("what"), over time ("when"), and space ("where") [11]. In humans, episodic memory is autobiographical in nature, a memory of personal experience instead of just the dry knowledge of a fact. The medial temporal lobe, which includes the hippocampus and adjacent cortical structures (see figure 4), is necessary for the acquisition of episodic memories, as demonstrated by the deficits of patients that have damage to this brain area [44], and by imaging studies [46,30].

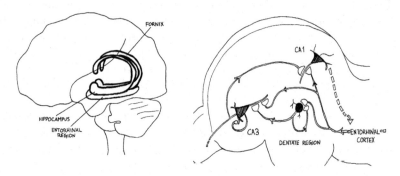

Fig. 4. *Left:* A cutaway of the medial temporal lobe including the hippocampus. Inputs from all over the brain converge onto the entorhinal region and from there enter the hippocampus. *Right:* A schematic of the hippocampal circuit. Input from the entorhinal cortex projects via the perforant path to dentate, CA3 and CA1 hippocampal subfields. Another pathway, the trisynaptic loop, projects from the entorhinal cortex to dentate to CA3 to CA1 and then back to the entorhinal cortex. Self-excitatory loops exists in both CA3 and dentate. The overall effect is that a great variety of signals pass through nested loops of various lengths — thus integrating different information streams over several time scales.

Navigation can be seen, in this sense, as a special case of a more general episodic memory process, where sequences (when) of egocentric perception (what) are converted into an allocentric map (where) of the environment [37]. The hippocampus is known to be important for navigation ability in both humans [21] and rodents [33], which suggests both that navigation and episodic memory rely on the same processes, and that animals may constitute a valid model for episodic memory research (on this topic, see also [5]).

One of the most interesting and well-studied neural correlates of episodic memory is the place cell. In the rodent hippocampus there are excitatory pyramidal neurons whose firing is highly correlated with the animal being at a particular location in the environment [32]. Such cells fire action potentials at rates up to 40Hz when near the location they respond to, tapering off their firing rates gradually while moving away from that location, often in a Gaussian-like manner.

This Gaussian-like relationship between place and firing rate of a cell is known as a place field (see figure 5). When the animal is far from the center of the place field the cell may be completely silent or at most fire a few handfuls of action potentials over very long periods of time.

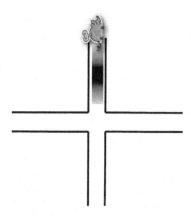

Fig. 5. A schematic example place field of a hippocampal neuron. When the animal is in a particular location in the maze the cell has high firing rate (dark). This firing rate drops away in a gradual fashion as the animal is further from the center of the place field (lighter color in the field map). Elsewhere the cell is totally silent.

Place cells also have several interesting characteristics from the standpoint of anticipatory learning. Such cells have a phase relationship with the theta rhythm that contains information about whether the animal is just approaching, at the center of, or departing from the place field [34]. The theta rhythm is an oscillation of 7-12 Hz observed in electro-encephalograms of the hippocampus. When a place cell in the hippocampus begins to fire action potentials as the rat enters the place field, it begins consistently at a particular phase relationship to the theta rhythm. Then as the rat traverses from the leading to the trailing edge of the place field, the phase shifts progressively forward on each theta cycle. It does so in a way that suggests it is actually location and not time that determines the phase relationship. Thus place cell firing contains not just information about how close the animal is to the center of a particular place cell, but also information about whether the animal is approaching the center of the field (a predictive encoding) or has already crossed the center and is departing the place field. There is evidence that phase relationships are indeed a fairly general mechanism used by several brain regions to encode the serial order of relationships [20]. Serial order encoding of stimuli is a necessary pre-condition before it is possible to produce behavior that anticipates a sequence.

Another interesting neural correlate of episodic memory that has a strong anticipatory flavor are hippocampal cells that have place fields only if the animal will take a particular path in the future, but have no firing field if the animal

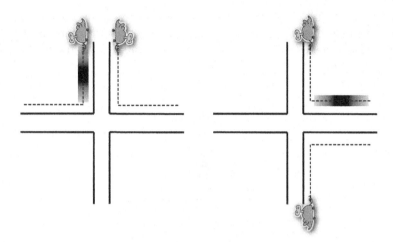

Fig. 6. A schematic example of a retrospective (left) and prospective (right) hippocampal place field. When the animal is in a particular location in the maze the cell has a place field only if the animal has taken (retrospective, left) or will take (prospective, right) a particular path. Otherwise the cell is silent.

will take a different path [7,52,10]. A place field whose activity is dependent on the future path in this fashion is said to have prospective coding. There are also cells whose place fields exist only if the animal has arrived at the field location via a particular path, but not when another path is used [7,10]. A place field whose activity is dependent on the past in this fashion is said to have retrospective coding. A schematic example of such a firing field can be seen in figure 6. Somewhere between half [7] and two-thirds [52] of hippocampal place cells have prospective or retrospective coding.

The existence of prospective and retrospective place fields implies that the hippocampus contains information not just about what, where and when, but also information that associates different wheres and whens together, which is an important ability for an anticipatory learning system. One of the most convincing elements establishing prospective activity as an observable correlate of an entire behavioral episode, is the discovery [7] of cells that fire in a prospective coding for one future path even if the animal makes a momentary error and begins traveling down the incorrect (non-coded) pathway before correcting itself and taking the correct path. This demonstrates that the prospective firing field is related to the whole path through the maze (the intended behavioral episode) and not the immediate decision that is about to be made at the next junction.

Why is it the hippocampus, and not some other brain region, that seems to be responsible for forming episodic memory? The unique anatomy and connectivity of the hippocampus is the likely mechanism by which episodic memories are formed [19,9]. The region receives many different inputs from both unimodal and polymodal sensory areas as well as higher association areas of the cortex. All this information is merged and flows through several nested loops of different time-scales in the hippocampus. The hippocampus then projects broadly back out to many regions of the cortex. This convergence, looping, and divergence of information, seen in figure 4, is a mechanism that can form associations across space and time.

Perhaps because of its central role in memory, the hippocampus is also involved in a variant of the eyeblink conditioning task known as trace conditioning [48]. Trace eyeblink conditioning is different from regular eyeblink conditioning in that there is a delay period between CS and US, such that the tone is off (between 500ms and tens of seconds) before the air puff is delivered. The lack of stimuli predicting the noxious US in the delay period makes this a more memory-intense version of eyeblink conditioning. Rabbits with hippocampal lesions are unable to acquire trace conditioning and rabbits that have been trained and then lesioned lose the conditioning. Additionally, there are strong neural correlates of trace conditioning in the hippocampus; around one quarter of hippocampal excitatory neurons fire maximally when the US should appear even if it is omitted [25].

5   Conclusion

This review has presented results demonstrating the involvement of three brain regions with anticipatory behavior. There are, however, other regions of the brain that are beyond the scope of this review that are probably involved with these same anticipatory behaviors. In addition, the regions covered in this review have neural correlates of other anticipatory behaviors that were not discussed here. The references discussed in the previous sections can provide the reader with an entry point into this literature.

Although each brain region was presented at the beginning of the review as being associated with a particular form of anticipatory behavior, there were in fact considerable overlaps that were pointed out at various points in the text. The hippocampus is involved in some forms of eyeblink conditioning, as well as being required for episodic memory. Basal ganglia are involved in timing, as well as producing a primary learning signal. The cerebellum is involved in learning, as well as being important for timing of motor actions. This should give the reader a feeling for behavioral neuroscience research and an understanding of why a point was made earlier about the need to accumulate evidence from multiple sources to formulate theories of function.

Although it is still not possible to fully understand the mechanisms that produce even some fairly simple behaviors in mammals, this review did illuminate a principle that could be of use in designing anticipatory learning systems. All three of these brain regions have unique anatomy that is very likely the mechanism by which they compute the functions proposed in this paper.

– The cerebellar circuit has a two-input, center excitation, surround inhibition, multi-layer structure that undoubtedly plays a large role in its ability to process information about millisecond level timings.

- The basal ganglia contains multiple segregated loops that may allow it to use the reward predictions it generates to select an appropriate action from many possibilities.
- The hippocampal circuit contains multiple nested loops of varying time-scales that help to integrate information over time to produce memories.

Hopefully computational theorists in the anticipatory learning community will be able to draw on these principles to create useful adaptive systems, and will in doing so develop an abiding interest in producing biologically-grounded models of how these brain structures produce adaptive behavior.

Acknowledgments

This work was supported by the Neurosciences Research Foundation. Many thanks to Amber McCartney for producing figures 1, 2, & 4, to John Iversen and Jeff McKinstry for valuable discussions that helped me organize this review, and to the anonymous reviewers for their useful suggestions.

References

1. Aldridge, J.W., Berridge, K.C.: Coding of serial order by neostriatal neurons: a natural action approach to movement sequence. J Neurosci 18, 2777–2787 (1998)
2. Bastian, A.J.: Learning to predict the future: the cerebellum adapts feedforward movement control. Curr. Opin. Neurobiol 16, 645–649 (2006)
3. Buhusi, C.V., Meck, W.H.: What makes us tick? functional and neural mechanisms of interval timing. Nat. Rev. Neurosci 6, 755–765 (2005)
4. Butz, M.V., Siguad, O., Gerard, P.: Anticipatory behavior: Exploiting knowledge about the future to improve current behavior. In: Butz, M.V., Sigaud, O., Gérard, P. (eds.) Anticipatory Behavior in Adaptive Learning Systems. LNCS (LNAI), vol. 2684, pp. 1–10. Springer, Heidelberg (2003)
5. Clayton, N.S., Bussey, T.J., Dickinson, A.: Can animals recall the past and plan for the future? Nat. Rev. Neurosci 4, 685–691 (2003)
6. Doya, K.: Complementary roles of basal ganglia and cerebellum in learning and motor control. Curr. Opin. Neurobiol 10, 732–739 (2000)
7. Ferbinteanu, J., Shapiro, M.L.: Prospective and retrospective memory coding in the hippocampus. Neuron 40, 1227–1239 (2003)
8. Fleischer, J., Marsland, S., Shapiro, J.: Sensory anticipation for autonomous selection of robot landmarks. In: Butz, M.V., Sigaud, O., Gérard, P. (eds.) Anticipatory Behavior in Adaptive Learning Systems.Foundations,Theories, and Systems. LNCS (LNAI), vol. 2684, Springer, Heidelberg (2003)
9. Fleischer, J.G., Gally, J.A., Edelman, G.M., Krichmar, J.L.: Retrospective and prospective responses arising in a modeled hippocampus during maze navigation by a brain-based device. Proc. Natl. Acad. Sci. USA 104, 3556–3561 (2007)
10. Frank, L.M., Brown, E.N., Wilson, M.A.: Trajectory encoding in the hippocampus and entorhinal cortex. Neuron 27, 169–178 (2000)
11. Griffiths, D., Dickenson, A., Clayton, N.: Episodic memory: what can animals remember about their past? Trends in Cognitive Science 3, 74–80 (1999)

12. Gurney, K., Prescott, T.J., Wickens, J.R., Redgrave, P.: Computational models of the basal ganglia: from robots to membranes. Trends in Neurosciences 27, 453–459 (2004)
13. Houk, J., Adams, J., Barto, A.: A model of how the basal ganglia generate and use neural signals that predict reinforcement. In: Models of Information Processing in the Basal Ganglia, MIT Press, Cambridge (1995)
14. Ivry, R.B., Keele, S.W., Diener, H.C.: Dissociation of the lateral and medial cerebellum in movement timing and movement execution. Exp. Brain Res. 73, 167–180 (1988)
15. Ivry, R.B., Spencer, R.M.C.: The neural representation of time. Curr. Opin. Neurobiol 14, 225–232 (2004)
16. Joel, D., Niv, Y., Ruppin, E.: Actor-critic models of the basal ganglia: new anatomical and computational perspectives. Neural Networks 15, 535–547 (2002)
17. Kandel, E., Schwartz, J., Jessell, T. (eds.): Principles of Neural Science, 4th edn. McGraw-Hill, New York (2000)
18. Kaplan, F., Oudeyer, P.Y.: Maximizing learning progress: An internal reward system for development. In: Iida, F., Pfeifer, R., Steels, L., Kuniyoshi, Y. (eds.) Embodied Artificial Intelligence. LNCS, vol. 3139, pp. 259–270. Springer, Heidelberg (2004)
19. Krichmar, J., Seth, A., Nitz, D., Fleischer, J., Edelman, G.: Spatial navigation and causal analysis in a brain-based device modeling cortical-hippocampal interactions. Neuroinformatics 3, 197–221 (2005)
20. Lisman, J.: The theta/gamma discrete phase code occuring during the hippocampal phase precession may be a more general brain coding scheme. Hippocampus 15, 913–922 (2005)
21. Maguire, E.A., Woollett, K., Spiers, H.J.: London taxi drivers and bus drivers: A structural MRI and neuropsychological analysis. Hippocampus 16, 1091–1101 (2006)
22. Marr, D.: A theory of cerebellar cortex. J. Phsysiol. 202, 437–470 (1969)
23. Mauk, M.D., Buonomano, D.V.: The neural basis of temporal processing. Annu. Rev. Neurosci 27, 307–340 (2004)
24. McCormick, D.A., Thompson, R.F.: Neuronal responses of the rabbit cerebellum during acquisition and performance of a classically conditioned nictitating membrane-eyelid response. J Neurosci 4, 2811–2822 (1984)
25. McEchron, M.D., Tseng, W., Disterhoft, J.F.: Single neurons in ca1 hippocampus encode trace interval duration during trace heart rate (fear) conditioning in rabbit. Journal of Neuroscience 23, 1535–1547 (2003)
26. McHaffie, J.G., Stanford, T.R., Stein, B.E., Coizet, V., Redgrave, P.: Subcortical loops through the basal ganglia. Trends Neurosci 28, 401–407 (2005)
27. Miall, R.C., Reckess, G.Z.: The cerebellum and the timing of coordinated eye and hand tracking. Brain Cogn. 48, 212–226 (2002)
28. Miall, R.C., Reckess, G.Z., Imamizu, H.: The cerebellum coordinates eye and hand tracking movements. Nat. Neurosci 4, 638–644 (2001)
29. Montague, P.R., Berns, G.S.: Neural economics and the biological substrates of valuation. Neuron 36, 265–284 (2002)
30. Moskovitch, M., Nadel, L., Winocur, G., Gilboa, A., Rosenbaum, R.S.: The cognitive neuroscience of remote episodic, semantic and spatial memory. Current Opinion in Neurobiology 16, 179–190 (2006)
31. O'Doherty, J., Dayan, P., Schultz, J., Deichmann, R., Friston, K., Dolan, R.J.: Dissociable roles of ventral and dorsal striatum in instrumental conditioning. Science 304, 452–454 (2004)

32. O'Keefe, J., Dostrovsky, J.: The hippocampus as a spatial map. preliminary evidence from unit activity in the freely-moving rat. Brain Research 34, 171–175 (1971)
33. O'Keefe, J., Nadel, L.: The Hippocampus as a Cognitive Map. Clarendon Press, Oxford (1978)
34. O'Keefe, J., Recce, M.L.: Phase relationship between hippocampal place units and the eeg theta rhythm. Hippocampus 3, 317–330 (1993)
35. O'Reilly, R.C., Frank, M.J.: Making working memory work: a computational model of learning in the prefrontal cortex and basal ganglia. Neural Computation 18, 283–328 (2006)
36. Redgrave, P., Prescott, T.J., Gurney, K.: Is the short-latency dopamine response too short to signal reward error? Trends in Neurosciences 22, 146–151 (1999)
37. Redish, A.D.: Beyond the cognitive map: From place cells to episodic memory. MIT Press, Cambridge (1999)
38. Rescorla, R., Wagner, A.: A theory of Pavlovian conditioning: Variations in the effectiveness of reinforcement and nonreinforcement. In: Classical Conditioning II. Appleton-Century-Crofts (1972)
39. Roberts, A., Robbins, T., Weiskrantz, L. (eds.): The Prefrontal Cortex: Executive and Cognitive Functions. Oxford University Press, Oxford, UK (1998)
40. Schmidhuber, J.: Curious model-building control systems. In: IEEE Intl Joint Conf. on Neural Networks. vol.2, pp. 1458–1463 (1991)
41. Schultz, W., Dayan, P., Montague, P.R.: A neural substrate of prediction and reward. Science 275, 1593–1599 (1997)
42. Schultz, W.: Getting formal with dopamine and reward. Neuron 36, 241–263 (2002)
43. Schultz, W.: Neural coding of basic reward terms of animal learning theory, game theory, microeconomics and behavioural ecology. Curr. Opin. Neurobiol 14, 139–147 (2004)
44. Scoville, W.B., Milner, B.: Loss of recent memory after bilateral hippocampal lesions. J. Neurochem, 20 (1957)
45. Shadmer, R., Wise, S.: The computational neurobiology of reaching and pointing: a foundation for motor learning. MIT Press, Cambridge (2005)
46. Squire, L.R., Stark, C.E.L., Clark, R.E.: The medial temporal lobe. Annu. Rev. Neurosci 27, 279–306 (2004)
47. Sutton, R.S., Barto, A.G.: Reinforcement learning: an introduction. MIT Press, Cambridge (1998)
48. Thompson, R.F.: In search of memory traces. Annual Review of Psychology 56, 1–23 (2005)
49. Thorndike, E.: Animal Intelligence. Macmillan Press (1911)
50. Waelti, P., Dickinson, A., Schultz, W.: Dopamine responses comply with basic assumptions of formal learning theory. Nature 412, 43–48 (2001)
51. Wolpert, D.M., Ghahramani, Z.: Computational principles of movement neuroscience. Nat. Neurosci 3, 1212–1217 (2000)
52. Wood, E.R., Dudchenko, P.A., Robitsek, R.J., Eichenbaum, H.: Hippocampal neurons encode information about different types of memory episodes occurring in the same location. Neuron 27, 623–633 (2000)

The Role of Anticipation in the Emergence of Language

Samarth Swarup[1] and Les Gasser[1,2]

[1] Department of Computer Science,
[2] Graduate School of Library and Information Science,
University of Illinois at Urbana-Champaign,
Urbana, IL 61801, USA
{swarup,gasser}@uiuc.edu

Abstract. We review some of the main theories about how language emerged. We suggest that including the study of the emergence of artificial languages, in simulation settings, allows us to ask a more general question, namely, *what are the minimal initial conditions for the emergence of language?* This is a very important question from a technological viewpoint, because it is very closely tied to questions of intelligence and autonomy. We identify anticipation as being a key underlying computational principle in the emergence of language. We suggest that this is in fact present implicitly in many of the theories in contention today. Focused simulations that address precise questions are necessary to isolate the roles of the minimal initial conditions for the emergence of language.

1 What is the Problem of Language Emergence?

It is very hard to imagine what life would be like without language. Before some point in our evolutionary history, however, our ancestors did not have language. How did language (and the capacity for it) evolve? This is the problem of language emergence.

The emergence of language is considered to be the last major transition in evolution [44]. It is one of the clearest distinctions between humans and other animals, and speculations about the origins of language go back to Plato's *Cratylus* dialogue, which discusses the connection between names and things. What makes the subject particularly difficult is the lack of data about the earliest languages. Despite this lack, the publication of Darwin's works on evolution lead to a great deal of speculation on possible scenarios for the evolution of language. This led the *Société de Linguistique de Paris*, when it was formed in 1865, to declare in its bylaws that it would not accept any communications dealing with the origin of language. A similar statement was made by the Philological Society of London in 1873 [31].

In the last fifty years or so, however, the question has again gained scientific validity due to relevant discoveries in archeology, anthropology, and neuroscience. A lot more is now known about the biology, environment, and lifestyles of the early *homo* species. This has lead to a renewed spate of theories about the origins

M.V. Butz et al. (Eds.): ABiALS 2006, LNAI 4520, pp. 35–56, 2007.

of language. It is the aim of this article to review some of the main contenders, and to address in particular the role of anticipation in the emergence of various aspects of language.

We start by describing the problem of the emergence of human language. Then we suggest that by expanding this question to ask what the minimal initial conditions for the emergence of language are, we can build a more general theory which will provide us with a better understanding of how to design systems that can create their own language. We lay out a space of communication systems, and analyze how the notion of anticipation can be used to build a framework to study the movement from simple to progressively more complex communication systems. After that we examine some of the main theories of the emergence of human language, and some of the work in artificial language evolution (through simulation). We find that these have been addressing different regions of the communication systems space. However, we can use these to infer some basic conditions for the emergence of various kinds of communication systems already, and by building an anticipatory framework we can provide the scaffolding for further simulations that will deal with more complex forms of language.

1.1   What Form Does an Answer to This Problem Take?

Theories of the origins of language address two questions: how language evolved, and why. Broadly, the answers to how language evolved consist of speculations on the mechanisms, or *preadaptations*, that made language possible, and the stages that lie between animal-like signaling and modern human language. The answers to why language evolved consist of speculations on the functional properties of language, environmental conditions, and selection pressures that gave language an adaptive advantage.

There are a couple of important points to remember here. First, the various proposals for why language evolved are not mutually exclusive. Indeed it is likely most of these contributed to the selection pressure for the evolution of language. Box 1 summarizes the main ideas about why language evolved.

Second, any postulated preadaptations for language must be selected for in their own right. This means that we cannot suppose that some preadaptation emerged in order to make language possible. Evolution does not proceed according to some pre-specified program, and therefore such a suggestion would violate causality.

An example of a preadaptation is the change in the shape and robustness of the jaw which made possible, as a side effect, the production of the range of speech sounds we enjoy today. This happened when *homo ergaster* moved from the arboreal habitats occupied by the australopithecines to a more open savannah habitat. This led to a change in diet from being predominantly vegetarian to incorporating more animal-based products. This in turn led to the change in the shape and robustness of the jaw [1].

Box 1: Functional Scenarios for the Evolution of Language
Johansson provides a nice overview of the various scenarios that have been proposed to have provided the selection pressure for the emergence of language [29]. We list them here.

1. Hunting, which leads to a pressure for a language to be able to cooperate.
2. Tool-making which, arguably, lead to an increase in intelligence, and provided the mental capabilities required for language (such as combinatoriality).
3. Sexual selection, such as a preference for more articulate mates, or sexual conflict as a driving force, or because better communicative ability can lead to social/political power.
4. Child-care and teaching, which leads to a pressure for a language for teaching.
5. Social relations in groups and tribes:
   - Predation, perhaps for coordination for group defense.
   - Inter-group competition.
   - Intra-group competition for resources.
   - Mating opportunities.
   - Intra-group aggression and politics, such as alliance-formation, negotiation, etc.
6. Children at play, where language may have appeared through mimicry, for example.
7. Music.
8. Story-telling.
9. Art.

A *complete* theory of the emergence of human language would need to answer at least the following questions:

- Why have only humans developed language?
- Is it due to a difference in degree, or a difference in kind?
- How much of language is innate, and how did it become so?
- Did language emerge gradually, and if so what did earlier forms of language look like?

Why have only humans developed language? Szathmáry has suggested that there can be two possible reasons for the uniqueness of an adaptation: it might be *variation-limited* or *selection-limited*. Being variation-limited means that the necessary mutations occur extremely rarely. Being selection-limited means that they only confer a selective advantage in extremely rare conditions [29, quoted]. Hurford has pointed out, however, that just because other species have not developed language does not mean they will not [27]. Language has only emerged in the last 100,000 to 500,000 years [1], which is a short while for evolution. Every major evolutionary transition must have had a vanguard - a species that was the first to achieve it, and solely enjoyed its benefits until the other species caught up.

Is it due to a difference in degree, or a difference in kind? A counterargument to many of the scenarios listed in box 1 is that other species also exhibit behavior of that kind. Why have they not developed language? Hunting, for example, is a very common activity in the animal kingdom. Even cooperative hunting, which is proposed to have provided the selective pressure for communication, is quite common. The question, then, is, are these viable propositions? Is a difference in the degree to which we engage in some activity, for example our increased period of childhood, or our increased social group size, sufficient to explain why language evolved? Or is there a different *kind* of activity we engage in, that other species do not, that led to the evolution of language? The same question holds for our cognitive capabilities. Does the emergence of language require some special cognitive capability that other animals lack, or is it that we are just better at (some aspects of) cognition?

How much of language is innate, and how did it become so? It is hard to argue that there are not at least some aspects of language which are innate. The capacity for symbolization is probably innate. Furthermore, children can acquire a grammatical language even if the linguistic input they receive is not grammatical, as in the emergence of creole languages from pidgins, and in the famous example of the Nicaraguan Sign Language, where a community of deaf children in a school in Nicaragua invented a grammatical sign language based on the pidgin-like *Lenguaje de Signos Nicaragüense* that they were exposed to at home [14]. This does not necessarily mean that grammar is innate, however. For one thing, the development of a creole seems to depend on the size of the community. If the community is not large enough, a grammatical language does not emerge.

The idea that we might have an innate language acquisition device (or a universal grammar), which appeared by means other than natural selection, was first proposed by Chomsky [12]. It has been extremely controversial [35], and in one of his most recent articles, he (with Hauser and Fitch) proposes that the only aspect of grammar that is innate is the ability to do recursion [25]. This proposal, also, has generated debate [36].

Did language emerge gradually, and if so, what did earlier forms of language look like? An idea that seems to find general agreement is Bickerton's proposal of a *protolanguage* [6]. A protolanguage is basically modern language without the rich syntax. It is compositional, that is, it consists of words that are strung together into sentences, but it does not have properties such as tense and aspect. It is also supposed to have a closed (that is, fixed) vocabulary. Bickerton has proposed that modern language was preceded by protolanguage, which may have existed for as many as a million years before modern language appeared, and further, that protolanguage still makes its appearance in pidgins, and in some aspects of language acquisition (a twist on "ontogeny recapitulates phylogeny"). Jackendoff has expanded on the idea of a protolanguage, suggesting several different stages. These are summarized in box 2. Johansson provides a nice summary of all the "protos" that make up protolanguage: proto-speech, proto-gestures, proto-semantics, and proto-syntax [30].

Box 2: Proto-language
Jackendoff has postulated the following stages in the evolution of the language capacity [28]. Bickerton's proposed protolanguage [6] is subsumed in this sequence.

1. The use of symbols in a non-situation-specific fashion.
2. An open, unlimited class of symbols.
3. A generative system for single symbols: proto-phonology.
4. Concatenation of symbols to build larger utterances.
5. Using linear position to signal semantic relationships.
6. Phrase structure.
7. Vocabulary for relational concepts.
8. Beyond phrase structure: inflection and further syntax.

Computer scientists have only recently become interested in the question of language emergence, partly because we believe that some of these issues can be addressed through agent-based simulation. However, we believe that in this case the appropriate question is slightly different, and more general.

2 A Modified Question: What Are the Minimal Initial Conditions for the Emergence of Language?

Part of the problem with trying to explain the emergence of language is that language is unique. No other species has evolved language, and so any explanation is going to be a "just so" story. However, when we include *artificial* language evolution[1] in the mix, we can ask the more general question, *what are the minimal initial conditions for the emergence of language?* One important thing to keep in mind is that the minimal conditions are not just cognitive, but also environmental. Another way of asking the same question is, what mechanisms/conditions do we need to design/provide to enable the emergence of language in a population of machines?

This is an important question because, besides being an important scientific problem, the study of the emergence of language is also very important from a technological perspective. Multi-agent systems are becoming increasingly widespread, being used in widely differing contexts such as spacecraft control, military mission scheduling, auctions, agent-based models of social networks and

---

[1] Note that by artificial languages, we do not mean those constructed by humans, such as Esperanto and Klingon. Rather, we are referring to attempts to evolve a language in a population of (simulated or real) agents.

organizations, etc. The general approach to communication and coordination in multi-agent systems is to pre-impose a designed language. However, such pre-defined languages are often found to be inadequate, especially as multi-agent systems increase in size and complexity, as they reflect the designer's viewpoint rather than the agents', and are unable to adapt to changing environmental conditions and task definitions. It is much more desirable for the agents to be able to create and maintain their own language.

The last decade has seen increasing application of computational and mathematical methods to the study of language evolution (see [55] for a recent review). This has led to important advances on questions such as how a shared language is established in a population [15], [32], [47], the emergence of syntax [34], and symbol grounding [41], [51]. However, we are still far from a general theory.

2.1   What Form Does an Answer to the Modified Question Take?

In order to construct such a general theory, we need to map out the space of possible communication systems, and analyze the factors that lead to the emergence of these. Figure 1 shows one possible way to lay out this space, and where some commonly considered communication systems in the language evolution literature would lie in this space.

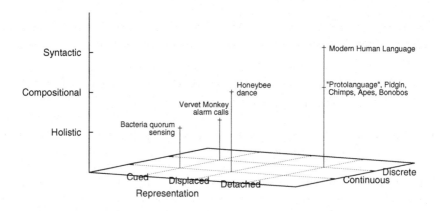

Fig. 1. The space of communication systems

Continuous and discrete communication systems are distinguished along the x-axis. A continuous communication system is one that uses the magnitude of some quantity to communicate information. For example, scout honeybees communicate the quality of a food-site they have discovered by the vigor of their waggle dance [20]. Bacteria do quorum sensing by producing molecules called autoinducers [5], [37]. Quorum sensing is the control of gene expression in response to cell density. This means, for example, that bacteria will often not express virulence factor until their colony is big enough to have a high probability of successfully infecting the host.

The y-axis in our graph distinguishes the kind of representation used by the communication system. The notion of cued and detached representations is due to Gärdenfors [22]. In his own words, "A *cued* representation stands for something that is present in the current external situation of the representing organism. When, for example, a particular object is categorized as food, the animal will then act differently than if the same object had been categorized as a potential mate... In contrast, *detached* representations may stand for objects or events that are neither present in the current situation nor triggered by some recent situation." Since the honeybee representation of food sources or nest sites seems to fall in-between cued and detached, we include *displaced* representations in our space of communication systems. By displaced representation, we mean representations which stand for objects or events that are not in the current situation, but have been triggered by some recent situation. The notion of displacement is one of Hockett's design features of language [26], but he uses the term to mean anything other than cued representations[2]. We think it is important to make these distinctions because they imply different computational properties of the underlying cognitive system. Deacon talks about a similar taxonomy of kinds of symbols: *iconic*, *indexical*, and *true symbols* [17]. Icons physically resemble that which they represent, for example onomatopoeic words like "pitter-patter". Indices involve correlations between the symbol and the referent, for example a symptom and a disease. True symbols, in contrast to the other types, are entirely arbitrary. For example, the word "chair" does not tell us anything about the (kind of) object to which it refers.

The z-axis in our graph distinguishes various levels of structure that might be present in the communication system. The simplest kind of communication in this sense is *holistic*, where every "meaning" or concept has an independent symbol associated with it. There is no relation between the symbols, and no internal structure to them. The classic example is the alarm call system of vervet monkeys [43]. Vervet monkeys have different calls for flying predators like eagles, and ground predators like snakes and leopards. These calls have no relation to each other. They do not, for example, have a common component that means "predator".

A significantly more complex form of structural organization is *compositional* language. This means that utterances are composed of meaningful parts, which then combine meaningfully. For example, "green ball" means not just that there is something green and something that is a ball, but that it is the ball that is green. The language capacity of chimps, bonobos, and apes seems to be at this level [42].

Finally, the most complex form of structural organization we know is modern human language with its rich syntax.

The communication space gets increasingly complex along each axis. Thus the simplest communication system is continuous, cued, and holistic, such as the quorum sensing of bacteria, and the most complex is discrete, detached, and

---

[2] Note that what we call displaced, Gärdenfors would probably include under cued, and what Gärdenfors calls detached, Hockett would include under displaced.

syntactic, of which the only known example is modern human language. We will see next that this space correlates well with different kinds of anticipation. This brings up the central question of this article.

3   What is the Role of Anticipation in the Emergence of Language?

Anticipation is widely considered to be a very important component of cognition. The notion of anticipation is closely related to prediction or expectation. To put it in a sentence, expectation is knowledge about the future, and anticipation is what you do with it. Robert Rosen has defined an anticipatory system to be one that has an internal model of itself and/or its environment, which it uses in planning (or action selection) [40]. The "model", of course, can be of varying degrees of complexity, from simple stimulus-response to systems with complex internal states. The most famous experimental demonstration of anticipatory behavior is Pavlov's dog, which learned to anticipate food at the sound of a bell, and showed this by starting to salivate.

We argue below that anticipation provides a very nice framework for studying cognitive requirements for language, because it is correlated with language: the more sophisticated the anticipatory behavior exhibited by a population, the more complex their communication system is.

Butz et al. have described four kinds of anticipation in relation to adaptive behavior [7]:

- Implicit anticipatory behavior
- Payoff anticipatory behavior
- Sensory anticipatory behavior
- State anticipatory behavior

Implicit anticipation corresponds to the situation where the agent is not explicitly computing expectations, but still exhibits some anticipatory behavior. The anticipation, in this case, has been carried out by evolution (or the designer, for artificial agents), by equipping the agent with a genome that will "work well" in its environment. There is no learning beyond that done through evolution, since learning is essentially equivalent to prediction. Bacteria are among the simplest kinds of implicitly anticipatory agents, though admittedly they blur the distinction between learning and evolution through horizontal gene transfer [16].

Payoff anticipation consists of forming expectations of rewards for states of the environment, and utilizing these expected rewards during planning. The simplest kind of reinforcement learning, called model-free reinforcement learning [49], is an example of payoff anticipation because it computes a value function which is simply the expected cumulative discounted future reward for each state, and then the agent chooses actions which take it to states with high values. Honeybees could be considered to be payoff anticipatory agents, because it is unlikely that they have a predictive model of the environment in their heads that they use for planning. They also exhibit associative learning (that is, classical conditioning), which again requires payoff anticipation. They are also capable of learning some other things, however, such as landmarks and other cues that they use for navigation on their foraging trips [24]. Vervet monkeys are probably capable of more sophisticated anticipatory behavior, but the kind of anticipation required for their alarm call system is only payoff anticipation.

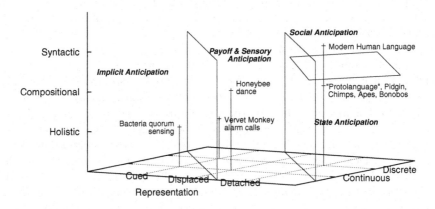

Fig. 2. The space of communication systems, partitioned by kinds of anticipation required

Sensory anticipation involves predictions that do not influence behavior directly, but only the sensory processing of the agent. It is strongly related to phenomena like priming, where a particular sensory input causes enhanced attention to a subsequent sensory input to the point where it can be hallucinated. This kind of anticipation is probably mediated by top-down connections in the sensory pathways, that carry predictions of expected sensory input. This kind of neural architecture is quite widespread, and probably honeybees as well as vervet monkeys are capable of sensory anticipation. It is thought that auditory anticipation is an important part of speech comprehension in humans, and so our linguistic capacity would probably be degraded if we were not capable of sensory anticipation. However, it probably is not one of the minimal initial conditions for language.

State anticipation involves having a detailed predictive model of the environment, which directly influences decision-making. Agents which have representations of goals, and perform mental simulations of actions to come up with a plan for reaching the goal from their current state are performing state anticipatory behavior. For example, a chimp that comes across a termite nest and then goes to find a stick from which it strips off the leaves, and then returns to the termite nest and uses the stick to fish out termites to eat is clearly performing state anticipatory behavior.

A fifth kind of anticipation, social anticipation, is now considered to be distinct from the other four. We believe that social anticipation is qualitatively different from other kinds of anticipation. In other kinds of anticipatory behavior, the environment is considered to be stationary. This means that, theoretically, environmental inputs are assumed to be generated by stationary distributions. Social anticipation, however, takes place in an "environment" consisting of other agents that are also performing social anticipation. This makes the environment non-stationary, which presents a much harder learning problem. We are not suggesting, however, that social anticipation could not have evolved from the previous kinds of anticipation. Unpacking how it might have done so however, is beyond the scope of this paper. It would involve, perhaps, anticipation in environments which are mostly stationary, for example where animals learn only occasionally or very slowly. It would also involve unpacking *degrees* of social anticipation. For instance, animals such as chimps and bonobos are capable of social anticipation to a degree perhaps, because they have to vie for dominance in their groups in order to achieve better feeding and mating opportunities. However, it is not clear to what degree they have a definite *theory of mind*, that is, a general understanding of how other members of their own species behave. It has been argued by several researchers that it is this theory of mind, or social intelligence, that sets us apart from other animals, and which brought about the emergence of language [2], [57].

Figure 2 shows how different kinds of anticipation partition the space of communication systems. In the figure, it should be understood that agents that are capable of a particular kind of anticipation are also capable of all the previous kinds of anticipation, but not subsequent ones. For example, honeybees are capable of implicit, payoff, and sensory anticipation, but not state or social anticipation.

Now, keeping this analysis in mind, we take a look at some of the theories and simulations of language emergence.

4   What Are Some of the Theories of the Emergence of Language?

There are three basic questions:

- Why/How did the capacity for symbolization evolve?
- Why/How did structured language (that is, compositionality and syntax) evolve?
- Why/How does a population converge upon a common language?

Various hypotheses have been put forward to answer one or more of these questions. We review some of the main ones below.

4.1 Natural Language Evolution

Anticipatory Planning. To our knowledge, Gärdenfors and his colleagues are the only ones who have explicitly invoked anticipation in theorizing about the emergence of language [22], [23], [4].

Gärdenfors and Osvath [23] have argued that Olduwan culture led to the evolution of anticipatory cognition, which in turn led to the emergence of symbolic communication. Olduwan culture is a term used to refer to the use of stone tools by pre-historic hominins[3], roughly in the period 2.6 to 1.5 million years ago. Furthermore, they claim that this sort of cognition is unique to humans. Once anticipatory cognition appeared, it made communication about future goals beneficial by enabling long-term planning.

Various animal species, particularly primates, are known to be capable of planning. However, they argue, the plans of other animals always address present needs, while humans are the only animals capable of planning for future needs, that is, *anticipatory* planning. For example, a chimp may go look for a stick on finding a termite nest, as mentioned earlier, however a chimp will not spend its time putting together a collection of sticks to carry around in anticipation of finding a termite nest.

The reason that Olduwan culture is presumed to have given rise to anticipatory planning is because it appeared at a time in our past when our hominin ancestors had to make the transition from a forest environment to the savannah. Food sources are much more scarce in the savannah, and this led to a number of changes. The dietary change and its effect on the jaw has already been mentioned. The scarcity of food also meant, however, that the hominins had to range farther in search of food. This made it beneficial to carry along tools (for dressing meat etc.) rather than trying to bring a carcass all the way back home. It might also have been a good idea to make caches of tools at various hunting locations, so that tools would not have to be carried everywhere. These kinds of behavior require anticipatory planning, though, because the hominin would have to fashion tools and carry them to caches without being cued by the need to hunt or scavenge. In other words, they would have to plan for anticipated *goals*.

This is a very sophisticated kind of planning. Butz et al. lay out their framework in the context of a partially-observable Markov decision process (POMDP). A POMDP contains a reward function, which essentially corresponds to a goal. An agent that is solving a POMDP is trying to maximize reward. The *anticipation* of goals can be interpreted in two ways.

The first approach is to think of the agent as having to solve several POMDPs, with different reward functions and possibly different state spaces. For example, one problem to solve would be finding food, another would be making tools, and so on. Over its lifetime, the agent encounters a series of POMDPs, which are related to each other. Anticipation of goals now corresponds to predicting what future reward functions will look like, based on the reward functions seen to date. This involves some kind of meta-learning or meta-cognition. In the

---

[3] *Hominin* refers to all the species of humans that ever evolved. The term *hominid* includes chimps and gorillas.

machine learning literature, this is most commonly referred to as transfer learning, and has been gaining increased attention [53], [50], though the focus is on using experience from past problems to solve new problems better (that is, more quickly, accurately, and robustly), rather than on predicting future goals and doing anticipatory learning.

The second approach is to think of the agent as having one *large* POMDP to deal with, where the goal is simply survival. The agent then has to decompose this POMDP into smaller POMDPs by discovering sub-goals. This is also a very hard problem and has been getting a lot of attention in the machine learning community [33], [48], [18].

It is not clear to what extent animals other than humans are capable of such sophisticated cognitive processes. It may be the case that other animals, such as the great apes, are capable of this *kind* of cognition, but to a lesser *degree*. They are, for example, capable of decomposing a relatively simple problem into a hierarchy of subgoals, such as pushing over a chair to a spot underneath a banana hanging from the ceiling so that they can climb up and reach the banana. But they certainly do not have the capabilities of humans, who can plan their entire careers and lives.

Gärdenfors et al. argue that once anticipatory planning appeared, it led to a selective pressure for evolving a means for cooperation about future goals, and that this led to the emergence of symbols. This is a very interesting hypothesis, which can be examined closely through simulations of agents with varying levels of anticipatory planning capabilities, perhaps by using some of the techniques cited above.

Social Intelligence. The key question for the emergence of structured language is, what cognitive preadaptation could provide the computational machinery for generating and processing highly structured language?

Several people have talked about social intelligence or "Machiavellian" intelligence as being the key factor in the emergence of structured language [57], [8], [11], [19].

Cheney and Seyfarth [11] point out that nonhuman primates do not seem to have a theory of mind. Their vocalizations may be intended to modify audience behavior, but are not intended to modify audience beliefs. They seem to be incapable of distinguishing their own knowledge from that of another individual's. As an example, they present an analysis of baboon contact barks. Baboons generally move through wooded areas in a group. Individuals that are separated from one another produce loud barks. These barks however do not seem to be produced with the intent of informing others of their location, rather they seem to be emotional responses to the stress of being potentially lost. If they had an informative intent, we would expect individuals that are securely in the center of the group to respond to contact barks in order to inform the others of the location of the group. However, playback experiments have shown this not to be the case. The baboons produce answering barks only if they themselves are separated from the group.

Despite this, nonhuman primates have a very structured understanding of social interactions, which they glean from both direct observation and through listening to vocalizations. Cheney and Seyfarth provide the following characterization of nonhuman primates' social knowledge.

- It is representational.
- It has discrete values.
- It is hierarchically structured.
- It is rule-governed and open-ended.
- It is propositional.
- It is independent of sensory modality.

They point out that these are very similar to the structural properties of human language, though they do not claim that all of the syntactic properties of human language are represented here. Recursion, for example, is not present in this list.

We hypothesize that, since nonhuman primates do not have a theory of mind, they are incapable of social anticipation. Instead, they are using state anticipation to keep track of social structure. This is computationally expensive, since it requires keeping track of each individual's responses to various kinds of cues, and therefore it sets an upper limit on the number of individuals and the number of different cues they can keep track of. When *homo ergaster* moved from the forest to the savannah, it would have faced a selection pressure for increasing group size, because in a more open landscape larger groups are more effective at repelling predators and protecting food. This probably led to adaptations such as a larger neo-cortex, and language, that can help maintain the cohesiveness of a larger group. This is also known as the social grooming theory of language [19]. Developing a theory of mind (that is, social anticipation) helps by allowing generalization, which reduces the computational burden of maintaining separate models of different individuals, and thus allows a larger group size. Further it requires the ability to do recursion, at least to a limited depth, because an individual must model another, who is in turn modeling him, and so on. These abilities, and the existing social knowledge system, probably got exapted (or recruited [46]) for the linguistic system.

The Mirror System Hypothesis. Arbib has suggested an alternate preadaptation for structured language: the mirror neuron system [3]. Mirror neurons were first discovered in the premotor area F5 of macaque monkeys [39]. Mirror neurons are observed to be active when the monkey executes a goal-directed arm movement, like picking up some food. They are also active when the monkey observes someone else (the experimenter, for example) perform the same movement. This observation has generated a lot of interest as to their functional role, and it has been suggested that they might form the precursor of a "mental simulation" system used to model internal states of conspecifics [21].

Interestingly, the homolog in the human brain of area F5 is Broca's area, which is critically involved in language production and comprehension. These findings led to the mirror system hypothesis of Arbib and Rizzolatti:

"The *parity requirement* for language in humans - that what counts for the speaker must count approximately the same for the hearer - is met because Broca's area evolved atop the mirror system for grasping, with its capacity to generate and recognize a set of actions." [38]

Box 3: The Mirror System Hypothesis for Language Evolution:
Arbib suggests the following stages in the evolution of the language capacity [3].

- S1: Grasping.
- S2: A mirror system for grasping shared with the common ancestor of human and monkey.
- S3: A simple imitation system for object-directed grasping through much-repeated exposure. This is shared with the common ancestor of human and chimpanzee.
- S4: A complex imitation system for grasping - the ability to recognize another's performance as a set of familiar actions and then repeat them, or to recognize that such a performance combines novel actions which can be approximated by variants of actions already in the repertoire.
- S5: *Protosign*, a manual-based communication system, breaking through the fixed repertoire of primate vocalizations to yield an open repertoire.
- S6: *Protospeech*, resulting from the ability of control mechanisms evolved for protosign coming to control the vocal apparatus with increasing flexibility.
- S7: *Language*, the change from action-object frames to verb-argument structures to syntax and semantics; the co-evolution of cognitive and linguistic capacity.

His criteria for language readiness are,

- LR1: *Complex imitation.*
- LR2: *Symbolization.*
- LR3: *Parity (mirror property).* What counts for the speaker must count for the listener.
- LR4: *Intended communication.*
- LR5: *From hierarchical structuring to temporal ordering.*
- LR6: *Beyond the here-and-now.*
- LR7: *Paedomorphy and sociality.* Paedomorphy is the prolonged period of infant dependency, which is especially pronounced in humans.

Arbib has developed a fairly detailed account of the evolutionary stages, starting with motor control (grasping), and proceeding through the development of the mirror system, that might have led to the emergence of language. His hypothesized stages are listed in box 3. He suggests that proto-language consisted of two stages, proto-sign and proto-speech. Proto-sign emerged first, by exaptation from a system for imitation of movements. This means that language was initially gestural. The compositional nature of movements provided the right computational machinery for developing structured language.

Note that there is quite a gap between having a mirror system, and having a (compositional) language. A mirror system does not provide the ability to do imitation learning. Gallese and Goldman point out that imitation behavior has never been observed with mirror neuron activity [21]. Arbib also points out that "further evolution of the brain was required for the mirror system for grasping to become an imitation system for grasping." He says, therefore, that stages S1 through S3 (see box 2) are pre-hominid. We believe that these stages do not require more than payoff and sensory anticipation.

Moving from simple imitation to complex imitation probably requires state anticipation, because it involves building a model of the behavior of conspecifics. It is not known, however, to what extent the mirror system is built (that is, learned) or inbuilt (that is, innate).

The crucial step for the development of language is the next one: from complex imitation to proto-sign. It is not clear how exactly this might have happened. Arbib points out that it must involve some neurological change. In fact he believes that proto-sign was preceded by pantomime, which is also qualitatively different from imitation. The key difference is intentionality. Imitation is performed for the purpose of reproducing a movement, whereas pantomime is performed with the intention of getting the other to think about what is being represented. Zlatev et al., similarly, posit "bodily mimesis" as the key transitionary phase from imitation to communication [58]. We believe that going from imitation to signing involves at least a rudimentary form of social anticipation, because it requires knowing that others have mental states that are distinct from one's own, and that they can be manipulated through one's actions. Going from mimesis to symbolicity might have resulted from a combination of goal-anticipation, as discussed in Section 4.1, and social anticipation. In other words, it may be more a difference of degree than kind, as evidenced by the fact that bonobos and chimpanzees are capable of understanding simple symbolic language, though only after a lot of training.

We look next at the attempts to explore the space of communication systems through computer simulation.

4.2   Artificial Language Evolution

Emergence of Signaling. Werner and Dyer did one of the earliest simulations of the emergence of language in a population of artificial organisms [56]. They simulated a population of "male" and "female" agents on a gridworld. Females stayed fixed in position (they were scattered over the grid), and males could move about. Furthermore, females could "see" males (up to a certain distance) and produce a "sound", whereas males could "hear" but not "see" females. The controllers for the agents were small recurrent neural networks that were updated by means of a genetic algorithm. Whenever a male succeeded in finding a female, their genomes were combined using crossover and mutation to produce two new individuals which were placed in random locations on the grid. The parents were removed to conserve the size of the population.

They showed that this simple setup was sufficient for the emergence of a communication protocol, which led to an increase in successful mating over time. The exceedingly simple cognitive architecture of the agents (and the evolutionary procedure) does not allow explicit model building, since the recurrent neural networks did not have their weights updated other than by the genetic algorithm. So this is an example of implicit anticipation. It should also be noted that females would have to give directions more or less continuously to the males. They did not develop any sense of compositionality. A female could not say, for example, "go straight and then turn left".

Tuci et al. have done some recent experiments on the emergence of signaling in very simple robots [52]. They had two kinds of robots, with different sensors, in a C-shaped maze. One type of robot could see the location of a goal (a light), using an ambient light sensor, and the other type could sense walls using infrared sensors. Furthermore, both kinds could produce and detect sounds. The goal was to navigate to the light without any collisions. Neither kind of robot could achieve this goal by itself, so they had to evolve a system of communication to help them to cooperate. The controllers for the robots were, again, small recurrent neural networks which were updated using a genetic algorithm, and weights were not updated other than through evolution. They showed that the robots were able to evolve a communication protocol for achieving the goal. The communication system was continuous (in time), because the robots were not capable of temporal abstraction and thus had to be constantly informing each other of their state. This, just like the Werner and Dyer simulation, is an example of implicit anticipation.

These two experiments show how easy it is to develop a signaling system, even in extremely simple cognitive agents, without any explicit anticipatory mechanism or any clear notion of symbols or language. In the space of communication systems, these systems would lie in the same location as bacteria. There have been several other simulations along these lines, some of which result in discrete communication systems, but none seem to go beyond cued representation. See [55] for a good overview of these as well as simulations of structured communication systems.

Emergence of Lexicons. Work on the emergence of shared symbol systems has focused more on how these systems come to be shared in a population (our third basic question), than the emergence of the capacity for symbolization.

The most famous of these experiments has been the series of Talking Heads experiments carried out by Steels et al. See [54] for a review of these and other experiments based on language games. A typical game consisted of a speaker and a hearer agent (robotic heads) that were presented with a collection of colored geometric shapes on a screen. This collection comprised the context for communication. The speaker agent would select one shape, and use its internal vocabulary to produce a symbol to communicate to the hearer which shape it had selected. A game was deemed a success if the hearer agent correctly picked out the shape based on the symbol it had heard. If it failed, it would be told what the correct shape was. There were two aspects to this game scenario. One was to show that a shared symbol system could emerge in the population of agents through these simple interactions. The other was to show that a shared conceptual space would emerge as well. The agents start without the ability to discriminate shapes, and learn to do it by building discrimination trees as needed. For example if the speaker picked the blue triangle, and the hearer had not yet learned to discriminate blue from other colors, it would extend its discrimination tree in order to do so.

It is difficult to put these experiments into our space of communication systems because they are not ecological. By this we mean that the agents are engaged in communication as their primary task. This makes it impossible to use these to conclude anything about the emergence of the communication *capacity*. However, for our purposes, they do make the important point that once the capacity for symbolization appears, the emergence of a shared symbolic system can happen through population dynamics alone, without the need for more complex cognitive mechanisms.

An example of an ecological simulation is the "mushroom-world" of Cangelosi and Parisi [10]. This world consists of two types of mushrooms: edible and poisonous. An agent, which is a feed-forward neural network, can learn to discriminate between these two types on the basis of the sensory impression they generate (differences in color, shape, size, etc.) Agents and mushrooms are distributed in a grid-world, and the agents move around and eat mushrooms. For each edible mushroom that they consume, they get 10 "energy points", and for each poisonous mushroom they consume, they get -11 energy points. At the end of a fixed lifetime, the 20 most energetic agents would be selected to form the next generation by replication and mutation. The weights of the neural networks only changed by mutation. They showed that when the agents were allowed to label mushrooms for each other (that is, communicate), eventually a shared stable communication system, consisting of just two symbols corresponding to "edible" and "poisonous", emerged. This is very similar to an alarm call system. Here again, once *capacity* to communicate was provided, a stable conventional symbol system eventually evolved.

Emergence of Structure. There have been very few ecological simulations of the emergence of structured language. Cangelosi extended the mushroom-world model to allow the production of two-word utterances [9]. In this version, there were six kinds of mushrooms, of which three were edible and three were poisonous. The edible ones needed to be "approached" in the right way in order to be eaten successfully. The right way to approach was determined by the color of the mushroom. The neural networks representing the agents had two clusters of linguistics units, which correspond to a two-word utterance. The agents were not forced to use two words, and in some cases, did not. However, in ten of eighteen simulations, the populations evolved compositional utterances, and in seven of those, the evolved language could be interpreted as having a verb-object structure. This is because one word in one cluster was consistently used to refer to the poisonous mushrooms ("avoid"), and another to the edible mushrooms ("approach"), and the words in the other cluster were used to distinguish types of edible mushrooms.

Interestingly, the simulation was carried out in two stages. In the first stage, the population was evolved to learn foraging, that is, how to distinguish the mushrooms and how to approach the edible ones. In the second stage, the population would consist of 20 parents from the previous generation along with 80 children from the new generation, and the children would learn from the parents by using the back-propagation algorithm. In other words, the children are forming a payoff and sensory anticipatory model. The communication was still cued, however. It is also not clear if compositional language emerged from anything other than the vagaries of the learning process.

Smith et al. have suggested that the emergence of grammar might, in fact, have more to do with the learning process than with ecological conditions [45]. Their Iterated Learning Model consisted of parents teaching language to children in a succession of generations. In such a situation, if the environment has structure, they showed that compositional language is more easily learnable, and therefore can be correctly passed through the *transmission bottleneck*. The transmission bottleneck refers to the fact that children have to learn the language from a finite sample, and therefore may not see all valid sentences. A rule-based language allows them to generalize correctly to unseen instances. Their model is quite abstract, however, and the agents do not have a cognitive architecture, so we cannot put it into our communication space.

5   What Are Some of the Sufficient Conditions for the Emergence of Language?

Based on the above examination of theories and examples of language emergence, we can start to infer some of the minimal conditions for the emergence of various kinds of language.

- **Adaptive value**: This is the most basic condition for the emergence of communication. If communication does not have an adaptive value, it will not evolve.
- **Memory**: Memory is a very large and complex part of cognition. Humans, e.g., have working memory, long-term memory, propositional memory, episodic memory, muscle memory, etc. We do not mean that all of these are necessary for language. What we mean is that the cognitive system must not be purely Markovian. This can be achieved by using a hierarchical planning system like a semi-Markov decision process, for example. This is necessary because our language is discrete. If the agents have no memory, then they would have to communicate continuously in order to cooperate in the achievement of some goal. See the experiments of Tuci et al. [52] about the evolution of signaling. Furthermore, if the agents have no memory, communication is forced to be cued.
- **Symbol generation**: This refers to the ability to generate new symbols when required. This is also known as having an *open* symbol system. In the absence of this ability, we cannot have detached representations.

- **Planning in non-stationary environments**: If social intelligence is one of the keys to developing rich syntactic language, then the agents must be capable of planning while taking into account that other agents are also planning. This makes the learning environment non-stationary, as we have observed earlier. It may have led to the recursion that is observed in modern language, such as center-embedding of clauses, e.g. "The cat *the dog chased* ran up a tree." In a recent article, Hauser et al. hypothesize that this may be the only innate aspect of grammar [25].

Anticipation provides a framework for analyzing the computational properties of cognitive systems. As we have seen, there has been very little simulation work examining how communication systems that have displaced or detached representations might have emerged. This means that most of the space of communication systems has not really been investigated through simulation. Theories of language emergence, however, deal mostly with this area of the space of communication systems. These theories are now acquiring sufficient detail so that the door to simulation studies has been opened. What we need are focused simulations that ask precise questions, for example, can a population that is capable of complex imitation develop the ability to communicate through mimesis? What exactly is required to make this transition? How complex does the cognitive system have to be to make the transition from compositional to syntactic language? And so on.

We are heading into an exciting period in the study of language evolution as we are beginning to see the emergence of a detailed understanding of the minimal conditions required for the emergence of language. When we succeed in answering these questions, not only will we have solved a very difficult scientific problem [13], but the technological possibilities will be tremendous. We will be able to design truly autonomous *populations* of agents which will be able to collectively accomplish feats that are currently beyond human or machine capabilities.

Acknowledgments

We thank Martin Butz, Kiran Lakkaraju, participants in the UIUC language evolution seminar in the fall semester of 2006, and an anonymous reviewer, for their feedback, which helped to improve the paper.

References

1. Aiello, L.C.: The foundations of human language. In: Jablonski, N.G., Aiello, L.C.(eds.) : The Origin and Diversification of Language. Wattis Symposium Series in Anthropology, Memoirs of the California Academy of Sciences, Number 24. University of California Press (1998)
2. Aiello, L.C., Dunbar, R.I.M.: Neocortex size, group size, and the evolution of language. Current Anthropology 34, 184–193 (1993)

3. Arbib, M.A.: From monkey-like action recognition to human language: an evolutionary framework for neurolinguistics. Behavioral and Brain Sciences 28, 105–167 (2005)
4. Balkenius, C., Gärdenfors, P., Hall, L.: The origin of symbols in the brain. In: Proceedings of the Conference on the Evolution of Language, Paris (2000)
5. Bassler, B.L.: How bacteria talk to each other: Regulation of gene expression by quorum sensing. Current Opinion in Microbiology 2, 582–587 (1999)
6. Bickerton, D.: Language and Species. University of Chicago Press, Chicago (1990)
7. Butz, M.V., Sigaud, O., Gérard, P.: Internal models and anticipations in adaptive learning systems. In: Butz, M.V., Sigaud, O., Gérard, P. (eds.) Anticipatory Behavior in Adaptive Learning Systems: Foundations, Theories, and Systems. LNCS (LNAI), vol. 2684, pp. 86–109. Springer, Heidelberg (2003)
8. Byrne, R., Whiten, A. (eds.): Machiavellian Intelligence. Social Expertise and the Evolution of Intellect in Monkeys, Apes and Humans. Clarendon Press, Oxford (1988)
9. Cangelosi, A.: Modeling the evolution of communication: From stimulus associations to grounded symbolic associations. In: Floreano, J.N.D., Mondada, F.(eds.): Proceedings of the European Conference on Artificial Life, Berlin, pp. 654–663 Springer-Verlag (1999)
10. Cangelosi, A., Parisi, D.: The emergence of a " language" in an evolving population of neural networks. Connection Science 10, 83–97 (1998)
11. Cheney, D.L., Seyfarth, R.M.: Constraints and preadaptations in the earliest stages of language evolution. Linguistic Review 22, 135–159 (2005)
12. Chomsky, N.: Rules and Representations. Columbia University Press, New York (1980)
13. Christiansen, M.H., Kirby, S.: Language evolution: The hardest problem in science? In: Christiansen, M., Kirby, S. (eds.) Language Evolution: The States of the Art, Oxford University Press, Oxford, UK (2003)
14. Comrie, B.: From potential to realisation: An episode in the origin of language. In: Trabant, J., Ward, S. (eds.) New Essays on the Origin of Language. Number 133 in Trends in Linguistics, Studies and Monographs. Mouton de Gruyter, Berlin - New York (2001)
15. Cucker, F., Smale, S., Zhou, D.X.: Modelling language evolution. In: Minett, J., Wang, W.S.Y. (eds.) Language Acquisition, Change and Emergence. City University of Hong Kong Press (2003)
16. Davidson, J.: Genetic exchange between bacteria in the environment. Plasmid 42, 73–91 (1999)
17. Deacon, T.: The Symbolic Species: The Co-Evolution of Language and the Brain. W. W. Norton and Co (1997)
18. Drummond, C.: Accelerating reinforcement learning by composing solutions of automatically identified subtasks. Journal of Artificial Intelligence 16, 59–104 (2002)
19. Dunbar, R.I.M.: Grooming, Gossip, and the Evolution of Language. Faber and Faber (1997)
20. Franks, N.R., Pratt, S.C., Mallon, E.B., Britton, N.F., Sumpter, D.J.T.: Information flow, opinion polling and collective intelligence in house-hunting social insects. Phil. Trans. R. Soc. Lond. B 357, 1567–1584 (2002)
21. Gallese, V., Goldman, A.: Mirror neurons and the simulation theory of mind-reading. Trends in Cognitive Sciences 2, 493–501 (1998)
22. Gärdenfors, P.: Cooperation and the evolution of symbolic communication. In: Oller, K., Griebel, U. (eds.) Evolution of Communicative Systems: A Comparative Approach, pp. 237–256. MIT Press, Cambridge, MA, USA (2004)

23. Gärdenfors, P., Osvath, M.: The evolution of anticipatory cognition as a precursor to symbolic communication. In: Proceedings of the Morris Symposium on the Evolution of Language, Stony Brook, NY, USA (2005)
24. Hammer, M., Menzel, R.: Learning and memory in the honeybee. The Journal of Neuroscience 15, 1617–1630 (1995)
25. Hauser, M.D., Chomsky, N., Fitch, W.T.: The faculty of language: What is it, who has it, and how did it evolve? Science 298, 1569–1579 (2002)
26. Hockett, C.F.: The origin of speech. Scientific American 203, 88–96 (1960)
27. Hurford, J.: The evolution of language and languages. In: Dunbar, R., Knight, C., Power, C.(eds.): The Evolution of Culture, pp. 173–193. Edinburgh University Press, (1999)
28. Jackendoff, R.: Possible stages in the evolution of the language capacity. Trends in Cognitive Science 3, 272–279 (1999)
29. Johansson, S.: Why Did Language Evolve? In: Origins of Language: Constraints on Hypotheses, John Benjamins Publishing Company, Amsterdam/Philadelphia (2005)
30. Johansson, S.: Protolanguage. In: Origins of Language: Constraints on Hypotheses, John Benjamins Publishing Company, Amsterdam/Philadelphia (2005)
31. Kendon, A.: Some considerations for a theory of language origins. Man, New Series 26, 199–221 (1991)
32. Komarova, N.L.: Replicator-mutator equation, universality property and population dynamics of learning. Journal of Theoretical Biology 230, 227–239 (2004)
33. McGovern, A., Barto, A.G.: Automatic discovery of subgoals in reinforcement learning using diverse density. In: Proceedings of the International Conference on Machine Learning (2001)
34. Nowak, M.A., Plotkin, J.B., Jansen, V.A.A.: The evolution of syntactic commmunication. Nature 404, 495–498 (2000)
35. Pinker, S., Bloom, P.: Natural language and natural selection. Behavioral and Brain Sciences 13, 707–784 (1990)
36. Pinker, S., Jackendoff, R.: The faculty of language: What's special about it? Cognition 95, 201–236 (2005)
37. Reading, N.C., Sperandio, V.: Quorum sensing: The many languages of bacteria. FEMS Microbiology Letters 254, 1–11 (2006)
38. Rizzolatti, G., Arbib, M.A.: Language within our grasp. Trends in Neurosciences 21, 188–194 (1998)
39. Rizzolatti, G., Camarda, R., Fogassi, L., Gentilucci, M., Lupino, G., Matelli, M.: Functional organization of inferior area 6 in the macaque monkey. ii. area f5 and the control of distal movements. Experimental Brain Research 71, 491–507 (1988)
40. Rosen, R.: Anticipatory Systems. Pergamon Press, New York (1985)
41. Roy, D.: Grounding words in perception and action: Computational insights. Trends in Cognitive Sciences 9, 389–396 (2005)
42. Savage-Rumbaugh, S., Rumbaugh, D.M., McDonald, K.: Language learning in two species of apes. Neuroscience and Biobehavioral Reviews 9, 653–665 (1985)
43. Seyfarth, R.M., Cheney, D.L.: Monkey responses to three different alarm calls: Evidence of predator classification and semantic communication. Science 210, 801–803 (1980)
44. Smith, J.M., Szathmáry, E.: The Major Transitions in Evolution. Oxford University Press, Oxford, UK (1995)
45. Smith, K., Kirby, S., Brighton, H.: Iterated learning: A framework for the emergence of language. Artificial Life 9, 371–386 (2003)

46. Steels, L.: The recruitment theory of language origins. In: Proceedings of the Alice, V. Morris, D.H.: International Symposium on Language and Cognition: The Evolution of Language, SUNY Stony Brook (2005)
47. Steels, L., Kaplan, F.: Collective learning and semiotic dynamics. In: Floreano, D., Mondada, F. (eds.) ECAL 1999. LNCS, vol. 1674, pp. 679–688. Springer, Heidelberg (1999)
48. Sutton, R., Precup, D., Singh, S.: Between MDPs and semi- MDPs: A framework for temporal abstraction in reinforcement learning. Artificial Intelligence 112, 181–211 (1999)
49. Sutton, R.S., Barto, A.G.: Reinforcement Learning: An Introduction. MIT Press, Cambridge (1998)
50. Swarup, S., Mahmud, M.M.H., Lakkaraju, K., Ray, S.R.: Cumulative learning: Towards designing cognitive architectures for artificial agents that have a lifetime. Technical Report UIUCDCS-R-2005-2514, Department of Computer Science, University of Illinois at Urbana-Champaign (2005)
51. Taddeo, M., Floridi, L.: Solving the symbol grounding problem: a critical review of fifteen years of research. Journal of Experimental and Theoretical Artificial Intelligence 17, 419–445 (2005)
52. Tuci, E., Ampatzis, C., Vicentini, F., Dorigo, M.: Operational aspects of the evolved signalling behavior in a group of cooperating and communicating robots. In: Vogt, P., Sugita, Y., Tuci, E., Nehaniv, C.L. (eds.) EELC 2006. LNCS (LNAI), vol. 4211, pp. 113–127. Springer, Heidelberg (2006)
53. Vilalta, R., Drissi, Y.: A perspective view and survey of meta-learning. Artificial Intelligence Review 18, 77–95 (2002)
54. Vogt, P.: Language evolution and robotics: Issues in symbol grounding and language acquisition. In: Loula, A., Gudwin, R., Queiroz, J. (eds.): Artificial Cognition Systems. Idea Group (2006)
55. Wagner, K., Reggia, J., Uriagereka, J., Wilkinson, G.: Progress in the simulation of emergent communication and language. Adaptive Behavior 11, 37–69 (2003)
56. Werner, G.M., Dyer, M.G.: Evolution of communication in artificial organisms. In: Langton, C.G., Taylor, C., Farmer, J.D., Rasmussen, S. (eds.) Artificial Life II, Reading, Mass., Addison-Wesley, London, UK (1991)
57. Worden, R.: The evolution of language from social intelligence. In: Hurford, J.R., Studdert-Kennedy, M., Knight, C. (eds.) Approaches to the Evolution of Language: Social and Cognitive Bases, Cambridge University Press, Cambridge (1998)
58. Zlatev, J., Persson, T., Gärdenfors, P.: Bodily mimesis as " the missing link" in human cognitive evolution. Technical Report LUCS121, Lund University Cognitive Science, Lund, Sweden (2005)

Superstition in the Machine

Alexander Riegler

Center Leo Apostel, Vrije Universiteit Brussel
Krijgskundestr. 33, B-1160 Brussels, Belgium
ariegler@vub.ac.be

Abstract. It seems characteristic for humans to detect structural patterns in the world to anticipate future states. Therefore, scientific and common sense cognition could be described as information processing which infers rule-like laws from patterns in data-sets. Since information processing is the domain of computers, artificial cognitive systems are generally designed as pattern discoverers.

This paper questions the validity of the information processing paradigm as an explanation for human cognition and a design principle for artificial cognitive systems. Firstly, it is known from the literature that people suffer from conditions such as information overload, superstition, and mental disorders. Secondly, cognitive limitations such as a small short-term memory, the set-effect, the illusion of explanatory depth, etc. raise doubts as to whether human information processing is able to cope with the enormous complexity of an infinitely rich (amorphous) world.

It is suggested that, under normal conditions, humans construct information rather than process it. The constructed information contains anticipations which need to be met. This can be hardly called information processing, since patterns from the "outside" are not used to produce action but rather to either justify anticipations or restructure the cognitive apparatus.

When it fails, cognition switches to pattern processing, which, given the amorphous nature of the experiential world, is a lost cause if these patterns and inferred rules do not lead to a (partial) reorganisation of internal structures such that constructed anticipations can be met again.

In this scenario, superstition and mental disorders are the result of a profound and/or random restructuring of already existing cognitive components (e.g., action sequences). This means that whenever a genuinely cognitive system is exposed to pattern processing it may start to behave superstitiously. The closer we get to autonomous self-motivated artificial cognitive systems, the bigger the danger becomes of superstitious information processing machines that "blow up" rather than behave usefully and effectively. Therefore, to avoid superstition in cognitive systems they should be designed as information constructing entities.

Keywords: Action-selection, anticipation, constructivism, decision-making, information-processing, pattern search, philosophy of science, schizophrenia, superstition.

Preliminary Remark

In his report, one of the reviewers wrote that the submitted version of this paper lead him "*initially* [...] into the wrong direction of thinking" [my emphasis]. *Voilà.* This is

M.V. Butz et al. (Eds.): ABiALS 2006, LNAI 4520, pp. 57–72, 2007.
© Springer-Verlag Berlin Heidelberg 2007

what this paper is about: picking up cues and running off in a direction that is determined by one's own experiential past. On a philosophical level the paper explores the relationship between rational thinking, anticipation, and superstition by building on the philosopher Kant's idea that "objects must conform to our knowledge" [32] rather than the other way around, which considers knowledge a mirror of the state of affairs. This paper is intended as a criticism of representationalist "third-person" modeling, of the attempt by humans to create intelligent artifacts in their own image.

1 Introduction

While early AI was mainly concerned with symbolic computation that assumed a readily structured propositional environment, more recent streams emphasize the embodied dynamical nature of cognition, e.g., [56]. In this paper I address one of the main consequences of the embodiment paradigm, i.e., the question of the relationship between the cognitive agent and its environment and the potential danger of the view that the agent is informed by the environment, i.e., that the agent processes input information in order to generate output, or as Ulric Neisser [48] puts it, that cognitive subjects are "dynamic information processing machines." Since in the context of human beings it can be shown that this leads to superstition and mental disorders, it seems reasonable to prevent machines from this destiny by carefully crafting alternative design principles.

This paper starts with the extreme case of a structureless (amorphous) world. This shifts the focus of attention from structures "out there" (entities, events, etc.) to what goes on *inside* a cognitive being. Research in adaptive behavior and cognitive systems is, after all, interested in creating cognitive artifacts rather than artificial worlds. I proceed with arguing that, based on experimental findings, there is a close relationship between pattern discovery and superstition since humans and animals alike excel at finding structures where there are none. How can this be explained?

At first sight it seems that the ability to find structures and compress them into rules is rather useful for anticipating future states. This ability is usually called "inductive reasoning." However, in many cases, instead of anticipating states that become actualized in the future, the cognitive systems merely exhibits wishful thinking, also referred to as "superstition." The point is that both induction and superstition are carried out by the same cognitive apparatus: from its point of view ("first-person perspective") there is no difference. But why is our information processing not always successful? We can identify two classes of reasons: (L1) our cognitive equipment is *very* limited by a small short-term memory, conservative bias in problem solving ("set-effect"), and the illusion of explanatory depth. (L2) The combinatorial explosion of different ways to account for the links between entities and events is such that the computational effort required to compute them becomes intractable or NP-complete in the sense that the time required to solve them grows exponentially with the number of components.[1] So how can complex situations be *processed* by limited cognition?

---

[1] This was one of the main reasons why attempts of AI failed to scale blockworld scenarios up to real world situations.

This suggests that the cognitive being does not map structures from the "outside world" onto its cognitive equipment but rather creates structures in the first place. The construction process may be triggered by sensorial, proprioceptive or other, "internal" cues. A reader, for example, would not be able to read a novel without the faculty of creating lively, rich mental pictures out of a few letters on a page. However these pictures are composed of parts that already exist and which were constructed previously. The parts are put together in a particular way, creating anticipatory "checkpoints" between them. The function of the checkpoints is to verify the viability of the constructed chain. By constructing this chain, cognition is canalized into a particular direction making it possible to effectively control the combinatorial explosion in the sense of L2 (i.e., to prune the vast search space – a very well-known problem in artificial intelligence and cognitive science). In this sense, cognitive structures are projected onto the "external" world.

Superstition occurs when the cognitive system is exposed to L2, i.e., when it leaves the construction mode and tries to find sense in the flood of incoming data (in humans this search for new rules may be accompanied by a feeling of anxiety). However the chances are that the newly constructed rule is nothing but a bad guess.

The remaining part of the paper is concerned with providing empirical and argumentative support for the thesis that cognition is about information construction rather than information processing. I start by presenting arguments that make it clear that the structure of the "world" must be almost infinitely rich so that we can speak of an amorphous world. Then I cite empirical results from the psychological literature that suggest a close link between pattern detection, anticipation and superstition. Furthermore, I discuss decision making from the perspective of both information processing and information constructing. I conclude that by following self-constructed information we do not fall prey to arbitrariness or insanity. Consequently, I suggest applying these insights to the design of genuinely autonomous artificial cognitive systems in order to prevent them from "blowing up."

2 The Search for Patterns

A generally accepted working hypothesis in artificial intelligence and cognitive science is that common-sense thinking and science both attempt to infer rule-like laws from the patterns in data sets, i.e., to perform some sort of data-mining.[2] In philosophy of science we find this claim expressed, for example, in Ernst Mach's *economy of thought* [41] which states that the goal of science is "the simplest and most economical abstract expression of facts" [42]. Similarly, Herbert Simon defined (scientific) discovery as "detecting the pattern information contained in the data, and using this information to recode the data in more parsimonious form" [60]. He argued that computer programs can discover the recursive rules generating sequences of letters. However, as the following examples demonstrate (from [28]), even in the case of seemingly simple sequences it can be hard to find the respective rule that specifies the criterion for separating the sets of letters in the following three sequences.

---

[2] Science may do so with a higher degree of systematicity, with the goal of uncovering underlying causes.

(A, E, F, H, I, K, L, M, N, T, V, W, X, Y, Z) versus (B, C, D, G, J, O, P, Q, R, S, U)
(A, B, D, O, P, Q, R) versus (C, E, F, G, H, I, J, K, L, M, N, S, T, U, V, W, X, Y, Z)
(A, B, C, D, E, F, G, H, I, J, K, L, M, N, O, P, Q, R, S, T, U, W, X, Y, Z) versus (V)

In order to establish rules in the first example you need to focus on whether all the segments of the respective letter are straight. In the second example the distinction is based on whether the respective letter has at least one enclosed area. That the criteria don't need to be based on topological characteristics is demonstrated in the third example in which the letters are separated based on whether they are part of the Polish alphabet, which does not include the letter V. With a little bit of imagination even more unusual letter sequences can be formulated. How could a heuristically working AI program possibly consider all possibilities to find the appropriate rule?

In his influential book *The logical structure of the world,* Rudolf Carnap presents the thought experiment of two geographers – a realist and an idealist – who travel to Africa to discover a certain mountain. After they have collected empirical data about its location, height and other physical characteristics upon which they agree, they find themselves in disagreement with regard to how to interpret the data. For the realist, the mountain "not only has the ascertained geographical properties, but is, in addition, also real" while the idealist claims that "the mountain itself is not real, only our perceptions and conscious processes are real" [10]. Carnap's original intention was to show that epistemological theses do not have any value if they go beyond experience in the sense that neither proponent can give an "indication of the design of an experiment through which his thesis could be supported" [10]. Historically, Carnap's arguments were meant to support logical positivism. However, the thought experiment also demonstrates the apparent arbitrariness of how to read sense into experiential data, which, in contrast to general belief, makes the scientist rather than "nature" responsible for decisions. The conclusions are twofold. For philosophy of science, theories are plausible at best; and for artificial intelligence, the structure of the environment seems to play only a marginal role.

3 The Amorphousness of the World

Let us investigate the role of the environment a little further. As early as 1906, Pierre Duhem [18] claimed that observational evidence can never conclusively disprove a theory (or thesis) as any seemingly disconfirming observational evidence can always be accommodated to it (the so-called "underdeterminism" of theories, in modified form later known as "Quine-Duhem thesis"). As a result, there will be many competing theories trying to explain a given set of experimental data.

Philosophy of science is full of examples that support the underdeterminism theorem, e.g., [35]. However, one can easily form an idea of how vast the range of possibilities is by considering the abstract concept of a black box, i.e., an entity whose inner mechanisms are unknown to the outside observer. It is already extremely difficult to make inferences for a black box with four input, four internal, and four output states, all of which can be wired in any way [21]. The total number of possible configurations is $4^{(4^4)} = 2^{32}$, i.e., about 4×10^9. In other words, starting from an observational protocol one needs to test 4 billion different models to find the one that reproduces a recorded behavior. Usually, empirical data contains much more extensive protocols.

Going one step further, James McAllister points out that there is an arbitrarily large number of ways to explain data points: "Any given data set can be interpreted as the sum of any conceivable pattern and a certain noise level. In other words, there are infinitely many descriptions of any data set as 'Pattern A + noise at m percent', 'Pattern B + noise at n percent', and so on"[3] [45].

> Asked in what the phenomenon of planetary orbits consists, Kepler would have replied "In the fact that, with such-and-such a noise level, they are ellipses," while Newton would have replied "In the fact that, with such-and-such a (lower) noise level, they are particular curves that differ from ellipses, because of the gravitational pull of other bodies." Thus, phenomena – understood as the patterns in data sets that investigators choose to model – vary from one investigator to another [45].

McAllister [46] continues to argue that since any given data set can be interpreted as the sum of any conceivable pattern and a certain noise level, all the rules and patterns that a data set displays have equal status and can, therefore, be said to equally correspond to structures in the "world." But if that world contains any structure, then it contains all possible structures, which is equivalent to exhibiting no structure at all, i.e., to being amorphous.

So in order to arrive at a decision, we let our choice be guided by pragmatic concerns such as simplicity, economy and elegance [55], or as Heinz von Foerster [22] put it,

> Only the questions which are principally undecidable, we can decide. Why? Simply because the decidable questions are already decided by the choice of the framework in which we are asked, and by the choice of rules of how to connect what we call 'the question' with what we may take for an 'answer'. ... [We] are under no compulsion, not even under that of logic, when we decide upon in principle undecidable questions.

This insight seems to be in sharp contrast to the self-understanding of the natural sciences, which aim at quantitative exactness through systematicity [29]. For example, in classical physics quantitative empirical data enters into a model in such a way that the mutual influence among single data items can be determined computationally. This principle assumes a homomorphism between the structure of the phenomenon and the structure of the model, i.e., that the phenomenon can be reduced to an isomorphic copy of the model through applying a many-to-one transformation [5]. However, for a given phenomenon an innumerable amount of homomorphic models can be found through many-to-one transformations. All these models will exhibit the same behavior. The reverse inference from a given model (i.e., the scientific image) to the "true" structures of the phenomenon (i.e., reality) is therefore impossible. Rather, it is the human (scientist or other) who decides which structural pattern to read in the data in order to explain the phenomena of interest in terms of a scientific law. In this way, quantitative completeness is replaced by *qualitative* schemata [3]. He understands the

---

[3] The term "noise" refers to the mathematical discrepancy between a particular pattern and a given data set.

behavior of the homomorphic model of the "real" phenomenon rather than the phenomenon itself. That is, he is able to cope with a "suitable" simplification of the system's states (which is his invention) in order to make predictions "in the head." As a result, people often overestimate their knowledge of facts and procedure, a phenomenon that Frank Keil [33] calls the "illusion of explanatory depth," so that despite the vast incompleteness of their knowledge, most people think that they know far more than they actually do.

4 Superstitious Anticipations

Given the amorphousness of the world one is tempted to caricature human attempts to systematically generate knowledge as nothing more than having one's fortune told from the coffee cup; any arbitrary and unconstrained interpretation seems possible. Still, as we will see in the following examples from the psychological literature, it would be wrong to say that superstitions are "just" erroneous modes of information construction. Rather, superstition is the active attempt by the subject to coerce order into structurelessness.

As is known from the psychological literature, the search for patterns in one's stream of experience leads to various conditions. Klaus Conrad [13] coined the term *apophenia* [*Apophänie*] to refer to the experience of "unmotivated seeing connections" in random or meaningless data. Similarly, *pareidolia* is the erroneous or fanciful perception of a pattern or meaning in something that is actually ambiguous or random. Examples are the Rohrschach test and the alleged face on Mars based on a photo made from the surface of the planet. In general, humans are prone to the so-called *clustering illusion*, i.e., the tendency to associate some meaning to certain types of patterns that must inevitably appear in any large enough data set. According to Scott Huettel, Peter Mack and Gregory McCarthy, even if patterns are generated randomly their recognition is "obligatory, in that it occurs without any conscious attentional effort" [30]. In their experiments they confronted test subjects with a random sequence of squares and circles. The subjects were asked to press a button in their right hand when they perceived a square, and a button in their left hand for a circle. Occasionally, brief periods of seemingly non-random patterns appeared, such as a series of alternating circles or squares. Even though the subjects were instructed that they were seeing random sequences, their unconsciousness reacted when such a series was violated. This was demonstrated by functional magnetic resonance imaging (fMRI) scans of their prefrontal cortex, which revealed the changing activity in a distributed set of regions that are highly sensitive to the presence of and deviations from patterns. It seems that the subjects' motor behavior was primed, based on the belief of having discovered a pattern that would continue. In other words, humans construct a belief in a genuine regularity where there is none: "the recognition of patterns is an obligatory, dynamic process that includes the extraction of local structure from even random sequences" [30].

Such *compulsive* pattern-perception is not limited to event patterns as was shown in the experiments of Frédéric Gosselin and Philippe Schyns [26]. The authors stimulated the visual system of test subjects with unstructured white noise[4] that superimposes on the contours of a face. In 20,000 trials the subjects were asked to determine whether the face was smiling, which according to the instruction was the case in 50% of the presentations, even though in none of the presentations whatsoever did the face have a mouth. Still, in many cases the subjects were certain that the face was indeed smiling. Clearly, the anticipated pattern was projected onto (partially correlated with) the perception of the white noise.[5]

A possible explanation for the built-in tendency to perceive patterns is that it enables the subject to better react to a sequence of cues that signal a potential threat or a source of food. Therefore, the obsessive search for patterns not only serves as a basis for human superstitious behavior, it can also be found in the animal kingdom. B. F. Skinner's article on *Superstition in the pigeon* [61] is a classical description of how birds react in situations beyond their cognitive capabilities and therefore beyond their control. Skinner presented food at regular intervals to hungry pigeons with no reference whatsoever to their current behavior. Soon the birds started to display certain rituals between the reinforcements, such as turning two or three times about the cage, bobbing their head, and incomplete pecking movements. As Skinner noticed, the birds happened to be executing some response when the food first appeared and they tended to repeat this response if the feeding interval was short enough. In some sense, pigeons associated their action with receiving food and started to believe that it caused the food to appear.

In the early 20th century Bronislaw Malinowski noticed that islanders in the Pacific who fished offshore beyond the coral reef displayed many superstitious rituals and ceremonies to invoke magical powers for safety and protection, while inshore fishermen carried out their job with a high degree of rational expertise and craftsmanship [43]. It reflects the desire of humans to find causal explanations and to organize their experience in a meaningful manner [25] in order to make predictions based upon them. Ellen Langer [36] accounted for the tendency to apply superstition as a response to uncertainty by introducing the notion of "illusion of control," i.e., the belief that one can control or at least influence outcomes in situations under which one has no control: "If there is a universal truth about superstition, it is that superstitious behavior emerges as a response to uncertainty – to circumstances that are inherently random and uncontrollable" [65].

Engaging in superstitious behaviors such as displaying patterns of stereotyped behavior is closely linked to an illusion of control since the people engaging in these patterns may actually believe that *they* are controlling an outcome [57]. A prominent example of this perspective is *feng shui*, the ancient Chinese superstitious practice of placement and arrangement of space, which is claimed to achieve harmony with the environment. Even today, Chinese managers resort to this superstitious practice when they have to make important decisions as many of them find it difficult to cope with the unknown [63]. It helps them to reduce uncertainty-induced anxiety. In this sense,

---

[4] White noise is a static bit pattern that has equal energy at all spatial frequencies and does not correlate across trials.

[5] The Italian painter Leonardo da Vinci had already recommended looking at blotches on walls as a means of initiating artistic ideas: "If you look at walls covered with many stains ... with the idea of imagining some scene, you will see in it a similarity to landscapes adorned with mountains, rivers, rocks, tree, plains, broad valleys, and hills of all kinds."

superstition provides an additional source of information which fills the void of the unknown or helps to deal with "information overload" by disentangling conflicting information to tip the scale: "superstition breaks the deadlock by indicating a superior alternative" [63]. One-third of Chinese managers rely so much on superstitious practices that they even neglect rational solutions. For others it is a stress-management tool, while for still others it has become an obsessive habit without which they will not feel at ease.

In the psychological literature, further links between superstitious thoughts and behaviors and obsessive-compulsive disorder (OCD) can be found, such as obsessive checking [23] and anxiety disorders in female test subjects [66]. Furthermore, as Peter Brugger [8] points out, the "ability to associate, and especially the tendency to prefer 'remote' over 'close' associations, is at the heart of creative, paranormal and delusional thinking." Again, a correlation between OCD and magical thinking, i.e., maintaining beliefs that defy culturally accepted laws of causality in general and the belief that certain thoughts or behaviors exert a causal influence over outcomes in particular, can be found [20].

The close connection between schizophrenia and creativity, i.e., the ability to think in terms of connections and links between entities belonging to different categories gives rise to the assumption that creativity has also close ties to superstition, which can be defined as the confusion of categories and core knowledge [37], i.e., building blocks that emerge early in human ontogeny and phylogeny [62]. This includes, among others, confusion of symbolic representations and the material objects they represent, the attribution of physical or animate entities to mental content, the ascription of independent existence to good and bad minds, which behave as animate entities by moving and initiating actions without external force, and thinking that badly placed furniture leads to crime and divorce.

The conclusion from these insights is that cognition, being the faculty to make decisions, depends on our ability to anticipate future events and states. This thesis, largely upheld and developed in cognitive science and AI, is not disputed. However, the origin of the anticipations is called into question: are they a result of information processing or alternatively do they emerge from information constructing? This question is legitimized by the fact that due to the amorphousness of the world any arbitrarily large number of underlying rules or laws for predicting events can be established. This leaves a cognitive system with uncertainty and even anxiety. In order to regain the feeling of control, cognition uses superstitious behaviors, which relieve the cognitive agent from the need to process the information overload from its environment. Therefore, for want of structure, superstition emerges as a response to uncertainty and may even transform into psychological disorders.

In the following section the consequences for decision-making and cognition are addressed.

5 Decision-Making and the Construction of Information

It has been argued that complex human thought and behavior are possible because cognition is able to simulate long sequences of responses and sensory consequences [27]. In a sense, cognition then consists of walking through a chain of decisions.

According to the traditional literature on decision-making, it is taken for granted that human cognition makes decisions based on careful considerations of the situational context. In economics, this idea has been condensed into "rational choice theory," which regards rationality as the only criterion a decision maker has to strive for. As pointed out above, there are reasons to assume that this is a misleading characterization, which is based on the assumption that human cognition basically functions as an information-processing device that extracts rules from incoming experience. This theory not only fails to take the limitations of (human and any other finite) cognition into account, it is also unable to deal with the amorphousness of the world. In artificial life research, this has been called the "what to do next" or action-selection problem [64, 31], i.e., formulating a mechanism that allows choosing an action in pursuit of a single coherent goal or several conflicting and heterogeneous goals.

Gerd Gigerenzer [24] argued that decisions must necessarily be adaptive, fast, and frugal if they are to ensure survival. In artificial intelligence, a similar paradigm has emerged focusing on behavior-based robotics. It does not rely upon *a priori* mathematical analysis of a given situation but rather on a *hic-et-nunc* strategy that takes *system-internal* drives into account rather than a sophisticated representation defined in terms of the programmer's semantic world, i.e., from the third-person perspective.

In order to understand how human and, consequently, artificial cognitive systems with limited processing capabilities can cognitively cope with an amorphous world, let us explore two examples from the animal kingdom which both question the notion of information-processing.

(1) Consider the behavior of an incubating goose that decides to use its bill to roll back the egg that has fallen out of its nest. Interestingly, it will continue its rolling behavior even if an ethologist takes away the egg [38]. It seems that the animal does not constantly screen its environment and filter out environmental changes. Rather, the environmental state becomes only important at certain, apparently evolutionarily important, *checkpoints* (which do not include the existence of ethologists). These checkpoints act as anticipations that determine whether an action that has been already started is on track with regard to a certain goal. As the psychological literature documents, human problem-solving, too, is dominated by a similar sort of conservative inflexibility that makes subjects repeatedly choose a once successful strategy irrespective of whether another, simpler, strategy might be better suited for new problems [19, 40]. Furthermore, human perception is determined by internal cognitive dynamics that only occasionally seek to verify certain anticipations about future input states, as shown in the sequential order of tactile object recognition [58]. Finally, it has been argued that the inability to ignore stimuli (i.e., low latent inhibition) can lead to mental illnesses such as schizophrenia [39].

(2) Consider a fly crawling over a painting of Rembrandt [44, 56]. It in no way *processes* the visual information presented in the painting, as from its perspective there is no painting whatsoever. Only the human observer may wonder which information *filters* the fly applies in order to *ignore* the rich informational input. The fundamental difference arises from the fact that human scientists and engineers from their third-person perspective (and lacking the first-order perspective of the observed systems, e.g., the fly) necessarily concentrate on the perceivable *output* of systems (such as the crawling of the fly). Cognitive systems, however, take actions in order to control and change their perceptive and proprioceptive *input*, e.g., they avoid the

perception of an obstacle or they drink to quench their thirst: "From the organism's point of view only actions which feed back to the organism's sensors can be observed [...] Any other action which simply disappears in the environment cannot be observed by the organism" [53].

In the latter sense a cognitive system needs to be defined as an information *creator* rather than as an information *processor*. The latter is defined in terms of an input–output relationship: a given input or perception yields (directly or after some "calculation") an action, i.e., the output is a function of the information in the input. Information-creating systems, however, control their input (perception) rather than their output (behavior), or as Powers [54] put it, "behavior is the process by which organisms control their input sensory data." This can only be achieved by first creating information (e.g., chains of cognitive elements connected via anticipatory checkpoints) internally and then allowing those checkpoints to be (occasionally) matched against the input.

Historically this idea can be traced back to the cybernetic concept of homeostasis [9]. It says that a living organism has to keep its intrinsic variables within certain limits in order to survive. These "essential variables" (which include body temperature, levels of water, minerals and glucose, and similar physiological parameters, as well as other proprioceptively or consciously accessible aspects in higher animals and human beings [4]) represent the purpose of a system. In order to account for a wider range of systems, we have to make the following (rough) distinctions here:

(a) *Man-made vs. genuinely autonomous systems.* In man-made systems goals are defined by the human designer, whereas in genuinely autonomous systems the goals are constructed by the system itself. They are defined before the system reads its inputs and tries to accommodate for any deviations, or as William Clancey [11] expressed it, "what constitutes information for an organism cannot be given by a teacher, but must arise from the organism's own organizing processes in interaction with its environment." Since the agent controls its inputs the output can become quite unpredictable for an *external* observer [53]. It is therefore useful to retain the distinction between a first-person perspective (the one of the input-controlling system) and the third-person perspective of the external observer who can only make guesses about the self-defined goals of the system. Susan Oyama [51] maintains that there is no preformed or a priori information, but rather that every system constitutes its own information: information is bound "inextricably to a point of view," i.e., the first-person perspective.

(b) *Simple vs. cognitive systems.* In simple systems (such as thermostats) the state of homeostasis is reached by the simple process of negative feedback. For example, in a thermostat, a given parameter, such as the temperature, is kept under control by appropriate counteractions, i.e., by turning the heating on or off depending on a certain reference value. Cognitive systems need to execute a certain sequence of actions in order to control and change their input state. This is a result of the fact that "perception and action arise together, dialectically forming each other" [12]. This renders the whole concept of representation doubtful as "we can walk through a room without referring to an internal map of where things are located, by directly coordinating our behaviors through space and time in ways we have composed and sequenced them before" [12]. Consequently sensory, cognitive, and motor functions can no longer be considered independent and sequentially working parts of the cognitive apparatus. And the latter can no longer be described as a computational device, the "task" of which is to acquire propositional knowledge about the mind-independent reality by processing information that is picked up from that reality.

Putting (a) and (b) together, *genuinely autonomous cognitive systems* can be characterized as systems that try to achieve self-determined goals. These goals are defined in terms of anticipatory checkpoints that glue together action sequences. In order to act as goals these action sequences must be constructed in the first place: the information they contain is constructed rather than representing processed input. The cognitive system then executes an action sequence as long as the checkpoints can be successfully met by the input. Any failure to do so causes the system either to find alternative chains of actions that more appropriately answer to the requirements of the checkpoints ("re-orientation") or to construct new chains that make more or less use of already existing elements. The lower the degree of reusability, the higher the degree of "perplexity" of the system. The vast number of different ways to form chains from components excludes random arrangements from being an option.

It is already clear in the case of simpler systems that action sequences have to be assembled in a hierarchical manner. While such a simple feedback loop may suffice for primitive intrinsic variables, higher order goals are accomplished in a hierarchical assemble of feedback loops in which each level provides the reference value for the next lower level: "The entire hierarchy is organized around a single concept: control by means of adjusting reference-signals for lower-order systems," as Powers [54] pointed out. So at higher levels the system controls the output of lower levels, at the bottom, however, it controls its perceptual input.

In cognitive systems action sequences must be arranged in a similar manner. Formally, the cognitive apparatus P may consist of schemata R that work over mental states S, $P = <R, S>$.[6] Each schema $r \in R$ is a chain of checkpoints c and actions a, $r = \{c \mid a\}^+$. We can extend this definition by allowing clusters of checkpoints and actions, respectively, $C = \{c\}^+$ and $A = \{a\}^+$, in order to account for multimodal perceptual entities and action sequences (e.g., the egg-rolling sequence in the example above), respectively. If, in addition, we allow recursions of the form $C = \{c \mid C\}^+$ and $A = \{a \mid A\}^+$ it follows that checkpoints and actions form *nested hierarchies* in which encapsulations are reused as building blocks on a higher level [1]. Elements can be functionally coupled as they make use of encapsulations and/or are part of encapsulations themselves. The encapsulation guarantees that, when new chains are being formed, partial solutions do not get lost, preventing the apparatus from making random guesses. By using a hierarchical arrangement the cognitive agent gains two advantageous aspects: speed and goals. According to Simon [59] "hierarchic systems will evolve far more quickly than non-hierarchic systems of comparable size" since the "time required for the evolution of a complex form from simple elements depends critically on the number and distribution of potential intermediate stable forms." In addition, the hierarchical composition of chains introduces a direction, a *goal*, "by the stability of the complex forms, once these come into existence." [59]

---

[6] Schemata are a popular explanatory vehicle for cognitive systems and have been widely discussed in the literature, e.g., Neisser [49], Drescher [17], Arbib [2], Bickhard [6].

In other words, goal-directed behavior can be explained as the execution of hierarchically nested chains composed of action sequences and checkpoints that glue them together. The checkpoint character of cognition defines decision-making and other cognitive acts as being based upon internal states rather than external states of affairs. For an observer, from her (third-person) perspective cognitive systems seem to make the proper decisions in response to the challenges of the environment. But from the (first-person) perspective of the organism, decisions are only the consequence of the internal cognitive dynamics. Since the observer is a cognitive system with her own internally defined goals, the respective goals of observer and observed do not necessarily coincide. This renders representational modeling difficult: how can the goal of the modeled cognitive system be defined in terms of modeler's goals? A design process like this requires a "God's Eye" perspective and is therefore, unless they trust in random guesses, out of reach for human scientists.

Reusing stable clusters of components, however, also has the disadvantage that the "right thing" may be done at the apparently "wrong moment." If we consider the superstitious behavior of pigeons or the rituals of the offshore fishermen, it is clear that exactly this characterizes superstitious behavior. None of the involved behaviors are new. Only the perceptual condition that triggers the execution of a certain sequence is different from the one that was originally associated with that behavior. In order words, superstition emerges when processed information is combined with already-constructed information (cf. the above discussion about the close link between superstition and creativity).

6 Conclusion

The arguments presented in this paper give rise to the assumption that artificial cognitive systems that are designed as information-processing devices are exposed to the danger of turning to superstitious behavior as their autonomy and cognitive competence increases. The arguments are based on the fact that in environments whose complexity transcends that of block- and microworlds[7] the number of possible rule-based explanations are NP-complete; any attempt to calculate them with the (very) finite means of the cognitive apparatus would render decision-making impossible. Making random guesses, however, leads to superstition and mental disorders.

Since Daniel Dennett's description of the frame problem [14], we know that robots that try to calculate every possible cause in their environment are doomed to remain inactive. In a sense, scientists are in no better of a situation. Johannes Kepler spent many years trying to find the regularity behind planetary movement by compressing the huge amount of position data collected by him and his teacher, Tycho Brahe [34]. Therefore it should not amaze us that "a scientific law or theory provides an algorithmic compression not of a data set in its entirety ... but only of a regularity that constitutes a component of the data set and that the scientist picks out in the data." [47] In other words, some of these patterns "are taken by investigators as corresponding to phenomena, not because they have intrinsic properties that other patterns lack, but because they play a particular role in the investigators' thinking or theorizing."

---

[7] The environments targeted by many action-selection solutions, e.g., [64,31].

[45] These are the checkpoints in cognitive chains. Instead of processing the structures of the world we generate this structure by applying our cognitive operators [16], an insight which was originally proposed by Immanuel Kant's Copernican Turn [32].

When checkpoints are not met and the cognitive agent is forced to switch to information processing in order to construct new cognitive chains, superstition and mental disorders may emerge. The bigger the information overload the longer it takes to get back to the normal mode of cognition (cf. the amount of time necessary to find one's way through new information) and the higher the chance of constructing superstitious information.

To sum up, by taking an input-oriented perspective, we no longer need to assume that information is being processed but, rather, that it is constructed. That is, cognition is safeguarded from cognitive overload, which could otherwise lead to the illusion of control and, consequently, the creation of superstitious routines. Such an information-constructing paradigm emphasizes the primacy of the internal cognitive dynamics over influences from the outside. Only if the cognitive artifact is in control of executing its structures rather than needing to cope with the computational costs of processing the flood of perceptual stimuli can it, like natural cognitive systems, effectively control its input.

Therefore, it would be useless to let autonomous cognitive machines data-mine the infinitely rich structures of the world. What we consider a regularity, a law, a rule or simply a heuristics for decision-making is an arbitrary component of the data set that only makes sense in the light of our thinking or theorizing, and is therefore an anticipatory construction based on aspects of the cognitive system's past experiences: "Things are a construction of ours, the function of which is to emphasize the resemblance between aspects of our immediate experience and aspects of our past experience." [7]

References

1. Angeline, P.J., Pollack, J.B.: Coevolving high-level representations. In: Langton, C., (ed.) : Artificial life III. Addison-Wesley, Reading, pp. 55–71 (1994)
2. Arbib, M.: Schema theory. In: Shapiro, S. (ed.) Encyclopedia of artificial intelligence, 2nd edn. vol. 2, pp. 1427–1443. Wiley, New York (1992)
3. Arthur, W.B.: Inductive reasoning and bounded rationality. American Economic Review 84, 406–411 (1994)
4. Ashby, W.R.: Design for a brain. Chapman & Hall, London (1952)
5. Ashby, W.R.: An introduction to cybernetics, 2nd edn. Chapman & Hall, London (1956)
6. Bickhard, M.H.: Function, anticipation and representation. In: Dubois, D.M. (ed.): Computing anticipatory systems (CASYS 2000). American Institute of Physics, Melville, pp. 459–469 (2001)
7. Bridgman, P.W.: The nature of physical theory. John Wiley & Sons, New York (1936)
8. Brugger, P.: From haunted brain to haunted science: A cognitive neuroscience view of paranormal and pseudoscientific thought. In: Houran, J., Lange, R. (eds.): Hauntings and poltergeists: Multidisciplinary perspective. McFarland, Jefferson, pp. 195–213 (2001)
9. Cannon, W.B.: The wisdom of the body. Norton, New Yorks (1932)
10. Carnap, R.: Der logische Aufbau der Welt. Felix Meiner Verlag, Leipzig (1928). English translation: Carnap, R.: The logical structure of the world. Pseudoproblems in philosophy. University of California: Berkeley (1967)

11. Clancey, W.J.: "Review of Rosenfield's 'The Invention of Memory'.". Artificial Intelligence 50, 241–284 (1991)
12. Clancey, W.J.: "Situated " means coordinating without deliberation. McDonnel Foundation Conference, Santa Fe (1992)
13. Conrad, K.: Die beginnende Schizophrenie. Versuch einer Gestaltanalyse des Wahns. Thieme, Stuttgart (1958)
14. Dennett, D.C.: Cognitive wheels: The frame problem of AI. In: Hookway, C. (ed.) Minds, machines, and evolution: Philosophical studies, pp. 129–151. Cambridge University Press, London (1984)
15. Dennett, D.C.: Consciousness explained. Little, Brown & Co, London (1991)
16. Diettrich, D.: A physical approach to the construction of cognition and to cognitive evolution. Foundations of Science 6, 273–341 (2001)
17. Drescher, G.L.: Made-up minds: A constructivist approach to artificial intelligence. MIT Press, Cambridge (1991)
18. Duhem, P.: The aim and structure of physical theory (French original published in 1906). Princeton University Press, Princeton (1954)
19. Duncker, K.: Zur Psychologie des produktiven Denkens. Springer, Berlin (1935). English translation: Duncker, K.: On problem solving. Psychological Monographs 58, 1–112 (1945)
20. Einstein, D., Menzies, R.: The presence of magical thinking in obsessive compulsive disorder. Behaviour Research and Therapy 42, 539–549 (2004)
21. Foerster von, H.: Molecular ethology. An immodest proposal for semantic clarification. In: Ungar, G. (ed.): Molecular mechanisms in memory and learning, pp. 213–248. Plenum Press, New York (1970) Reprinted in Foerster von, H.: Observing systems. Intersystems Publications, Seaside, pp. 149–188 (1982)
22. Foerster von, H.: Ethics and second-order cybernetics. Cybernetics & Human Knowing 1, 9–19 (1992)
23. Frost, R., Krause, M., McMahon, M., Peppe, J., Evans, M., McPhee, A., Holden, M.: Compulsivity and superstitiousness. Behaviour Research and Therapy 31, 423–426 (1993)
24. Gigerenzer, G.: Adaptive thinking. Oxford University Press, Oxford (2000)
25. Glasersfeld von, E.: Radical constructivism. Falmer Press, London (1995)
26. Gosselin, F., Schyns, P.G.: Superstitious perceptions reveal properties of internal representations. Psychological Science 14, 505–509 (2003)
27. Hesslow, G.: Conscious thought as simulation of behaviour and perception. Trends in Cognitive Sciences 6, 242–247 (2002)
28. Hong, F.T.: Deciphering the enigma of human creativity: Can a digital computer think? IPSP-2003 VIP Forum, October 4-11, 2003, Sveti Stefan, Montenegro (2003)
29. Hoyningen-Huene, P.: The nature of science. Nature & Resources 35, 4–8 (1999)
30. Huettel, S.A., Mack, P.B., McCarthy, G.: Perceiving patterns in random series: Dynamic processing of sequence in prefrontal cortex. Nature Neuroscience 5, 485–490 (2002)
31. Humphrys, M.: Action selection methods using reinforcement learning. PhD Thesis Trinity Hall, Cambridge (1997)
32. Kant, I.: Kritik der reinen Vernunft. Zweite Ausgabe. Reclam jun, Leipzig, English translation: Critique of pure reason, Second edition.(1781)
33. Keil, F.C.: Folkscience: Coarse interpretations of a complex reality. Trends in Cognitive Sciences 7, 368–373 (2003)
34. Kozhamthadam, J.: The discovery of Kepler's Laws. University of Notre Dame Press, Notre Dame (1994)

35. Lakatos, I.: Falsification and the methodology of scientific research programmes. In: Lakatos, I., Musgrave, A. (eds.) Criticism and the growth of knowledge, pp. 91–195. Cambridge University Press, London (1970)

36. Langer, E.J.: The illusion of control. Journal of Personality and Social Psychology 32, 311–328 (1975)

37. Lindeman, M., Aarnio, K.: Superstitious, magical, and paranormal beliefs: An integrative model. Journal of Research in Personality (in press)

38. Lorenz, K.Z., Tinbergen, K.: Taxis und Instinkthandlung in der Eirollbewegung der Graugans. Zeitschrift für Tierpsychologie 2, 1–29 (1939)

39. Lubow, R.E., Gewirtz, J.C.: Latent inhibition in humans: Data, theory, and implications for schizophrenia. Psychological Bulletin 117, 87–103 (1995)

40. Luchins, A.S.: Mechanization in problem solving. The effect of einstellung. Psychological Monographs 54/248 (1942)

41. Mach, E.: Knowledge and error. Sketches on the psychology of enquiry. (German original was published in 1905). Reidel, Dordrecht (1976).

42. Mach, E.: Popular scientific lectures. The Open Court, La Salle (Originally published in 1893) (1986)

43. Malinowski, B.: Magic, science and religion and other essays. Free Press, Glencoe (Originally published in 1925) (1948)

44. Maturana, H.R.: Autopoiesis: reproduction, heredity and evolution. In: Zeleny, M. (ed.) Autopoiesis, dissipative structures and spontaneous social orders, pp. 48–80. Westview Press, Boulder (1980)

45. McAllister, J.W.: Phenomena and patterns in data sets. Erkenntnis 47, 217–228 (1997)

46. McAllister, J.W.: The amorphousness of the world. In: Cachro, J., Kijania-Placek, K. (eds.): IUHPS 11th International Congress of Logic, Methodology and Philosophy of Science. Jagiellonian University: Cracow, 189 (1999)

47. McAllister, J.W.: Algorithmic randomness in empirical data. Studies in the History and Philosophy of Science 34, 633–646 (2003)

48. Neisser, U.: Cognitive psychology. Meredith, New York (1967)

49. Neisser, U.: Cognition and reality. W. H. Freeman, San Francisco (1976)

50. O'Regan, J.K.: Solving the " real " mysteries of visual perception: The world as an outside memory. Canadian Journal of Psychology 46, 461–488 (1992)

51. Oyama, S.: The ontogeny of information: Developmental systems and evolution (Republished in 2000). Cambridge University Press, Cambridge (1985)

52. Pörksen, B.: The certainty of uncertainty (German original appeared in 2001). Imprint, Exeter (2004).

53. Porr, B., Wörgötter, F.: Inside embodiment. What means embodiment to radical constructivists? Kybernetes 34, 105–117 (2005)

54. Powers, W.T.: Behavior. The control of perception. Aldine de Gruyter, New York (1973)

55. Quine, W.V.: Word and object. MIT Press, Cambridge (1960)

56. Riegler, A.: When is a cognitive system embodied? Cognitive Systems Research 3, 339–348 (2002)

57. Rudski, J.M.: The illusion of control, superstitious belief, and optimism. Current Psychology 22, 306–315 (2004)

58. Sacks, O.: An anthropologist on Mars. Alfred A. Knopf, New York (1995)

59. Simon, H.A.: The architecture of complexity. In: Simon, H.A.: The sciences of the artificial, pp. 192–229. MIT Press, Cambridge (1969)

60. Simon, H.A.: Does scientific discovery have a logic? Philosophy of Science 40, 471–480 (1973)

61. Skinner, B.F.: 'Superstition' in the pigeon. Journal of Experimental Psychology 38, 168–172 (1948)
62. Spelke, E.S.: Core knowledge. American Psychologist 55, 1233–1243 (2000)
63. Tsang, E.W.K.: Superstition and decision-making. Contradiction or complement? Academy of Management Executive 18, 92–104 (2004)
64. Tyrrell, T.: Computational mechanisms for action selection. PhD thesis, University of Edinburgh, Centre for Cognitive Science (1993)
65. Vyse, S.A.: In: The psychology of superstition. Oxford University Press, New York (1997)
66. Zebb, B.J., Moore, M.C.: Superstitiousness and perceived anxiety control as predictors of psychological distress. Anxiety Disorders 17, 115–130 (2003)

From Actions to Goals and Vice-Versa: Theoretical Analysis and Models of the Ideomotor Principle and TOTE

Giovanni Pezzulo[1], Gianluca Baldassarre[1], Martin V. Butz[2],
Cristiano Castelfranchi[1], and Joachim Hoffmann[2]

[1] Istituto di Scienze e Tecnologie della Cognizione,
Consiglio Nazionale delle Ricerche,
Via San Martino della Battaglia 44, I-00185 Roma, Italy
{giovanni.pezzulo,gianluca.baldassarre,
cristiano.castelfranchi}@istc.cnr.it
[2] University of Würzburg, Röntgenring 11, 97070 Würzburg, Germany
{mbutz,hoffmann}@psychologie.uni-wuerzburg.de

Abstract. How can goals be represented in natural and artificial systems? How can they be learned? How can they trigger actions? This paper describes, analyses and compares two of the most influential models of goal-oriented behavior: the ideomotor principle (IMP), which was introduced in the psychological literature, and the "test, operate, test, exit" model (TOTE), proposed in the field of cybernetics. This analysis indicates that the IMP and the TOTE highlight complementary aspects of goal-orientedness. In order to illustrate this point, the paper reviews three computational architectures that implement various aspects of the IMP and the TOTE, discusses their main peculiarities and limitations, and suggests how some of their features can be translated into specific mechanisms in order to implement them in artificial intelligent systems.

Keywords: Teleonomy, goal, goal selection, action triggering, feedback, anticipation, search, robotic arms, reaching.

1 Introduction

Intelligence of complex organisms, such as humans and other apes, resides in the capacity to solve problems by working on internal representations of them, that is by acting upon "images" or "mental models" of the world on the basis of simulated actions ("reasoning"). These capabilities require that internal representations of world states, goals and actions are intimately related. With this respect, accumulating evidence in psychology and neuroscience is indicating that anticipatory representations related to actions' outcomes and goals play a crucial role in visual and motor control [16]. As suggested by the discovery of mirror neurons [38], representations are often action-related and are thus grounded on the representations sub-serving the motor system. Barsalou [2] and Grush

M.V. Butz et al. (Eds.): ABiALS 2006, LNAI 4520, pp. 73–93, 2007.

[14] try to provide unitary accounts of these phenomena respectively proposing *perceptual symbol systems* and *emulation* theories of cognition. In a similar vein, Hesslow [16] proposes a *simulation hypothesis* according to which cognitive agents are able to engage in simulated interactions with the environment in order to prepare to interact with it. According to Gallese [11]: "To observe objects is therefore equivalent to automatically evoking the most suitable motor program required to interact with them. Looking at objects means to unconsciously 'simulate' a potential action. In other words, the object-representation is transiently integrated with the action-simulation (the ongoing simulation of the potential action)".

Recently anticipatory functionalities have been started to be explored from a conceptual point of view [6,7,40] as well as from a computational point of view [8,5,32,47]. This paper contributes to this effort by analyzing two important now "classic" frameworks of goal-oriented behavior, namely the *ideomotor principle (IMP)*, and the *test operate test exit* model *(TOTE)*. The IMP and the TOTE can be dated back in their origin for decades if not centuries. The IMP, which was proposed multiple times during the 19th century within the psychological literature [15,22], hypothesizes a bidirectional action-effect linkage in which the desired (perceptual) effect triggers the execution of the action that previously caused it. The TOTE, introduced within the field of cybernetics [27], proposes that goal-oriented action control is based on an internal representation of the desired world's state(s) with which the current world's state is repeatedly compared in order to direct action.

The first goal of the paper is to provide a comprehensive introduction to both the IMP and the TOTE and to highlight their similarities, differences, and drawbacks in explaining anticipatory goal-oriented behavior (sections 2-4).

The second goal of the paper is to analyze, at an abstract level, three computational architectures, which implement various different features of the IMP and the TOTE in distinct ways (Section 5). The architectures are only reviewed here, while the reader is referred to specific papers for details). This analysis aims at exemplifying and clarifying the principles underlying the IMP and the TOTE. it is intended to serve as a starting point for future research on the investigation of anticipatory goal-oriented behavioral mechanisms.

A final discussion concludes the paper with an outlook of the most important challenges that the two principles pose to cognitive science (Sec. 6).

2   The Ideomotor Principle

According to the IMP [17,20,22], *action planning takes place in terms of anticipated features of the intended goal*. Greenwald [13] underlines the role of anticipation in action selection: *a current response is selected on the basis of its own anticipated sensory feedback*. The *Theory of Event Coding* [21] proposes a common coding in perception and action, suggesting that the motor system plays an important role in perception, cognition and the representation of goals. The theory focuses on learning action-effect relations which are used to reverse the "linear stage theory" of human performance (from stimulus to response) supported by the sensorimotor view of behavior. Neuroscientific evidence suggesting common mechanisms in organisms' perception and action is reported in [23,37]. With this respect, Gallese [12] suggests that "the goal is represented as a goal-state, namely, as a successfully terminated action pattern". Recently Fogassi et al. [10] discovered that inferior parietal lobule neurons coding an observed specific act (e.g., grasping) show markedly different activations when the act is part of different courses of actions leading to different distal goals (e.g., for eating or for placing). Since the activation begins before the course of action starts, they postulate that those neurons do not only code the observed motor act but also the anticipation (in an ideomotor coding, we would say) of the distal goal, that is the understanding of the agent's intentions.

In the ideomotor view, in a sense, *causality, as present in the real world, is reversed in the inner world.* A mental representation of the intended effect of an action is the cause of the action: here it is not the action that produces the effect, but the (internal representation of the) effect that produces the action. Minsky [28, par. 21.5] describes an "automatic mechanism" that can be considered as an example of how to realize this principle (cf. Figure 1): when the features of, say, an apple are endogenously activated, an automatic mechanism is oriented toward seeing or grasping apples teleonomically.

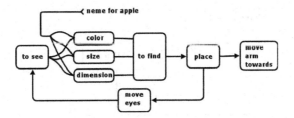

Fig. 1. The "automatic mechanism" proposed by Minsky [28, par. 21.5]

The main constituents of the IMP. The comparison of the presentations of the IMP by these various authors allows identifying three main constituents of the principle. These form the core of the principle and abstract over minor details and different aspects stressed by the various authors. The three constituents are now analyzed in detail (cf. Figure 2):

1. *Perceptual-like coding of goals.* An important characteristic of the IMP is that it has been developed within a vision of intelligence seen as closely related to the sensorimotor cycle (for an example drawn from the psychology literature see [25], whereas for an example drawn from embodied artificial intelligence see [14]). As a consequence, the authors proposing the IMP usually stress the fact that the system's internal representations of goals are similar, or the same, as the internal representations activated by perception. This feature of the principle has also an important "corollary": the source of goals is usually

assumed to be experience, that is, goals tend to correspond to previously perceived (eventually represented in an abstract way) states.

2. *Learning of action-effect relations.* Another important constituent of the principle is that experience allows the system to create associations between the execution of actions (e.g., due to exploration, "motor babbling", etc.) and the perceived consequences resulting from it. This requires a learning process that is based on the co-occurrence of actions and their effects observed in the environment [20]).

3. *Goals are used to select actions.* Another core constituent of the principle is the fact that the system exploits the learned association between actions and the resulting perceived states of the world to select actions. According to Greenwald [13]: *For the ideomotor mechanism, a fundamentally different state of affairs is proposed in which a current response is selected on the basis of its own anticipated sensory feedback.* The idea is that the activation of the representation of a previously experienced state allows the system to select the action that led to it. When this process occurs, the representation of the state assumes the function of goal both because it has an anticipatory nature with respect to the states that the environment will assume in the future, and because it guides behavior so that the environment has higher chances to assume such states.

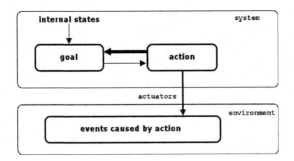

Fig. 2. A scheme that represents the main features of the IMP. Thin arrows represent information flow, whereas the bold arrow represents the direction of the internal association between goal and action corresponding to physical causality. See text for further explanations.

It is important to note that the selection of actions with this process requires an "inversion" of the direction of the previously learned action-effect association, from "actions → resulting states" to "resulting states → actions". This inversion is particularly important because it implies that the system passes from the causal association that links the two elements, as resulting from experience, to the teleonomic association between them, as needed to guide behavior. It is only thanks to this inversion that the system can use the effects as goal states.

3   TOTE and Cybernetic Principles

TOTE was introduced by Miller, Galanter and Pribram [27] as the basic unit of behavior, opposed to the stimulus-response (SR) principle. The TOTE was inspired by cybernetics [39], that however focused on homeostatic control and not on *goals*. In a TOTE unit, firstly a goal is tested to see if it has been achieved: if not, an operation is executed until the test on the goal's achievement is successful. One of the examples of a TOTE unit is a plan for hammering a nail: in this case, the test consists in verifying if the nail's head touches the surface and the operation consists in hitting the nail. In this case, the representation used for the test is in sensory format, and the operation is always the same, even if the TOTE cycle can involve many steps. TOTE units can be composed and used hierarchically for achieving more complex goals, and can include any kind of representation for the test and any kind of action. The TOTE inspired many subsequent theories and architectures such as the General Problem Solver (GPS) [29].

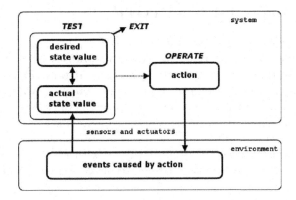

Fig. 3. A scheme that represents the main features of the TOTE. Words in *Italics* represent the main processes composing the principle. Thin arrows represent information flows. The double-headed arrow represents a process of comparison between the desired and the actual state value. The dashed arrow represents the fact that an action is selected and executed in the case the *Test* fails, but not how it is selected. The bold arrow represents a switch in the sequence of processes implemented by the system. See text for further explanations.

The main constituents of the TOTE. The three main constituents of the TOTE (cf. Figure 3) are now analyzed in detail on the basis of the comparison of the various formulations proposed in the literature.

1. *Test.* A first fundamental constituent of the principle is the internal representation of the desired value(s) of the state of the environment. The representation of this value is a key element of the *Test* sub-process composing the principle. This sub-process implies that the system repeatedly checks if the current state of the environment matches the goal.

2. *Abstract goal.* Another important feature of TOTE is that the desired state value of the system, that is the goal, can be abstract. Indeed, the TOTE is underspecified with this respect, and the literature has used several different types of encodings for goals, from perceptive-like encodings to more abstract symbolic ones. The principle can manage this type of goals as the Test sub-process can be as complex as needed, from simple pattern-matching to more sophisticated processes of logical comparison of several features. This (possibly) abstract nature of the definition of goals has also an important implication for the origin of goals themselves, which can derive from previous experience but also from other sources such as other systems (communication or external setting) or "imagination" processes (internally generated goals).

3. *Multiple steps.* An important aspect of the TOTE is the fact that it is naturally suited to implement a course of action formed by multiple steps, as suggested by the repetition of the "Test" sub-process in its acronym. In these steps sensory feedback might be used for chaining actions.

4 Comparison of IMP and TOTE

From the descriptions of the IMP and the TOTE reported in the previous sections, it should be apparent that the two frameworks specify rather general behavioral and learning principles and abstract over details. Thus, designing an artificial adaptive learning system according to them requires to integrate the indications that they give with many implementation details. The guidelines that the two frameworks give for the implementation of systems will now be presented and compared in detail highlighting the respective strengths and weaknesses. In particular, the two frameworks will be analyzed with respect to goal selection and representation, action selection, action execution, context dependencies, and learning.

4.1 Origin of Goals and Their Selection

If goals/effects have to trigger actions, they need to be generated and selected in the first place. However, none of the two approaches gives suggestions on how such goal selection process might be implemented. Certainly, strong links with motivational and emotional mechanisms might be called into play to tackle this problem. For example undesired low values of variables controlled homeostatically may trigger a goal that previously caused the variable to increase in value (e.g., empty stomach leads to the search and consumption of food). However, literature on IMP simply assumes that some events internal to the system eventually trigger the (re-)activation of an internal representation of action-consequences, that hence assume the function of a pursued goal, without specifying the mechanisms that might lead to this. On the other side, the literature on TOTE tends to generically assume that goals derive from experience or that they originate form outside the system (e.g., other intelligent systems, other modules, con-specifics, etc.). Thus, how goal generation and selection could be implemented lies outside the scope of IMP and TOTE.

4.2 Goal Representation

Regardless of how goals are generated and selected, goals may be represented in multiple ways. In the IMP, goal representations are encoded perceptually. As a consequence anything that can be perceived might give origin to a goal representation. These goal representations can then trigger linked action codes or action programs that previously led to the activated goal representation.

The IMP does not specify which perceptual goal representations may be used and how concrete or abstract they might be. However, two aspects are usually emphasized in the literature: the role of experience in the formation of potential goals and the perceptual basis of goal representations. With these restrictions in mind it seems hard to generate some kinds of abstract goals within the IMP, in particular, goals that are defined in terms of qualitative or quantitative comparisons, such as: "find the biggest object in the scene", or "find the farthest object". In fact, in these cases the goal cannot be a template or a sensory prototype to be matched with percepts, but corresponds to complex processes such as "find an object, store it in memory, find a second object, compare it with the previous one".

An interesting additional problem arises in the implementation of the IMP in that it does not specify how the system may distinguish between the current perceptual input and the pursued goal. In fact the IMP postulates that the goal is represented in the same format as the percepts generated from the sensation of the state the world. With this respect, authors usually claim that the physical machinery used to represent the goals and the one used at the higher levels of perceptual processing are the same (e.g. [38]). This raises a problem of how the system can distinguish between the activation of the representation corresponding to the pursued goal and the activation of such representation caused by the perception of the world. Indeed this information is needed by the system to control actions, but the IMP framework does not indicate how this can be done.

The TOTE, on the other hand, explicitly assumes "abstract" goal representations, and the level of such abstraction can be decided by the designer. This gives to the TOTE much freedom with respect to the IMP: goals could be encoded in abstract forms, but they could even be perceptually specified. However, even when abstract encodings are used, the TOTE needs to be perceptually grounded since the "Test" sub-process of the mechanism needs to compare goals with environmental states, and these can only be derived from the perceptual input. Thus, differently than the IMP, the TOTE stresses the importance of abstract goal representations but its goals' representations need ultimately to be grounded in perceptual input to test if they have been achieved.

4.3 Action Selection and Initialization

Once a goal representation is invoked, the next question arises: how the corresponding motor program or action is selected and triggered. Both principles remain silent on when the invocation of a goal actually triggers an action, assuming that this is always the case. However, in an actual cognitive system it can be expected that the invocation of a goal representation may not always lead to an actual action trigger, for example when the goal is currently not achievable or too hard to achieve.

The IMP stresses that the perceptual goal representations directly trigger actions or motor programs that previously led to that goal. In contrast to TOTE, though, the IMP does not specify how long this goal is pursued. In particular, it does not specify what happens if the selected goal is already achieved, nor it specifies how the system checks if the currently pursued goal has been achieved. This information is important for the successive selection of actions depending on the fact that the pursued goal has been achieved or not. On the contrary, TOTE contemplates an explicit test, applied repeatedly, that allows the system to check when the selected goal has been achieved.

On the other side, whereas IMP suggests the existence of bidirectional links between goal representations and motor programs or actions that achieve them, TOTE is silent on how specific actions are triggered on the basis of the activated goal. For example, the origin of the knowledge needed to select the suitable actions in correspondence to goals is not specified. This is in line with the fact that the literature on TOTE tends to overlook the role that learning and experience might have in goal directed behavior. Given this underspecification, the models working on the basis of TOTE have adopted various solutions. For example a common solution (e.g. sometimes used in the General Problem Solver) assumes that the controlled state is quantitative and continuous, and uses a mechanism that selects and executes actions so as to diminish the difference between the current and the desired values of the state. Another example of solution is presented in [34] where an explicit representation of a causal/instrumental link between the actions and the resulting consequences is used to trigger actions.

4.4 Action Execution

The TOTE is thus an explicit closed loop framework, which by definition takes the initial state and feedback into account. However, it does not specify if the system should only check for the final goal or for intermediate perceptual feedback, as suggested for example in the emulation framework of Grush [14] or also in the closed-loop theory by Adams [1]. Moreover, the authors of the TOTE do not furnish any specific indication about the specific mechanisms used for control, such as the overall architecture of the system (e.g., hierarchical, modular, etc.). Also the IMP is silent with regard to the question on how the execution of the "selected" action is carried out, in particular whether or not feedback is used. Finally, none of the two frameworks distinguishes between different types of perceptual feedback such as proprioceptive versus exteroceptive feedback.

4.5 Context Dependence

Both approaches do not make any suggestion on how goal selection and action selection may be dependent on the current context. The IMP approach considers merely the relation between the desired goal and the "action" to reach it, without taking into account that the required action almost always depends on the given initial state of the system. Although modulations of action-effect links are certainly imaginable dependent on currently available contextual information, these are not specified in any form. Also the TOTE is silent on this issue as the link between activated goal and corresponding operation is not specified. However, the TOTE is context dependent at least in a sense: it explicitly takes the current state into account in order to determine the action.

4.6 Learning

The IMP assumes the learning of action-effect associations with a *bidirectional* nature, contrasting the view that the learning of "forward models" and "inverse models" should take place as distinct learning mechanisms. However, how such bidirectional learning is actually accomplished is not specified. If one assumes that the connections between actions and effects are mutually formed by Hebb-like mechanisms ("what fires together wires together"), one has to face the problem that sensory and motor parameters have to be represented in a way that allows the system to "wire" together different types of representations. This assumption leads to the "common code" hypothesis [35].

The TOTE stays completely silent on how "operator modules" for pursuing goals might be learned or acquired. Indeed, probably because of its historical origin within the cognitive psychology literature, the TOTE does not consider learning at all but rather expects that the system designer creates appropriate operator modules for the goals that may be pursued.

In general, both frameworks remain underspecified with respect to other important issues related to learning. For example, they do not address important challenges such as learning generalization over different control programs or the problem for which goals may be achieved in multiple ways. This underspecifications with respect to learning represent some of the most crucial challenges for the application of both frameworks.

4.7 Goal Orientedness, the IMP and the TOTE

After having compared in detail the strengths and weaknesses of the IMP and the TOTE, it is now important to consider their relations with goal-orientedness. To this purpose, one can distinguish between three kinds of teleonomic mechanisms:

1. *Stimulus determined*, in which some relevant final states are reached thanks to learned regularities (e.g. stimulus-response associations), without any explicit representation of the final states to be achieved.
2. *Goal determined*, in which there is an explicit representation of the expected effect which also triggers an action, via previously learned action-effect links. Notice that, as discussed in Sec. 2, an effect can be used as a goal state because there is an "inversion" of the direction of the previously learned action-effect association.

3. *Goal driven*, in which there is an explicit representation of the states to be achieved (goals): the system compares these states with the current state and activates a suitable action if there is a mismatch between them. Note that this third type of mechanism is a sub-case of the second one.

The IMP can be considered an instance of the second kind of mechanisms and the TOTE an instance of the third kind. The main difference is that in the IMP the goal is causally reached but not pursued as such. In other terms, the IMP is functionally able to reach a state which is represented in an anticipatory way, but the state is not treated as a goal that as is something motivating and to be pursued.

On the contrary, the TOTE is goal driven: it is based on an explicit goal representation which serves to evaluate the world (in particular, to be matched against the current state). With this respect, the Test sub-process has both the functions of action trigger and stopping condition. More precisely, the mismatch serves to select and trigger the rule whose expectation minimizes the discrepancy. Moreover, differently from the IMP, the TOTE "knows" if and when a goal is achieved. Another related point is that in the IMP desired results (motivating the action) are not distinguished from expected results of actions, the latter including the former.

The comparison has shown that both frameworks are rather underspecified under many aspects. Whereas the TOTE stresses the test-operate cycle, the IMP stresses the linkage between action and contingently experienced effects and the reversal thereof to realize goal-oriented action triggering. With this regards, it seems possible that both principles might be combined into a unique system whose goals are perceptually (but possibly very abstractly) represented, and in which these perceptual goal representations trigger the associated action commands. The triggered goal may then be continuously compared to the current perceptual input enabling the recognition of current goal achievement. To realize this, goal-related perceptual codes need to be distinguished from actual

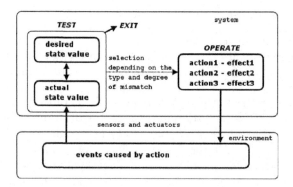

Fig. 4. An example of model integrating some functionalities of both IMP and TOTE. Actions, as in the TOTE, are selected and triggered by the mismatch produced by the test. The action-effect rules are the same used in the IMP.

perceptual codes, by, for example, a tag-based mechanism, a difference-based representation, or a simple duplication of perceptual codes. As an example of such a combination, Figure 4 indicates how the TOTE can exploit action-effect rules as in the IMP, still retaining the test component and using the mismatch for selection and triggering. Of course, the functioning of many processes such as matching, selection and triggering are left unspecified here, because they can be implemented in different ways. Next Section presents some implemented architectures that provide concrete examples of possible models that can be obtained by merging different aspects drawn from the two principles, and that show some of the elements composing them "in action".

5   Implementations of IMP and TOTE in Artificial Systems

After having analyzed the IMP and TOTE at a theoretical level, this section reviews and discusses some computational models, presented in detail elsewhere, that on one side represent concrete implementations of some important features of such frameworks, and on the other side offer concrete answers to the issues left open by both frameworks.

5.1   Case Study I: An Architecture for Visual Search

A hierarchical architecture [33] inspired by the IMP and by the "automatic mechanism" proposed by Minsky [28] was tested in a *Visual Search* task [46]. The goal the system was to find the a red T in a picture containing also many distractors, namely green Ts and red Ls. The system could not see all the picture at once, but had a movable spotlight with three concentric spaces characterized by a good, mild and bad resolution.

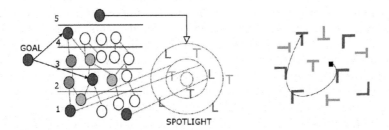

Fig. 5. *Left*: the components of the simulation: the goal, the spotlight and the modules, whose layers are numbered. Light and dark nodes represent more or less active modules. Modules learn to predict the activity level of some modules in the lower layer, which they receive in input (dotted lines). *Right*: a sample trajectory in the visual field, starting from the center (red letters are dark Grey, green letters are light Grey).

The architecture performs the visual search task on the basis of many feature-specific modules, such as color-detectors and line-detectors. Like in *pandemonium* models [41], modules are organized hierarchically and include increasingly complex representations (cf. the left part of Figure 5). According to [9, p. 444] "search" consists in *matching input descriptions against an internal template of the information needed in current behavior*: each module is composed by an input template and a behavior. Modules have a variable level of activation: more active modules can act more often and, as we will see, influence more strongly the overall computation. Modules in layers 1 and 2 obtain an input from a simulated fovea. The other modules have no access to the fovea, but use as input the activation level of some modules in the immediately lower layer (dotted lines in Figure 5). The architecture has five layers:

1. **Full Points Detectors** receive input from portions of the spotlight, for example the left corner, and match full or empty points. Modules are more numerous in the inner spotlight than in the central and outer spotlight.
2. **Color Detectors** monitor the activity of Full Points Detectors and recognize if full points have the color they are specialized to find (red or green).
3. **Line Detectors** categorize sequences of points having the same color as lines: they do not store positions and can only find sequences on-the-fly.
4. **Letter Detectors** categorize patterns of lines as Ls or Ts: they are specialized for letters having different orientations.
5. **The Spotlight Mover** is a single module: as explained later, it receives asynchronous motor commands from all the other ones (e.g. *go to the left*) and consequently moves the center of the spotlight.

In the **learning phase**, by interacting with a simulated environment, each module learns *action-expectation pairs*. Modules learn the relations between their actions and their successive perceptions (the activation level of some modules in the lower layers), as in *predictive coding* [36]. In this way they also learn which actions produce successful matching. For example, a line-detector learns that by moving left, right, up or down the fovea its successive pattern matching operation will be successful (i.e. it will find colored points, at least for some steps), while by moving in diagonal its matching will fail. In this way the line-detector implicitly learns the form of a line by learning how to "navigate" images of lines. In a similar way, a T-detector learns how to find Ts by using as inputs the line-detectors . There is also a second kind of learning: modules *evolve links* toward the modules in the lower layer, whose activity they use as input and can successfully predict. For example, T-detectors will link some line-detectors[1]. These top-down, *generative* links are used for spreading activation across the layers.

The **simulation phase** starts by setting a Goal module (e.g. *find the red T*) that spreads activation to the red-detector(s) and the T-detector(s). This introduces a strong goal directed pressure: at the beginning of the task some

---

[1] By learning different sets of action-prediction rules, modules can also specialize: for example, there can be *vertical lines detectors* and *horizontal lines detectors*.

modules are more active than others and, thanks to the top-down links, activation propagates across the layers. During the search, each module in the layers 2, 3, and 4 tries to move the spotlight where it anticipates that there is something relevant for its (successive) matching operation, by exploiting their learned action-perception associations. For example, if a red-detector anticipates something red on the left, it tries to move the spotlight there; a green-detector does the opposite (but with much less energy, since it does not receive any activation from the Goal module). Line- and letter- detectors try to move the spotlight for completing their "navigation patterns". Modules which successfully match their expectations (1) gain activation, and thus the possibility to act more often and to spread more energy; and (2) send commands to the Spotlight Mover (such as *move left*); the controller dynamically blends them and the spotlight moves, as illustrated in the right part of Figure 5. In this way the fovea movements are sensitive to both the goal pressures and the more contextually relevant modules, i.e. those producing good expectations, reflecting attunement to actual inputs. The simulation ends when the Goal module receives simultaneous success information by the two modules it controls; this means that the Goal module has only two functions: (1) to start the process by activating the features corresponding to the goal state and (2) to stop the process when the goal is achieved. As reported in [33], this model accounts for many evidences in the Visual Search literature, such as sensitivity to the number of distractors and "pop-out" effects [46].

The IMP and the TOTE in Play. According to the IMP, activity is preceded and driven by an endogenous activation of the anticipated (and desired) goal state. In this case, the goal "find the red 'T'" can be reformulated as "center the fovea in a position in which there is a red T"; and the process starts by pre-activating the features of the desired state, i.e. the modules for searching the color red (red-detector) and the letter T (T-detector); the "finding machine", once activated, can only search for an object having these features. The key element of the model is the fact that modules embed action-expectation rules and are self-fulfilling; when a module is endogenously activated, its effect becomes the *goal* of the system. It is worth noting that this system does not use any map of the environment, but only sensorimotor contingencies [31] and a close coupling between perception and action.

This system can achieve only two kinds of goals: (1) *goal states that were experimented during learning*, such as "find the red T"; and (2) *goal states that are a combination of features*; for example, by combining a green-detector and an L-detector, the system can *find a green L* even if it has never experimented green Ls during learning, but only green Ts and red Ls. On the contrary, this system cannot achieve other kinds of goals such as: (1) *The red T on the left*, since locations are not encoded; (2) *The biggest red T*, since there is no memory of past searches and different Ts can not be compared; (3) *The farthest red T*, since temporal features are not encoded. These goals require a more sophisticated procedure for testing and a more abstract encoding: two of the features of the TOTE. The system uses a feature of TOTE: a stopping condition, consisting in a matching between the goal and the activation level of the corresponding features.

5.2   Case Study II: An Architecture for Reaching

The second system used to illustrate the IMP and the TOTE in play has been used to control a simple 2D two-segment arm involved in solving sequential reaching tasks by reinforcement learning. Here we present only the features of the system useful for the purposes of the paper and refer the reader to [30] for details.

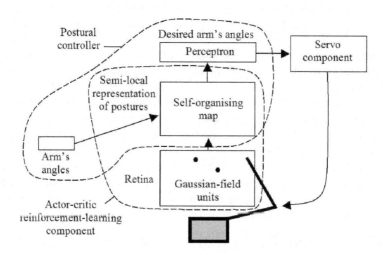

Fig. 6. The architecture of the model of reaching. Rectangular boxes indicate neural-network layers. Text in boxes indicates the type of neural-network model used. Text near boxes indicates the type of information encoded in the layers. Enclosed areas indicate the two major components of the system. The graph also shows the controlled arm and two targets activating the retina (black dots). See text for further explanations.

The system is mainly formed by two components, a *postural controller* and a *reinforcement-learning component* ("RL component" for short). In a first learning phase, the postural controller learns how to execute sensorimotor primitives that lead the arm to assume certain postures in space. In order to do so, while the system performs random actions (similarly to "motor babbling" in infants, cf. [26]), the postural controller learns to categorize the perceived arm's angles in a 2D self-organizing map [24]. At the same time a two-layer network is trained, by a supervised learning algorithm [44], to associate the arm's angles (desired output pattern) with the map's representation of them (input pattern). This process allows the system: (a) to develop a population-code representation of sensorimotor primitives within the self-organizing map, encoded in terms of the corresponding "goals" (i.e. postures); (b) to develop weights between the map and the desired arm's angles that allow selecting sensorimotor primitives by suitably activating the corresponding goals within the map.

In a second learning phase, the RL component learns to select primitives to accomplish reward-based reaching-sequence tasks, for example in order to reach two visible dot targets in a precise order (cf. Figure 6 for an example; the RL component is an "actor-critic model", cf. [42]). Each time the RL component selects an action (i.e. the achievement of a "desired posture"), the desired arm's angles produced by it are used to perform detailed movements (variations of the arm's angles) through a hardwired servo-component that makes the arm's angles to progressively approach the desired angles (postures): when this happens, control is again passed to the RL component that selects another action.

The IMP and the TOTE in Play. The system has strong relations with both the IMP and the TOTE, and in so doing it emphasizes their complementarity. In line with the IMP, in the first phase of learning (motor babbling) the system performs exploratory random actions, and learns to associate the resulting consequences, in terms of the proprioception of the arm's angles, to them. In the second phase of learning, the system uses the expected consequences of the actions as goals ("expected" in terms of final postures), to trigger the executions of the actions themselves so as to pursue rewarding states. This feature of the system is in line with two core features of the IMP, namely learning action-effect relations and using them in a reversed fashion to select actions. However, notice how it encodes the action-effects relations (that is the relations "current posture angles seen as action - internal posture representations ") and the effects/goals-action relations (that is the "internal posture representations - desired posture angles seen as action" relations) in two separated sets of connection weights. With respect to this feature of the model, recall that the IMP does not furnish any specific mechanisms on how the action-effects/effects-action associations should be learned.

A first important departure of the model from the IMP is that the "goals" of the actions (i.e. the corresponding postures perceived in the first learning phase), through which the system selects and triggers the actions themselves, are not encoded in a "pure" perceptual-like format, but in terms of more abstract representations generated by the self-organizing map. This might represent a first step toward a more abstract representation of goals in the spirit of the TOTE. A second important departure from the IMP is that the system incorporates a "stop" mechanism on the basis of which control passes again to the RL component when the execution of an action achieves the goal for which it was selected. As we have seen, this is a typical feature of the TOTE. Note how this "stopping" condition had to be introduced to allow the system to accomplish a task that required the execution of more than one action in sequence (two actions in this case).

From an opposite perspective, it is interesting to notice how, by using some of the core ideas behind the IMP, the system overcomes some limitations of the TOTE. In particular, first it uses experience both to create goals' representations and to associate them to actions, two issues that, as we have seen, are not specified by the TOTE. Second, it uses motor babbling to create an association between goals and actions, overcoming the TOTE's underspecification about how specific actions are selected and triggered in correspondence to a given activated goal.

5.3   Case Study III: Anticipatory Classifier Systems

The *anticipatory learning classifier system* ACS2 [3] learns anticipatory representations in the form of condition-action-effect schemata, similar to Drescher's schema system [8]. However, ACS2 learns and generalizes these schemata online using an interactive mechanism that is based on Hoffmann's theory of anticipatory behavioral control [17,18,19] and on genetic generalization [3]. Similar to the described arm-control approach, ACS2 executes some form of motor babbling. It consequently learns a generalized model of the experienced sensorimotor contingencies of the explored environment. In difference to the above system, though, ACS2 learns purely symbolic schema representations, in difference to the dynamically abstracted real-valued sensory information. Generally, though, such an abstraction mechanism might be linked with the ACS2 approach. More importantly, though, ACS2 makes sensorimotor contingencies explicit: The systems learns a complete but generalized predictive model of the environment.

ACS2 was combined with an online generalizing reinforcement learning mechanism, based on the XCS classifier system [45]. The resulting system, XACS [4], learns a generalized state value function using XCS-based techniques in combination with the model learning techniques of ACS2. Figure 7 sketches the resulting architecture. The reinforcement component is intertwined with the model learning component using the model information for both predictive reinforcement learning and action decision making. For learning, XACS iteratively updates its reinforcement component using a Q-learning-based [43] update mechanism—testing all possible reachable situations and using the maximum reward value to update the currently corresponding reward value. For action decision making, XACS uses the model to activate all immediately reachable future situations and then uses the reinforcement learning component to decide on which situation to reach and consequently which action to execute. It was also proposed that XACS may be used in conjunction with a motivational module representing different drives. The reinforcement module would then consist of multiple modules that work in parallel, each module influencing decision making according to its current importance [4] (cf. Figure 7).

The IMP and the TOTE in Play. XACS plays a hybrid role being situation-grounded but goal-oriented. In this way, goals that cannot be achieved currently will not have any influence on behavior. Vice-versa, goals that are easily achievable currently will be pursued first. Due to the generalization in the predictive model and in the reinforcement component, abstract generalized goal representations can be reached within differing contexts.

XACS realizes ideomotor principles in that actions are directly linked to their action effects. Initially, XACS learns such schemata during random exploration. Goals are coded using the given perceptual input, which is symbolic. XACS, however, does not start from the goal itself but interactively activates potential goals (that is, future situations), then chooses the currently most desirable one, which finally triggers action execution. In this way, the system is goal-driven—but it is grounded in the current situation.

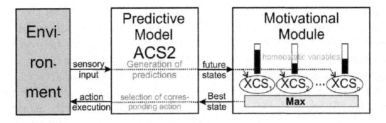

Fig. 7. XACS realizes the IMP in that it selects actions according to their associated perceptual effects. A desired effect is selected using the developed motivational module that is designed to maintain the system in homeostasis. The TOTE is realized in that each iteration currently possible effects are compared with currently desired effects.

Goal selection is integrated in XACS by the separate reinforcement component that links to the behavioral component. Thus, XACS proposes a goal selection mechanism realized with reinforcement learning techniques. In difference to the TOTE, there is never an explicit test that controls if a goal was reached. This mechanism is implicitly handled by the reinforcement learning component in conjunction with the proposed motivational module. Once a goal is reached, a motivation will become satisfied and thus another drive will control behavior.

6   Conclusions

This paper has investigated the implications of the Ideomotor Principle (IMP) and the Test Operate Test Exit framework (TOTE) for adaptive behavior and action selection. The paper showed that the frameworks are actually rather closely related as both stress the importance of anticipation as goal-oriented action selection. Whereas goals are represented perceptually and are bidirectionally linked to associated actions in the IMP, the TOTE emphasizes the interactive cycle of triggering actions by desired goals while iteratively testing if such goals are achieved. Overall the two frameworks enlighten important aspects of the anticipatory nature of goal-driven systems. However, neither of them get concrete enough to pinpoint specific actual implementations.

The paper also reviewed three implementations that not only exemplify the power and interest of the guidelines proposed by the IMP and TOTE, but also represent important attempts to give possible answers to the problems left unresolved by them. The lessons learned by trying to implement the theoretical principles suggested by the IMP and the TOTE in the three architectures can be summarized as follows:

1. The first architecture accomplished a visual search task. It has a "goal node", which contains a test condition (similarly to the TOTE) having a sensorimotor encoding of two conditions: color and shape. Like the IMP, actions are preceded and triggered by a pre-activation of the desired goal state, but like the TOTE this happens as a consequence of a mismatch between the

pursued goal and the percepts. The search proceeds thanks to the learned action-expectation links, which are encoded in both the modules, which are procedures that attempted to "self-realize", and the links between them. Interestingly, to allow the architecture to function we had to design a mechanism for which the goal to pursue was selected through an activation with a level above zero (in order to trigger the search), but below the activation achieved when the state corresponding to it was actually achieved. In fact, had the pre-activation and the activation the same level, the test would have a positive outcome and the search would have immediately stopped. In several experiments it was also found that different initial amounts of pre-activation lead to different response times in finding a solution and can also lead to different search strategies. The interpretation of this was that such pre-activation encoded a measure of *urgency*. The IMP and the TOTE do not specify any mechanism to encode quantitative aspects of teleonomic behavior such as urgency: this is surely an important limitation of the two frameworks pointed out by the attempt to translate them into efficient computational systems.

2. The second architecture was a neural-network system directed to tackle reaching tasks with a simulated robotic arm. This architecture is based on the central idea of IMP related to the creation of the association between actions and their effects through exploratory experience and learning, and to the use of such effects as goals to suitably trigger actions. With this aspect the model implemented the "inversion" required by the IMP, from "actions to effects" to "effects/goals to actions" by actually creating two distinct neural mappings (even if on the basis of a common learning process). The implementation of the architecture also highlights the importance of testing the achievement of goals, similarly to what is suggested by the TOTE, to assign the responsibility of control either to the (reinforcement-learning) selector of goals/actions or to the components executing the actions themselves. In this respect, the implementation of the architecture again highlighted the necessity to have distinct representations of goals to pursue and of current states in order to perform such tests.

3. The last architecture, the XACS architecture, was a symbol-based architecture that can pursue different goals. It implements the IMP by directly forming a forward model of the environment, and by using this forward model to trigger action execution. In the TOTE it remains underspecified how goals may emerge and how they may trigger actions. Also the IMP does not specify how desired perceptual states are triggered, nor how the bidirectional sensorimotor knowledge can activate appropriate actions. XACS proposes an interlinked process that (1) activates all reachable (currently immediate) future states and (2) selects the action that leads to the currently most desirable one. Multiple goals may thus be active concurrently, leading to the pursuance of the currently most relevant and most reachable goal.

With this conceptualization and characterization of the IMP and the TOTE in hand, the next step along this line of research is to further investigate the many questions left open by the two principles, as well as to further identify the specific advantages and disadvantages stemming from actual implementations of them. Hereby, it will be important to use real-world simulations, or actual robotic platforms, to both identify the issues left unresolved, and to crystallize the true potential of the two anticipatory principles.

Acknowledgments. This work was supported by the EU funded projects *MindRACES - from Reactive to Anticipatory Cognitive Embodied Systems*, FP6-STREP-511931, and *ICEA - Integrating Cognition Emotion and Autonomy*, FP6-IP-027819.

References

1. Adams, J.A.: A closed-loop theory of motor learning. Journal of Motor Behavior 3, 111–149 (1971)
2. Barsalou, L.W.: Perceptual symbol systems. Behavioral and Brain Sciences 22, 577–600 (1999)
3. Butz, M.V.: Anticipatory learning classifier systems. Kluwer Academic Publishers, Boston, MA (2002)
4. Butz, M.V., Goldberg, D.E., Tharakunnel, K.: Analysis and improvement of fitness exploitation in XCS: Bounding models, tournament selection, and bilateral accuracy. Evolutionary Computation 11, 239–277 (2003)
5. Butz, M.V., Hoffmann, J.: Anticipations control behavior: Animal behavior in an anticipatory learning classifier system. Adaptive Behavior 10(2), 75–96 (2002)
6. Butz, M.V., Sigaud, O., Gérard, P.: Internal models and anticipations in adaptive learning systems. In: Butz, M.V., Sigaud, O., Gérard, P. (eds.) Anticipatory Behavior in Adaptive Learning Systems. LNCS (LNAI), vol. 2684, pp. 86–109. Springer, Heidelberg (2003)
7. Castelfranchi, C.: Mind as an anticipatory device: For a theory of expectations. In: BVAI 2005, pp. 258–276 (2005)
8. Drescher, G.L.: Made-Up Minds: A Constructivist Approach to Artificial Intelligence. MIT Press, Cambridge, MA (1991)
9. Duncan, J., Humphreys, G.W.: Visual search and stimulus similarity. Psychological Review 96, 433–458 (1989)
10. Fogassi, L., Ferrari, P., Chersi, F., Gesierich, B., Rozzi, S., Rizzolatti, G.: Parietal lobe: From action organization to intention understanding. Science 308, 662–667 (2005)
11. Gallese, V.: The inner sense of action: Agency and motor representations. Journal of Consciousness Studies 7, 23–40 (2000)
12. Gallese, V., Metzinger, T.: Motor ontology: The representational reality of goals, actions, and selves. Philosophical Psychology 13(3), 365–388 (2003)
13. Greenwald, A.G.: Sensory feedback mechanisms in performance control: With special reference to the ideomotor mechanism. Psychological Review 77, 73–99 (1970)
14. Grush, R.: The emulation theory of representation: Motor control, imagery, and perception. Behavioral and Brain Sciences 27(3), 377–396 (2004)
15. Herbart, J.: Psychologie als Wissenschaft neu gegründet auf Erfahrung, Metaphysik und Mathematik. Zweiter, analytischer Teil. August Wilhem Unzer, Königsberg, Germany (1825)

16. Hesslow, G.: Conscious thought as simulation of behavior and perception. Trends in Cognitive Sciences 6, 242–247 (2002)

17. Hoffmann, J.: Vorhersage und Erkenntnis: Die Funktion von Antizipationen in der menschlichen Verhaltenssteuerung und Wahrnehmung [Anticipation and cognition: The function of anticipations in human behavioral control and perception]. Hogrefe, Göttingen, Germany (1993)

18. Hoffmann, J.: Anticipatory behavioral control. In: Butz, M.V., Sigaud, O., Gérard, P. (eds.) Anticipatory Behavior in Adaptive Learning Systems. LNCS (LNAI), vol. 2684, pp. 44–65. Springer, Heidelberg (2003)

19. Hoffmann, J., Stöcker, C., Kunde, W.: Anticipatory control of actions. International Journal of Sport and Exercise Psychology 2, 346–361 (2004)

20. Hommel, B.: Planning and representing intentional action. TheScientificWorld JOURNAL 3, 593–608 (2003)

21. Hommel, B., Musseler, J., Aschersleben, G., Prinz, W.: The theory of event coding (tec): A framework for perception and action planning. Behavioral and Brain Science 24(5), 849–878 (2001)

22. James, W.: The Principles of Psychology. Dover Publications, New York (1890)

23. Knoblich, G., Prinz, W.: Higher-order motor disorders, chapter Linking perception and action: An ideomotor approach, pp. 79–104. Oxford University Press, Oxford, UK (2005)

24. Kohonen, T.: Self-Organizing Maps, 3rd edn. Springer-Verlag, Berlin Heidelberg, New York (2001)

25. Kosslyn, S.: Image and Brain: The Resolution of the Imagery Debate. MIT Press, Cambridge (1994)

26. Meltzoff, A.N., Moore, M.K.: Explaining facial imitation: A theoretical model. Early Development and Parenting 6, 179–192 (1997)

27. Miller, G.A., Galanter, E., Pribram, K.H.: Plans and the Structure of Behavior. Rinehart and Winston, New York (1960)

28. Minsky, M.: The Society of Mind. Simon & Schuster (1988)

29. Newell, A.: Unified Theories of Cognition. Harvard University Press, Cambridge, MA (1990)

30. Ognibene, D., Rega, A., Baldassarre, G.: A model of reaching integrating continuous reinforcement learning, accumulator models, and direct inverse modeling. In: Nolfi, S., Baldassarre, G., Calabretta, R., Hallam, J.C.T., Marocco, D., Meyer, J.-A., Miglino, O., Parisi, D. (eds.) SAB 2006. LNCS (LNAI), vol. 4095, pp. 381–393. Springer, Heidelberg (2006)

31. O'Regan, J., Noe, A.: A sensorimotor account of vision and visual consciousness. Behavioral and Brain Sciences 24(5), 883–917 (2001)

32. Pezzulo, G., Calvi, G.: A schema based model of the praying mantis. In: Nolfi, S., Baldassarre, G., Calabretta, R., Hallam, J.C.T., Morocco, D., Miglino, O., Meyer, J.-A. (eds.) SAB 2006. LNCS (LNAI), vol. 4095, pp. 211–223. Springer, Heidelberg (2006)

33. Pezzulo, G., Calvi, G., Ognibene, D., Lalia, D.: Fuzzy-based schema mechanisms in akira. In: CIMCA '05, Proceedings of the International Conference on Computational Intelligence for Modelling, Control and Automation and International Conference on Intelligent Agents, Web Technologies and Internet Commerce, vol. 2, pp. 146–152. IEEE Computer Society Press, Washington, DC, USA (2005)

34. Pollack, M.E.: Plans as complex mental attitudes. In: Cohen, P.R., Morgan, J., Pollack, M.E. (eds.) Intentions in Communication, pp. 77–103. MIT Press, Cambridge, MA (1990)

35. Prinz, W.: An ideomotor approach to imitation. In: Hurley, S., Chater, N. (eds.) Perspectives on imitation: From neuroscience to social science, vol. 1, pp. 141–156. MIT Press, Cambridge, MA (2005)
36. Rao, R.P., Ballard, D.H.: Predictive coding in the visual cortex: A functional interpretation of some extra-classical receptive-field effects. Nature Neuroscience 2(1), 79–87 (January 1999)
37. Rizzolatti, G., Arbib, M.A.: Language within our grasp. Trends in Neurosciences 21(5), 188–194 (1998)
38. Rizzolatti, G., Fadiga, L., Gallese, V., Fogassi, L.: Premotor cortex and the recognition of motor actions. Cognitive Brain Research, 3 (1996)
39. Rosenblueth, A., Wiener, N., Bigelow, J.: Behavior, purpose and teleology. Philosophy of Science 10(1), 18–24 (1943)
40. Roy, D.: Semiotic schemas: A framework for grounding language in action and perception. Artificial Intelligence 167(1-2), 170–205 (2005)
41. O. Selfridge. The Mechanisation of Thought Processes, chapter Pandemonium: A paradigm for learning, vol. 10, pp. 511–529. National Physical Laboratory Symposia. Her Majesty's Stationary Office, London (1959)
42. Sutton, R., Barto, A.: Reinforcement Learning: An Introduction. MIT Press, Cambridge MA (1998)
43. Watkins, C.J.C.H., Dayan, P.: Q-learning. Machine Learning 8, 279–292 (1992)
44. Widrow, B., Hoff, M.: Adaptive switching circuits. In: IRE WESCON Convention Record, pp. 96–104, Part 4. (1960)
45. Wilson, S.W.: Classifier fitness based on accuracy. Evolutionary Computation 3, 149–175 (1995)
46. Wolfe, J.M.: Visual search. In: Pashler, H. (ed.) Attention, University College London Press, London, UK (1996)
47. Wolpert, D.M., Kawato, M.: Multiple paired forward and inverse models for motor control. Neural Networks 11(7-8), 1317–1329 (1998)

Project "Animat Brain": Designing the Animat Control System on the Basis of the Functional Systems Theory

Vladimir G. Red'ko[1], Konstantin V. Anokhin[2], Mikhail S. Burtsev[3],
Alexander I. Manolov[4], Oleg P. Mosalov[4], Valentin A. Nepomnyashchikh[5],
and Danil V. Prokhorov[6]

[1] Center of Optical Neural Technologies, Scientific-Research Institute for System Studies,
Russian Academy of Sciences,
Vavilova Str., 44/2, Moscow, 119333, Russia
vgredko@gmail.com
[2] P.K. Anokhin Research and Development Institute of Normal Physiology, Russian Academy
of Medical Sciences, Mokhovaya Str., 11/4, Moscow, 103009, Russia
k_anokhin@yahoo.com
[3] M.V. Keldysh Institute for Applied Mathematics, Russian Academy of Sciences,
Miusskaya Sq., 4, Moscow, 125047, Russia
mbur@narod.ru
[4] Moscow Institute of Physics and Technologies,
Institutsky per., 9, Dolgoprudny, Moscow region, 141700, Russia
olegmos_@mail.ru, paraslonic@yandex.ru
[5] I.D. Papanin Institute for Biology of Inland Waters, Russian Academy of Sciences,
Borok, Yaroslavl region, 152742, Russia
nepom@ibiw.yaroslavl.ru
[6] Toyota Technical Center in Ann Arbor, MI, USA
dvprokhorov@gmail.com

Abstract. The paper proposes the framework for an animat control system (the Animat Brain) that is based on the Petr K. Anokhin's theory of functional systems. We propose the animat control system that consists of a set of functional systems (FSs) and enables predictive and purposeful behavior. Each FS consists of two neural networks: the actor and the predictor. The actors are intended to form chains of actions and the predictors are intended to make prognoses of future events. There are primary and secondary repertoires of behavior: the primary repertoire is formed by evolution; the secondary repertoire is formed by means of learning. This paper describes both principles of the Animat Brain operation and the particular model of predictive behavior in a cellular landmark environment.

Keywords: Animat control system, predictive behavior, learning, evolution.

1 Introduction

This paper proposes the framework for an animat control system (the Animat Brain) that is based on the biological theory of functional systems. This theory was proposed and developed in the period 1930-1970s by Russian neurophysiologist Petr K.

M.V. Butz et al. (Eds.): ABiALS 2006, LNAI 4520, pp. 94–107, 2007.

Anokhin [1] and pays special attention to prediction and anticipation of a final needful result of a goal-directed action.

There are a number of researches that analyze prediction and anticipation in animat control systems [2,3]. Tani investigated recurrent neural network (RNN) approach implementing predictive models for mobile robots [4,5]. Witkowski proposed the expectancy model that is based on a set of heuristic rules [6]. Butz et al [7] developed anticipatory learning classifier systems (ALCSs) that incorporate methods of reinforcement learning, genetic algorithm and earlier versions of classifier systems [8,9].

The main goal of our work is to propose the neural network (NN) animat control system that enables explicit models of predicted states. The architecture of the NN control system is formed by biologically plausible self-organizing processes. We also propose simple cellular environments that can be used in both biological and computer simulation experiments.

The ideas for our work are similar to that developed in Tani's research [4,5], however, we propose more distributed NN architecture as compared with RNN. The explicit NN models of predicted states in our approach are similar to sign-action-sign (SAS) relations in Witkowski's dynamic expectancy model [6]. A more detailed comparison of our approach with other works will be given at the end of the paper.

The paper is organized as follows. Section 2 outlines Anokhin's theory of functional systems. Section 3 describes principles of animat control system operation. A particular example of the proposed model is described in Section 4. Section 5 contains the discussion and conclusion.

2 Anokhin's Theory of Functional Systems

Functional systems were put forward by Petr K. Anokhin in the 1930s as an alternative to the predominant concept of reflexes [1]. Contrary to reflexes, the endpoints of functional systems are not actions themselves but adaptive results of these actions. According to the functional systems theory, initiation of each behavior is preceded by the stage of afferent synthesis (Figure 1). It involves integration of neural information from a) dominant motivation (e.g., hunger), b) environment (including contextual and conditioned stimuli), and c) memory (including innate knowledge and individual experience). The afferent synthesis ends with decision making, which results in selection of a particular program of an action.

A specific neural module, acceptor of the action result, is formed before the action itself. The acceptor stores an anticipatory model of the needful result of a goal-directed action. Such a model is based on a distributed neural assembly that includes various parameters (i.e., proprioreceptive, visual, auditory, olfactory) of the expected result. Execution of every action is accompanied by a backward afferentation. If parameters of the actual result are different from the predicted parameters stored in the acceptor of action result, a new afferent synthesis is initiated. In this case, a new functional system is formed and all operations of the functional system are repeated. Such processes take place until the final needful result is achieved.

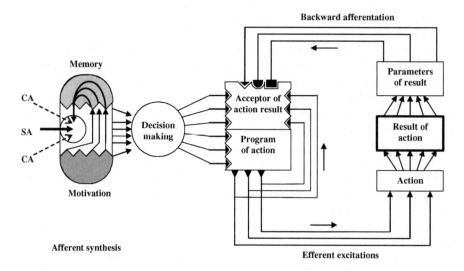

Fig. 1. General architecture of a functional system. SA is starting afferentation, CA is contextual afferentation. Operation of the functional system includes: 1) preparation for decision making (afferent synthesis), 2) decision making (formation of a program of an action), 3) prognosis of the action result (generation of acceptor of action result), 4) backward afferentation (comparison between the result of action and the prognosis).

A separate branch of the general functional system theory is the theory of systemogenesis that studies mechanisms of functional systems formation during 1) evolution, 2) individual or ontogenetic development, and 3) learning. In the current paper we consider two of these mechanisms: evolution and learning.

3 Architecture and Principles of Operation of the Animat Brain

It is supposed that the animat control system consists of neural network (NN) blocks and is analogous to an animal control system. Each block is a formal functional system (FS). At any moment in time ($t = 1,2,...$), only one FS is active, in which the current action is formed. There are connections between FSs; the active FS can transmit activation to every FS through these connections.

Each FS consists of two NNs: the actor and the predictor. Operation of the active FS can be described as follows. The state vector $S(t)$ characterizing the current external and internal environment is fed to the FS input. The actor forms the action $A(t)$ in accordance with given state $S(t)$, i.e. the actor forms the mapping $S(t) \rightarrow A(t)$. The predictor makes prognosis of the next state for given vectors $S(t)$ and $A(t)$, i.e. the predictor forms the mapping $\{S(t), A(t)\} \rightarrow S^{pr}(t+1)$. So, the predictor stores a model of casual relation between the current state $S(t)$, action $A(t)$ and the next state $S(t+1)$. The prediction $S^{pr}(t+1)$ of the next state corresponds to the acceptor of action result in the functional system theory. The mappings $S(t) \rightarrow A(t)$ and $\{S(t), A(t)\} \rightarrow S^{pr}(t+1)$ are stored in NN synaptic weights.

Activation is transmitted from one FS to others in accordance with connectivity matrix C_{ij}. The value C_{ij} characterizes the probability that the j-th FS is activated by the i-th FS.

The animat receives reinforcements (rewards and punishments) which are related to animat needs.

It is supposed that there are primary and secondary repertoires of behaviors. The primary repertoire is formed by evolution: there is a population of animats and a set of FSs. The corresponding synaptic weights of the involved NNs and the connectivity matrix C_{ij} are adjusted during the evolutionary processes.

The secondary repertoire of behavior is formed by learning. There are two regimes of learning: 1) the extraordinary mode and 2) the fine tuning mode.

The extraordinary mode is a rough search of behavior that is adequate to the current situation. This mode comes, if the predicted state $S^{pr}(t+1)$ in the active FS strongly differs from the real state $S(t+1)$. In terms of the functional system theory, a large difference between $S^{pr}(t+1)$ and $S(t+1)$ means that parameters of the result differ essentially from parameters stored in the acceptor of action result.

In the extraordinary mode, a random search for new behaviors takes place; namely, the connectivity matrix C_{ij} is substantially changed and new FSs can be randomly generated and selected. This mode is similar to neural group selection in Edelman's theory of Neural Darwinism [10].

In the fine tuning mode, learning is adjustment of NN weights in the FS that is active at the current moment of time and in the FSs that were active in several previous steps of time. As synaptic weights are updated in those NNs, which were active in previous time steps, this learning mode allows forming chains of consecutive actions. Synaptic weights in predictors are modified to minimize prediction errors (e.g. by means of error back-propagation [11]). Synaptic weights in actors are adjusted by a Hebbian-like rule: the synaptic weights in actors are modified to make the mappings $S(t) \rightarrow A(t)$ stronger/weaker for positive/negative reinforcements.

We introduce two modes of learning, that is, the extraordinary and the fine tuning mode, for the following reasons:

1) We believe that learning by means of these two modes (rough search in a new situation and fine tuning in a partially known situation) is more effective as compared with one mode.

2) Analogies to the fine tuning and extraordinary modes can be seen in animal behavior. For example, bees, butterflies and other insects are able to crop pollen or nectar from various flower species, but individual insects tend to choose flowers of a particular species, while ignoring others. The reason for this so called "flower constancy" is that different flower species differ by their structure, so insects should learn a structure of particular flowers to extract food efficiently. Learning requires numbers of visits to the same flower species, resulting in a gradual decline in handling time on successive visits [12,13]. This learning is analogous to our fine tuning mode. If a production of food by preferred flower species falls low, then an insect starts to sample various other species [12] (an extraordinary mode).

An extraordinary mode also can be seen under artificial conditions unfamiliar to an animal. For example, certain species of jumping spiders live on trees and have no experience with water spaces. When faced with a water-filled tray in the laboratory for the first time, they may choose one of only two solutions available for them: jump over the tray or "swim" (in fact, walk) across it. Individuals, which attempt to swim first and fail to reach the opposite side of the tray, switch to jumping, if allowed a next trial. Those animals, which jump first and fail, switch to swimming (the extraordinary mode). If a spider reaches the opposite side, it repeats a successful behavior (swimming or jumping) in the course of next trials. Once an appropriate behavior is chosen, only minute quantitative details of this behavior are varied [14]. One may speculate that these minute variations help to improve spider's performance (the fine tuning mode of learning).

Existence of extraordinary and fine tuning modes in various animal species suggests that a learning based on two very different modes could be of adaptive value.

A particular version of the Animat Brain model is described in the next section. It should be underlined that we propose only a possible version of the model. In order to ensure that all components of the model are consistent with each others, we describe concrete possible mechanisms of NNs operation, leaning, and evolution. In the current work we consider simplest variants of these mechanisms; in further research these mechanisms could be replaced by similar ones. We propose also a certain landmark environment that can be used to compare behavior of simulated and real animals in the same model "world".

4 Particular Model of Animat Brain Operation

4.1 Animat Environment and Features

Environment. We assume simple 2D cellular landmark environment (figures 2 and 3). Any marked cell A, B, C, D, G has its own landmark. The modeled "world" is restricted by impenetrable barriers. The animat sensory system is able to perceive the state of a marked cell (5 different signals), an unmarked cell and a cell of the barrier. So, there are 7 different possible signals from cells. The goal cell is G.

Animat Features. An animat senses its local environment and executes some actions. Actions are executed in accordance with the commands of the active FS of the animat control system. At any moment in time, the animat executes one of the following five actions:

1- 4) to move one cell up/down/right/left,
5) to wait.

The animat has internal energy resource R. Performing actions, an animat spends its resource. We suppose that at every movement (actions 1-4), the animat resource is decreased by r_1, and when waiting (action 5), the decrease of the resource is negligible. Reaching the goal cell G, the animat increases its resource by r_2.

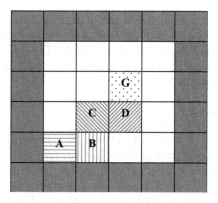

Fig. 2. Simple cellular environment. The landmarks A, B, C, D, G are in adjacent cells. The goal cell is G. The "world" consists of 4x4 cells; it is surrounded by impenetrable barriers (grey cells)

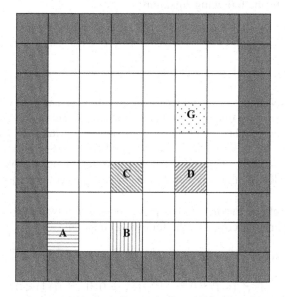

Fig. 3. The cellular environment that is similar to the "world" in Figure 2, but the landmarks A, B, C, D, G are separated by one cell distance

Animat Sensory System. The animat perceives the states of five cells: its own cell and the four surrounding cells (up/down/right/left). In each of the four cells, the animat estimates one of 7 signals (5 kinds of landmarks, unmarked cell or barrier cell); in its own cell, the animat estimates one of 6 signals (5 kinds of landmarks or unmarked cell). For definiteness we suppose that every such signal is a binary component (1 or 0) of the state vector $S(t)$. Also the animat perceives its resource $R(t)$ and the resource change for the last time step $R(t) - R(t-1)$. As other signals take values 0 or 1, it is convenient to characterize resource and resource change by binary values too. So, we assume that the animat estimates the binary values S_R and S_{DR}, where $S_R = 1$ if $R(t) > R_T$ and $S_R = 0$ if $R(t) < R_T$ (R_T is a predetermined threshold resource value), and $S_{DR} = 1$ if $R(t) > R(t-1)$ and $S_{DR} = 0$ if $R(t) < R(t-1)$. Thus, the animat perceives 36 binary parameters, characterizing its external and internal environment. These 36 values form the current state vector $\mathbf{S}(t)$.

4.2 Animat Control System

The animat control system is a set of FSs; each FS consists of two neural networks: the actor and the predictor. At any moment in time, only one FS is active, in which the current action is formed; after performing its operation, this FS transmits activation to other FSs. The new active FS is chosen probabilistically. The probability that the j-th FS is activated by the i-th FS is equal to $C_{ij} / (\sum_k C_{ik})$, with C_{ij} as the element of the connectivity matrix ($C_{ij} > 0$).

Neural Network of the Actor. The actor is a two layer NN. The operation of the actor is described by the following equations:

$$\mathbf{x}^A = \mathbf{S}(t), \quad y^A_j = \text{th} \left(\sum_i w^A_{ij} x^A_i \right), \quad z^A_k(t) = F\left(\sum_j v^A_{jk} y^A_j \right), \tag{1}$$

$$F(a) = 1/[1+\exp(-a/b)], \tag{1a}$$

where $\mathbf{x}^A$ is the NN input vector, which is equal to the current state vector $\mathbf{S}(t)$, $\mathbf{y}^A$ is the vector of hidden layer outputs, $z^A_k(t)$ are signals of output layer neurons, w^A_{ij} and v^A_{jk} are NN synaptic weights, F(.) is the sigmoid activation function, and parameter b regulates the slope of this function. The probability that m-th action is selected is equal to $z^A_m(t) / \sum_k z^A_k(t)$. The action vector $\mathbf{A}(t)$ is determined as follows: $A_m(t) = 1$, if m-th action is selected, all other components of $\mathbf{A}(t)$ are set to be equal to 0.

Neural Network of the Predictor. The predictor is also a two layer NN. The operation of the predictor is described by the following equations:

$$\mathbf{x}^P = \{\mathbf{S}(t), \mathbf{A}(t)\}, \quad y^P_j = \text{th}\left(\sum_i w^P_{ij} x^P_i \right), \quad z^P_k(t+1) = F_1\left(\sum_j v^P_{jk} y^P_j \right), \tag{2}$$

$$S^{pr}_k(t+1) = 1 \text{ if } z^P_k(t+1) > 0.5, \quad S^{pr}_k(t+1) = 0 \text{ if } z^P_k(t+1) < 0.5, \tag{2a}$$

$$F_1(a) = 1/[1+\exp(-a)], \tag{2b}$$

where $\mathbf{x}^P$ is the NN input vector, which is the compound vector $\mathbf{x}^P = \{\mathbf{S}(t), \mathbf{A}(t)\}$, $\mathbf{y}^P$ is the vector of hidden layer outputs, w^P_{ij} and v^P_{jk} are NN synaptic weights, $z^P_k(t+1)$ are signals of output layer neurons, $S^{pr}_k(t+1)$ are components of the predicted state vector $\mathbf{S}^{pr}(t+1)$.

Equations (1,2) describe the simple version of NNs operation that ensures both natural schemes of learning (see below) and binary components of vectors $\mathbf{A}(t)$ and $\mathbf{S}^{pr}(t+1)$.

4.3 Learning Mechanism

There are two regimes of learning: 1) the extraordinary mode and 2) the fine tuning mode.

The extraordinary mode occurs, if there is a strong mismatch between the expected and real results: the predicted state $S^{pr}(t+1)$ in the active FS essentially differs from the real state $S(t+1)$. A strong mismatch means the difference in essential components of vectors $S^{pr}(t+1)$ and $S(t+1)$: for example, the increase of the animat resource was expected, but really the resource was reduced.

In order to define essential components, we introduce a mask for every block. The mask is the vector M of dimension 36; this vector has components that are equal to 0 or 1. The unit components of the vector M define essential components of the state vector $S(t+1)$. Namely, the component $S_k(t+1)$ is determined as essential, if $M_k =1$. If $M_k = 0$, the component $S_k(t+1)$ is considered as inessential. The essential components determine, which causal relation between the current state $S(t)$, current action $A(t)$ and next state $S(t+1)$ is checked by the given predictor.

In the current version of our model it is supposed that the extraordinary mode of learning occurs as follows: a) activation is returned back to the i-th FS, that activated the current j-th FS, b) the element of the connectivity matrix C_{ij} corresponding to the link between these two FSs is changed.

The change of connection value C_{ij} occurs as follows. First, this value C_{ij} strongly decreases in the next time moment $t+1$, at which the i-th FS repeats activation of other FSs. At this moment, the temporary value of connection C_{ij}^{Temp} is used, and then there is a return to the usual connection value C_{ij} :

$$C_{ij}^{Temp}(t+1) = K_1\, C_{ij}(t)\,. \tag{3a}$$

Secondly, the connection value C_{ij} slightly decreases in long-term manner:

$$C_{ij}(t+2) = K\, C_{ij}(t)\,, \tag{3b}$$

where $0 < K_1 < K < 1$. For example, we can set $K_1 = 0.1$, $K = 0.9$.

The described scheme of adjusting the connection value C_{ij} suggests that activation is transferred with high probability from the j-th FS that performed "unsatisfactory" action at the moment t to some other FS and then the probability to activate the j-th FS in future is slightly reduced.

Learning in extraordinary mode means that there is certain reorganization of animat control system operation. It is also possible to implement random generation and selection of new FSs in the extraordinary mode of learning; we intend to consider this option in further versions of the Animat Brain.

During *the fine tuning mode*, learning occurs by adjusting NN synaptic weights. This learning takes place when there is no strong mismatch between the expected and obtained result. Learning in actors and predictors occurs in different ways.

Learning in actors occurs according to reinforcements. Synaptic weights are adjusted in the FS that is active at the current moment of time t, and in FSs, that were active in several previous steps of time. These synaptic weights are modified as follows:

$$\Delta W_{ij} = \alpha_A \, \gamma^k \, X_i(t-k) \, Y_j(t-k) \, [R(t) - R(t-1)] \, , \tag{4}$$

where W_{ij} is the weight of the considered synapse, $X_i(t-k)$ is the signal on the synapse input, $Y_j(t-k)$ is the output of the neuron corresponding to the given synapse, α_A is learning rate of actors, γ is the discount factor ($0 < \gamma < 1$), k is the difference between the current moment of time and the time of operation of the considered FS, $[R(t) - R(t-1)]$ is the value of the current reinforcement.

As learning occurs in those actors, which were active in several previous steps of time, this type of training allows the formation of chains of actions.

Learning in the predictor occurs, if there is a mismatch between the prediction $S^{pr}(t+1)$ and the result $S(t+1)$ in any components of these vectors.

Learning in the predictor is carried out by the usual method of error back-propagation [11]. At this learning the target vector is $S(t+1)$, and the NN output vector (that is compared with the target vector) is the vector $z^P(t+1)$, that is formed at the output layer of the predictor NN, see formulas (2).

In addition to fine tuning mode, we consider *learning upon achievement of the final needful result* (see description of the functional system theory in Section 2). We suppose that, upon achievement of the final needful result, there is strengthening connections between several FSs, which were active immediately before achievement of this result. In the current model the final needful result corresponds to reaching the goal cell G. For this type of learning connections between FSs are modified as follows:

$$\Delta C_{ij} = \alpha_L \, (\gamma_L)^k \, r_2 \, , \tag{5}$$

where C_{ij} is the connection between considered FSs, α_L and γ_L are learning rate and the discount factor for this type of learning, k is the difference between the reward time and the time of considered activation transfer, r_2 is the value of the reinforcement in the cell G.

4.4 Evolution Mechanism

We consider a simple genetic algorithm (GA) [15,16] that can be described as follows. An evolving population consists of n animats. Evolution passes through a number of generations, $n_g = 1,2,...$ At any generation, each animat is tested during T time steps independently of other animats of the population. At the beginning of the test, the animat resource $R(t)$ is set to certain predetermined value R_0 and the animat itself is set into the cell A. Then the animat acts in accordance with its control system and its resource is changed according to reinforcements. When the animat reaches the goal cell G and receive the reward r_2, it is returned to the start cell A. Such process is repeated, until the time T is over. After testing all n animats, the transition to the new generation occurs. At this moment, the animat having the maximum resource $R_{max}(n_g)$ is determined. This best animat gives birth to n children that constitute a new (n_g+1)-th generation.

The initial architecture of the animat control system (the set of FSs and the connectivity matrix C_{ij}) as well as initial synaptic weights of NNs form the animat genome **G**. The genome **G** is received at animat birth and is not changed during animat life. It is transferred (with small mutations) from the parent (the best animat of n_g-th generation) to descendants (all animats of (n_g+1)-th generation). Temporary architecture and synaptic weights of the NNs are changed during animat life via learning described in section 4.3.

At the beginning of (n_g+1)-th generation, the genome **G** of each newborn animat is determined: the offspring genomes are obtained from the genome of the parent through mutations that include:

1) duplication (with certain probability P_D) of every existing FS;
2) forming of elements of the connectivity matrix C_{ij}, corresponding to new FSs;
3) removing (with certain probability P_R) of every existing FS;
4) small random variations of elements of the connectivity matrix C_{ij} and synaptic weights of all NNs;
5) small random variations of the mask vector **M** for every predictor.

The described evolution mechanism is the simple version of the GA that takes into account all compounds of the current Animat Brain model. Similar and more sophisticated versions of the GA [15,16] could be used in future research.

4.5 Interaction Between Selection of Actions and Predictions

In the current model, we pay special attention to predictions of future states. We suppose that essential learning takes place in the extraordinary mode, when there is a large difference between predictions and results of action. This implies that chains of actions (formed by actors) should correspond to predictions (formed by predictors).

For example, consider the "world" shown in Figure 3. When the animat placed in the cell A moves two times right, it should be able to predict the movement into an unmarked cell after the first step and into the landmark cell B after the second step. Moving further two times upwards, it should predict the displacement into an unmarked cell and into the landmark cell C after the first and second steps, respectively. Then it should be able to predict movements to the landmark cells D and G. In principle, the animat can find an alternative path to the goal cell G, however, using landmarks, it is able to find the reliable path. Chains of actions and predictions should be in agreement with each other for reliable behavior.

Thus, we plan to analyze, how the agreement between chains of actions and predictions can be formed through learning and evolution in the current model.

5 Discussion and Conclusion

Comparison with Other Approaches. As was stated in the introduction, our approach is similar to models by Tani [4,5], who models predictive behavior of mobile robots using RNNs. As compared with Tani's works, our model provides more explicit representation of states $S(t)$, actions $A(t)$ and predictions $S^{pr}(t+1)$.

Referring to Witkowski work [6] and research by Butz et al [7], we can note that our NN approach is based on the biological theory of functional systems [1] and we believe that it will be more flexible as compared with rule-based methods used in [6,7].

We can also compare our approach with works by Edelman et al, who investigate adaptive behavior that is controlled by huge NN control systems [17,18]. Our approach is at intermediate positions between small NN control system investigated in [4,5] and very large NN "brains" simulated in [17,18].

Our model includes two types of learning: 1) the extraordinary mode and 2) the fine tuning mode; and this can provide additional advantages as compared with similar models [4-7,17,18].

In our previous work, we designed Animat Brain architecture that is based on the reinforcement learning (RL) and consists of a set of hierarchically linked FSs [19]. Every FS is a simple adaptive critic design (ACD) that consists of two NNs: the model (predictor) and the critic. The model is intended to predict the next state $S(t+1)$ for given current state $S(t)$ and all possible actions a_i (the number of actions a_i is supposed to be small). The critic is intended to estimate state value function $V(S(t))$. Actions are chosen in accordance with ε-greedy rule [8] ensuring selection of those actions that maximize state values V. However, analyzing evolution and learning in populations of such adaptive critics [20], we observed that ACD operation can be evolutionary unstable. This is due to the necessity to estimate state value function $V(S)$; these estimations impose too strong a restriction on adaptive agent functioning. In the current work we introduce Hebbian-like learning modulated by rewards and punishments instead of the usual RL scheme. A similar viewpoint on RL and evolution was expressed by Stanley, Bryant and Miikkulainen [21], who emphasized that discovering complex NN control systems of adaptive agents by means of evolution is more effective than RL. In contrast to neuroevolution method [21], our schemes of search for adaptive behavior by both evolution and learning correspond to the biologically inspired concept of primary (formed by evolution) and secondary (formed by learning) repertoire of behaviors.

Our approach is similar to works by Wolpert, Kawato et al [22,23] on multi-modular NN systems for motor control. The architectures investigated in [22,23] include multiple pairs of inverse (controller) and forward (predictor) models. The inverse model is similar to the actor in our architecture; the forward model plays the same role as the predictor in our schemes. It should be noted that we consider the control system of an autonomous animat, whereas Wolpert, Kawato et al analyze learning at human motor control that corresponds to psychological experiments on movement of different objects at different conditions by an arm.

It should be underlined that the simulation of adaptive behavior in landmark "worlds" proposed in this work (figures 2 and 3) can be used to compare different approaches, such as RNN [4,5], adaptive critic designs [19], brain-inspired NN control system [17,18], ALCSs [7], and distributed NN-based FSs.

Possible Variations on the Proposed Model. One of the difficulties of the current model is the too large dimension of state vectors $S(t)$ that include 36 components. To overcome this difficulty we can consider more specialized FSs. A particular FS can perceive only a small subset of parameters from the local environment. For example, the FS that is responsible for movement from cell A to cell B (Figure 2) can perceive only landmarks A and B, only in left and right cells. Such specialization can be implemented by means of mask vectors M^* that have components 0 or 1. Parameters corresponding to zeros ($M^*_k = 0$) are not included into state vectors for the considered FS. This option can provide a distributed animat control system, in which many small specialized FSs constitute the whole Animat Brain. The specialized FSs can be formed through evolution and extraordinary mode of learning. It should be noted that this scheme of small specialized FSs is similar to the multi-modular architecture that was proposed and investigated in [22,23]. We can also consider the concept of module responsibility from [22,23] in order to organize a flow of FS activity throughout the Animat Brain architecture.

Figures 2 and 3 show simple landmark "worlds". Obvious generalizations and variations are possible: several different goals can be introduced; the landmark distribution can be unstable, noisy, etc.

Biological Aspects. We propose to investigate animat behavior in landmark environments (simple examples of which are shown in figures 2 and 3). This is interesting from a biological viewpoint for the following reasons:

-    It is possible to design the cellular "world" with exactly the same structure for real biological experiments. Namely, we can construct the 2D array of cells with nontransparent walls between cells, color floor in certain cells by different landmarks and make a door between every neighboring pair of cells. Any door is automatically closing but it can be opened by an investigating animal.
-    Landmarks are really used by animals in adaptive behavior. For example, honey bees use landmarks for efficient goal navigation [24].
-    In some biological experiments, such as investigations of rat orientation in a Morris water maze [25], animals seem to be able to select and use landmarks to find a goal.

So, we can state that it is possible to compare the goal-directed behavior of simulated and real animals in proposed landmark environments.

Conclusion. We proposed the biologically inspired Animat Brain architecture that consists of a set of functional systems (FSs). Every FS includes two NNs: the actor and the predictor, and provides action selection and predictions of action results. In the case of unexpected events, considerable learning takes place and animat behavior is reorganized. We intend to study conditions for which predictions of future events (formed by predictors) and generations of action chains (formed by actors) are consistent with each other. We also propose to investigate the predictive animat behavior in landmark environments that ensure comparison of behavior of simulated and real animals in the same model "world".

Acknowledgments. This work is supported in part by the Russian Foundation for Basic Research, Grants No 05-07-90049 and No 04-01-00510, the Program No 14 of the Presidium of the Russian Academy of Science on "Intelligent Technologies and Mathematical Modeling", and "Russian Science Support Foundation". The authors thank Martin Butz, Stewart Wilson and anonymous reviewers for valuable comments on the earlier versions of this paper.

References

1. Anokhin, P.K.: Biology and Neurophysiology of the Conditioned Reflex and Its Role in Adaptive Behavior. Pergamon, Oxford (1974)
2. Butz, M.V., Sigaud, O., Gérard, P. (eds.) Anticipatory Behavior in Adaptive Learning Systems. LNCS (LNAI), vol. 2684, Springer, Heidelberg (2003)
3. Butz, M.V., Sigaud, O., Gérard, P.: Internal Models and Anticipations in Adaptive Learning Systems. In: Butz, M.V., Sigaud, O., Gérard, P. (eds.) Anticipatory Behavior in Adaptive Learning Systems. LNCS (LNAI), vol. 2684, pp. 86–109. Springer, Heidelberg (2003)
4. Tani, J.: Model-Based Learning for Mobile Robot Navigation from the Dynamical Systems Perspective. IEEE Trans. on Systems, Man, and Cybernetics. Part B: Cybernetics 26, 421–436 (1996)
5. Paine, R W., Tani, J.: How Hierarchical Control Self-organizes in Artificial Adaptive Systems. Adaptive Behavior 13, 211–225 (2005)
6. Witkowski, M.: Towards a Four Factor Theory of Anticipatory Learning. In: Butz, M.V., Sigaud, O., Gérard, P. (eds.) Anticipatory Behavior in Adaptive Learning Systems. LNCS (LNAI), vol. 2684, pp. 66–85. Springer, Heidelberg (2003)
7. Butz, M.V., Goldberg, D.E.: Generalized State Values in an Anticipatory Learning Classifier System. In: Butz, M.V., Sigaud, O., Gérard, P. (eds.) Anticipatory Behavior in Adaptive Learning Systems. LNCS (LNAI), vol. 2684, pp. 282–302. Springer, Heidelberg (2003)
8. Sutton, R.S., Barto, A.G.: Reinforcement Learning. An Introduction. In: A Bradford Book, MIT Press, Cambridge, MA (1998)
9. Holland, J.H., Holyoak, K.J., Nisbett, R.E., Thagard, P.: Induction: Processes of Inference, Learning, and Discovery. MIT Press, Cambridge, MA (1986)
10. Edelman, G.M.: Neural Darwinism: The Theory of Neuronal Group Selection. Oxford University Press, Oxford (1989)
11. Rumelhart, D.E., Hinton, G.E., Williams, R.G.: Learning Representation by Back-Propagating Error. Nature 323, 533–536 (1986)
12. Chittka, L., Thomson, J.D., Waser, N.M.: Flower Constancy, Insect Psychology, and Plant Evolution. Naturwissenschaften 86, 361–377 (1999)
13. Goulson, D., Stout, J.C., Hawson, S.A.: Can Flower Constancy in Nectaring Butterflies Be Explained by Darwin's Interference Hypothesis? Oecologia 112, 225–231 (1997)
14. Jackson, R.R., Carter, C.M., Tarsitano, M.S.: Trial-and-error Solving of Confinement Problem by Araneophagic Jumping Spiders, Portia Fimbriata. Behaviour 138, 1215–1234 (2001)
15. Holland, J.H.: Adaptation in Natural and Artificial Systems. The University of Michigan Press, Ann Arbor, MI (1975). 2nd edn. MIT Press, Boston, MA (1992)
16. Goldberg, D.E.: Genetic Algorithms in Search, Optimization and Machine Learning. Addison-Wesley, London, UK (1989)
17. Krichmar, J.L., Edelman, G.M.: Machine Psychology: Autonomous Behavior, Perceptual Categorization and Conditioning in a Brain-Based Device. Cerebral Cortex 12, 818–830 (2002)
18. Krichmar, J.L., Seth, A.K., Nitz, D.A., Fleischer, J.G., Edelman, G.M.: Spatial Navigation and Causal Analysis in a Brain-Based Device Modeling Cortical-Hippocampal Interactions. Neuroinformatics 3, 197–221 (2005)
19. Red'ko, V.G., Prokhorov, D.V., Burtsev, M.S.: Theory of Functional Systems, Adaptive Critics and Neural Networks. In: Proc. International Joint Conference on Neural Networks (IJCNN 2004), Budapest, pp. 1787–1792 (2004)

20. Red'ko, V.G., Mosalov, O.P., Prokhorov, D.V.: A Model of Evolution and Learning. Neural Networks 18, 738–745 (2005)
21. Stanley, K.O., Bryant, B.D., Miikkulainen, R.: Evolving Neural Network Agents in the NERO Video Game. In: Proceedings of the IEEE, Symposium on Computational Intelligence and Games (CIG'05), Essex University, Colchester, Essex, UK, pp. 182–189 (2005)
22. Wolpert, D.M., Kawato, M.: Multiple Paired Forward and Inverse Models for Motor Control. Neural Networks 11, 1317–1329 (1998)
23. Haruno, M., Wolpert, D.M., Kawato, M.: MOSAIC Model for Sensorimotor Learning and Control. Neural Computation 13, 2201–2220 (2001)
24. Fry, S.N., Wehner, R.: Look and Turn: Landmark-Based Goal Navigation in Honey Bees. The. Journal of Experimental Biology 208, 3945–3955 (2005)
25. Morris, R.G.M., Garrud, P., Rawlins, J.N.P., O'Keefe, J.: Place Navigation Impaired in Rats with Hippocampal Lesions. Nature. 297, 681–683 (1982)

Cognitively Inspired Anticipatory Adaptation and Associated Learning Mechanisms for Autonomous Agents

Aregahegn Negatu, Sidney D'Mello, and Stan Franklin

Department of Computer Science and the Institute for Intelligent Systems
The University of Memphis
Memphis, TN 38152, USA
{asnegatu,sdmello,franklin}@memphis.edu

Abstract. This paper describes the integration of several cognitively inspired anticipation and anticipatory learning mechanisms in an autonomous agent architecture, the Learning Intelligent Distribution Agent (LIDA) system. We provide computational mechanisms and experimental simulations for variants of payoff, state, and sensorial anticipatory mechanisms. The payoff anticipatory mechanism in LIDA is implicitly realized by the action selection dynamics of LIDA's decision making component, and is enhanced by importance and discrimination factors. A description of a non-routine problem solving algorithm is presented as a form of state anticipatory mechanism. A technique for action driven sensational and attentional biasing similar to a preafferent signal and preparatory attention is offered as a viable sensorial anticipatory mechanism. We also present an automatization mechanism coupled with an associated deautomatization procedure, and an instructionalist based procedural learning algorithm as forms of implicit and explicit anticipatory learning mechanisms.

1 Introduction

It is widely acknowledged that adaptive behavior, an essential component of intelligence, is enhanced by anticipatory activities, where predictions of the future modulate and influence current decision making. Simply put, organisms survive by anticipating the future. While the role of anticipations on deliberation, memory, attention, behavior, and other facets of cognition has been well studied in cognitive psychology, neuropsychology, and ethology, the literature on explicit mechanisms to realize anticipations in artificial agents is considerably more sparse and scattered (Blank, Lewis, & Marshall, 2005; Butz, Sigaud, & Gerard, 2002; Kunde, 2001; Rosen, 1985; Schubotz & von Cramon, 2001). Since anticipations have been acknowledged to be an influential component of the cognitive facilities of humans (and other animals), the need to model and integrate theories of anticipations in our artificial systems becomes essential.

Over the last decade a variety of mechanisms that realize anticipations in artificial systems have been proposed. These include reinforcement learning systems that are model-free (Watkins, 1989) as well as (predictive) model based such as Drescher's

M.V. Butz et al. (Eds.): ABiALS 2006, LNAI 4520, pp. 108–127, 2007.

Schema mechanism (1991), Sutton's dynamical architecture (1991), and the expectancy model proposed by Witkowski (2002). Learning classifier systems, that make explicit predictions of future states, have also been widely used as anticipatory mechanisms (e.g., Stolzmann's, 1998 - ACS system, the YACS system - Gerard, Stolzmann, and Sigaud, 2002). Artificial neural network based anticipatory systems include an attention mechanism proposed by Baluja and Pomerleau (1995) and Tani's recurrent neutral network based model learning and planning mechanism (1996). More recently, anticipatory mechanisms are being implemented in developmental and evolutionary robots (Blank, Lewis, & Marshall, 2005; Hartland & Bredeche, 2006). Butz, Sigaud, and Gerard (2002) have provided a useful nomenclature for a variety of functional anticipatory processes found in humans and animals and computationally realized by artificial systems. They recognize four fundamental types of anticipatory systems that include *payoff, sensorial, state, and implicitly* anticipatory systems. The fundamental difference between implicit and the other three anticipatory systems is that in implicitly anticipatorial systems no explicit predictions about the future are made, even though the structure of the action selection component must contain certain anticipatory elements. Sensorial anticipation differs from payoff and state anticipatory mechanisms in that the predictions influence both early and later stages of sensory processing without having a direct impact on action selection. Finally, the main difference between payoff and state anticipatory mechanisms is that in payoff anticipatory systems anticipations play a role as payoff predictions only and explicit predictions of future states are not made. On the other hand state anticipatorial mechanisms make explicit predictions of future states during decision making processes.

Our interest in anticipation and anticipatory learning mechanisms emerges from a desire to model several facets of human (and animal) cognition in an autonomous agent the Learning Intelligent Distribution Agent (LIDA). LIDA is the partially conceptual, learning extension, of the original IDA system implemented computationally as a software agent (D'Mello et al., 2006). The original IDA system was designed as an autonomous agent, and performed personnel work for the US Navy in a human-like fashion (Franklin, 2001). Although the design of IDA was inspired by several theories of human and animal cognition, it did not learn. The LIDA system adds three fundamental forms of learning to IDA: perceptual, procedural, and episodic learning.

As mentioned above, over the last few years there has been a sustained effort to devise computational mechanisms for anticipations and to incorporate these algorithms into existing or new animats. Although our approach shares several similarities to well established research along this avenue, we can point out two substantial differences. First, we describe the manner in which computational mechanisms for several anticipatory facilities may be integrated into a large working model of cognition. While several of the existing cognitive models do incorporate some sort of anticipatory element, a degree of uniqueness in our approach ensues due to the fundamental difference between LIDA and other cognitive models. Most of the cognitive models developed by AI researchers and cognitive scientists are designed around some unified theory of cognition (Newell, 1990). Some of these well known models include SOAR (Laird, Newell, & Rosenbloom, 1987), ACT-R (Lebiere & Anderson, 1993), and Clarion (Sun, 1997). Most of these are based on some extension of Post production systems (Post, 1965). In contrast, the LIDA model is based on a number of mostly psychological theories of cognition. These include: situated or embodied cognition (Varela, Thompson, & Roach 1991, Glenberg, 1997), Barsalou's theory of perceptual symbol systems (1999), working memory (Baddeley & Hitch, 1974), Glenberg's theory (1997) of the importance of affordances to understanding, Baars' global workspace theory (GWT) (1988), and Sloman's architecture for a human-like agent (1999). We argue that basing our anticipatory mechanisms on functional aspects of human cognition may serve dual purposes by yielding both engineering and scientific gains. We expect engineering improvements because we are basing our computational mechanism on the best known example of intelligence, i.e. humans. Scientific gains can be achieved by using computer systems to test and perhaps augment psychological theories of anticipatory adaptation.

One potential pitfall of relying on psychological theories is that they typically model only small pieces of cognition. In contrast, by its very nature the control system of any autonomous agent or cognitive robot must be fully integrated. That is, it must chose its actions based on real world sensation and perception along with incoming endogenous stimuli utilizing *all* needed internal processes. Once again the use of the LIDA system helps to divert this problem, while at the same time highlighting our second major contribution to anticipatory research in artificial systems. Due to its very breadth we are able to develop and experiment with a variety of anticipatory mechanisms. In this paper we describe computational algorithms for several anticipatory mechanisms, as used in the LIDA model. Sensory anticipation is accomplished in LIDA via a preafferent signal (Kay et al., 1996), and a preparatory attentive process (LaBerge, 1995), sent upon the decision to take an action, that biases LIDA's perceptual and attentive mechanisms in favor of the anticipated sensory information (see 3.3 below). Payoff anticipation is implemented implicitly by LIDA's action selection mechanism, as the next action to be taken is chosen (see 3.1). The non-routine problem solving algorithm (see 3.2 below) produces state anticipation in LIDA. An automatization and associated deautomatization mechanism, along with a procedural learning mechanism, are also presented as anticipatory learning mechanisms (see 4 below).

2 Architectural Support for Anticipation

The LIDA architecture is partly symbolic and partly connectionist with all symbols being grounded in the physical world in the sense of Brooks (1986). The fundamental computational mechanism of the LIDA system is the *codelet* (Hofstadter & Mitchell, 1994), a small piece of code executing as an independent thread that is specialized for some relatively simple task. The components of the LIDA system that are related to anticipations and anticipatory learning include perceptual associative memory, selective attention, procedural memory, and action selection. Other components, being only peripheral to this paper, are not described here.

2.1 Perceptual Associative Memory

The perceptual knowledge-base takes the form of a semantic net with activation (called the slipnet) motivated by Hofstadter and Mitchell's Copycat architecture (1994). Nodes of the slipnet constitute the agent's perceptual symbols (Barsalou, 1999), representing individuals, categories and simple relations. The perceptual symbols are grounded in the real world by their ultimate connections to various primitive feature detectors having their receptive fields among the sensory receptors. An incoming stimulus, say a visual image, is descended upon by a hoard of perceptual codelets. Perceptual codelets respond to specific features from the various sensory streams and perform perceptual tasks such as recognition and identification. Each of these codelets is looking for some particular feature (a certain color, an edge at a particular angle, etc) or more complex features (a T junction, a red line). Upon finding a feature of interest to it, the codelet will activate an appropriate node or nodes in the slipnet. Activation is passed. The network will eventually stabilize. Nodes with activations over threshold, along with their links, are taken to provide the constructed meaning of the stimulus, the percept (see Figure 1).

2.2 Selective Attention

Selective attention in LIDA is an implementation of Global Workspace Theory (Baars, 1988) with hosts of attention codelets, each playing the role of a daemon, watching for an appropriate condition under which to act. Each attention codelet watches for some particular situation that might call for selective attention (i.e. novelty, changes, etc). Upon encountering such a situation, the attention codelet is associated with a few nodes (from the slipnet) carrying a description of the situation. A coalition of codelets (collection of related codelets) is thus formed. During any given cycle one of these coalitions with the highest average activation is considered relevant and broadcasts its information to every other codelet (Baars, 1988). This broadcast is used to recruit schemes (see below) and perform various types of learning (D'Mello, et al., 2006).

2.3 Procedural Memory

Procedural memory in LIDA is a modified and simplified form of Drescher's schema mechanism (1991), the scheme net. The scheme net is a directed graph whose nodes are (action) schemes and whose links represent the 'derived from' relation. Built-in primitive (empty) schemes directly controlling effectors are analogous to motor cell assemblies controlling muscle groups in humans. A scheme consists of an action, together with its context and its result (see Figure 1). The context and results of the schemes are represented by perceptual symbols (Barsalou, 1999) for objects, categories, and relations in perceptual associative memory. The action of a scheme consists of one or more behavior codelets (discussed next) that execute the actions in parallel.

2.4 Action Selection

The LIDA architecture employs an enhancement of Maes' behavior net (1989) for high-level action selection in the service of drives (primary and internal motivators) (Cañamero, 1997). The behavior net is a digraph (directed graph) composed of behavior codelets (a single action), behaviors (multiple behavior codelets operating in parallel), and behavior streams (multiple behaviors operating in some partial order) and their various links. These three entities all share the same representation in procedural memory (i.e., a scheme). As in connectionist models, this digraph spreads activation. The activation comes from three sources: from pre-existing activation stored in the behaviors, from the environment, and from drives. To be acted upon, a behavior must be executable (preconditions satisfied), must have activation over threshold, and must have the highest such activation.

LIDA's action selection mechanism incorporates five major enhancements over Maes' behavior net: (i) *Variables* – While Maes' behavior net operates on the basis of boolean propositions only, LIDA's mechanism supports variables that get bound during the instantiation of procedural schemes; (ii) *Restricted search space* – During the action selection phase Maes' mechanism performs a global search over all the available competency modules while the enhanced behavior net restricts its search to relevant (instantiated) goal hierarchies, which are a subset of the available competencies; (iii) *Failure handling* - Maes' mechanism assumes that the result of a selected action is deterministic in that every action produces its expected outcome. Therefore, this mechanism is unable to handle execution failures which frequently occur in any real system. On the other hand LIDA's enhanced behavior net is endowed with a degree of fault tolerance via its expectation mechanism; (iv) *Priority control* – Maes' mechanism modulates the priorities of competing goals by building *static* causal links among competence modules while LIDA's mechanism provides parametric control to *dynamically* change goal priorities at run time; (v) *Planning and subgoaling* – Maes' mechanism does not support classic AI planning and subgoaling but LIDA's mechanism, as a collection of goal structures, supports both (see Negatu, 2006).

2.5 LIDA's Cognitive Cycle

Since the LIDA architecture is composed of several specialized mechanisms a continual process that causes the functional interaction among the various components is essential. We offer the cognitive cycle as such an iterative, cyclical, continually active process that brings about the interplay among the various components of the architecture. A complete description of the cognitive cycle can be found in Franklin et al. (2005). We restrict our discussion to the four major components described above as follows.

The meaning of an incoming stimulus is constructed in perceptual associative memory and is taken to be nodes that are above a certain threshold. The attention codelets build coalitions among these nodes and compete for attention. The contents of a winning coalition are broadcast to procedural memory to instantiate action schemes. Instantiated schemes compete for execution in the behavior net as behaviors. The dynamics of the behavior net select an action and the agent then directs its focus to perception.

3 Anticipatory Mechanisms

The LIDA architecture includes payoff, state, and sensorial anticipatory mechanisms as outlined by Butz, Sigaud, and Gerard (2002). At this stage, the payoff and sensorial anticipatory mechanisms have been computationally implemented. A graphical depiction of these anticipatory mechanisms embedded with LIDA's cognitive cycle is presented as Figure 1.

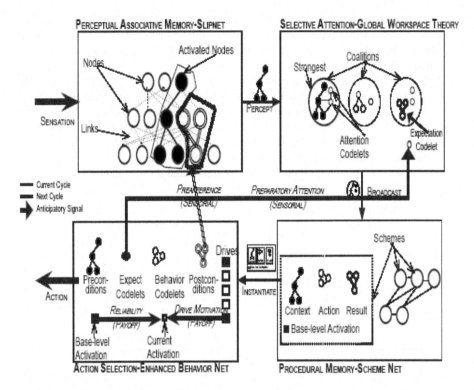

Fig. 1. LIDA's Cognitive Cycle and Payoff and Sensorial Anticipatory Mechanisms

3.1 Payoff Anticipatory Mechanisms

In a payoff anticipatory mechanism no explicit predictions of future states are made, with the role of anticipations being restricted to some form of payoff, or utility, or reinforcement signal. In the LIDA model the payoff for a behavior is calculated on the basis of predictive assessments by its *current activation* (i.e., relevance to the current goals or drives and environmental conditions) and its *base-level* activation (i.e., reliability in past situations).

LIDA's motivational system to influence goal-directed decision making is implemented on the basis of *drives*. Drives are built-in or evolved (in humans or animals) primary and internal motivators. All actions are chosen in order to satisfy one or more drives, and a drive may be satisfied by different goal structures. A drive has an importance parameter (real value in [0,1]) that denotes its relative significance or priority compared to the other drives. Each drive has a preconditional proposition that represents a global goal. A drive spreads goal-directing motivational energy, which is weighted by the importance value, to behaviors that directly satisfy its global or deep goal. Such behaviors in turn spread activation backward to predecessor behaviors. Although external activation spreading includes situational motivation, in this discussion of anticipation, we will attend only to the action selection dynamics that are tuned to goal-end motivation. From this point of view, the current activation of a behavior at a given time represents the motivation level for its execution to satisfy sub-goals, which in turn contributes towards satisfying one or more global goals at some future time. In other words, anticipating the predictive payoff in satisfying a goal influences the selection of the current action.

This payoff anticipatory mechanism was tested as a controller of a Khepera robot in a simulated environment (a warehouse) with tasks of differing priorities (drive levels). Our results indicate that the enhanced behavior net was found to correlate highly with a simulated human operator (Negatu, 2006) using GOMS task analysis (Card, Moran, & Newell, 1983). However, in order to maximize the effect of the payoff two enhancements had to be made to establish sufficient priority control in executing tasks. These included the use of importance and discrimination factors.

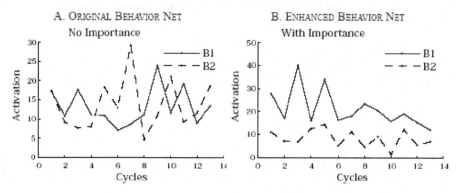

Fig. 2. Variation of motivation levels of competing behaviors: (a) without priority control (Maes' mechanism); (b) with the effect of the importance parameter to control priorities (LIDA's mechanism)

As described above, in order to maintain priorities for the competing tasks (goals), behavior streams (goal structures or partial plans of action) that satisfy the various goals have executable behaviors with motivation levels that reflect the priorities of the goals. In general, maintenance of priorities is possible if and only if, when considering all behavior streams with specified priorities (obtained from the importance parameter of the motivating drive), the selected behavior should belong to the one with the highest priority. All other things being equal between two behavior streams, the one motivated by a drive with a high importance parameter is expected to have a higher average activation level than the other motivated by a drive with a lower importance parameter. Figure 2 shows how goal-driven motivation levels serving as payoff anticipations vary with time. The test data was obtained by taking a snapshot of the activation/motivation levels of two behaviors that were parts of two competing instantiated behavior streams (a behavior stream is instantiated by the selective attention mechanism as described above). Figure 2a shows that in the absence of the importance parameter for the drives (as in the case of Maes' original mechanism), the payoff anticipatory mechanism could not be effectively tuned to control the priorities of the competing tasks as evidenced by the interleaving of the activation levels of the two behaviors. However, as illustrated in Figure 2b, LIDA's use of the importance parameter (for the drives) consistently increases the activation or payoff of the behavior with the higher priority.

While the use of the importance parameter improves the payoff anticipatory mechanism of Maes' behavior net, in some situations the inclusion of this parameter is not sufficient to guarantee priority control. This occurs when behaviors in two competing behavior streams interact, i.e., they mutually spread activation between themselves. In such cases, a behavior in one behavior stream of given importance could have a lower motivation level than a behavior under the second behavior stream with less importance. That is, the net effect of the intra behavior stream activation passing over time could be to increase the motivation of a low-importance behavior stream at the expense of the high-importance behavior stream. As a result, and as shown in Figure 3a, priority of tasks could be violated since a behavior under a low-importance behavior stream could eventually obtain a higher activation level than a behavior under a high-importance behavior stream. In order to address this issue we include an additional tunable global parameter in LIDA's enhanced behavior net. This parameter, called discrimination-factor, determines the level of spreading activation among interacting behavior streams. By weighing the intra behavior streams activation passing, the discrimination factor modulates the independence of a goal from the effects of other competing/cooperating goals. Considering a goal with high importance, we conjecture that a *high discrimination-factor parameter value determines the concentration level or persistence in performing the associated task and vice versa.* Figure 3b shows the effect of both the importance and discrimination-factor parameters in LIDA's behavior net in controlling priorities of interacting behavior streams that underlie the execution of two tasks performed by a simulated warehouse robot. These run-time tunable parameters together provide an unattended mechanism to control the motivation dynamics of the action selection system that enable an animat to execute tasks with strict prioritization.

The third factor that influences the payoff in selecting an action involves the use of the base-level activation of a scheme, an uninstantiated behavior template in procedural

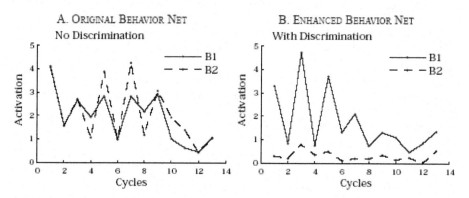

Fig. 3. Variation of motivation level of competing behaviors in setting priorities of competing tasks. (a) Importance parameter is not enough in setting priority. (b) Importance and discrimination-factor parameters together guarantee priority setting.

memory. The base-level activation is a measure of the scheme's overall reliability in the past, and is computed on the basis of the procedural learning mechanism described in the next section. It estimates the likelihood of the result of the scheme occurring upon taking the action in its given context. When a scheme is deemed somewhat relevant to the current situation as a result of the attention mechanism, it is instantiated from the scheme template as a behavior into the action selection mechanism (see Figure 1), and allowed to compete for execution. This behavior shares the base-level activation of the scheme which, when aggregated with its current activation, produces a two-factor assessment of the anticipated payoff in selecting this behavior for execution. That is, goal-end motivation and past reliability produce anticipation value such that the satisfaction of deep goal(s) in the future and likelihood of success biases what action is to be executed during the current cycle.

3.2 State Anticipatory Mechanism

In the design of a state anticipatory mechanism we are concerned with explicit predictions of future states influencing current decision making. In LIDA, state anticipations come to play when the agent is confronted by a novel situation in which it fails to converge upon an existing plan of action (behavior stream). In these situations, LIDA utilizes its non-routine problem solving (NRPS) process to generate solutions, usually the adaptation of existing ones, to handle an encountered novelty (Negatu, 2006). This is on par with the solution finding strategy called meshing (Glenberg, 1997), which in humans, is typically accomplished by putting together bits and pieces of knowledge and techniques that have been stored and, perhaps, used in the past to help solve other problems.

Based on the premise that non-routine problem solving is a deliberative state anticipatory behavior system, the NRPS module in LIDA is a special behavior stream

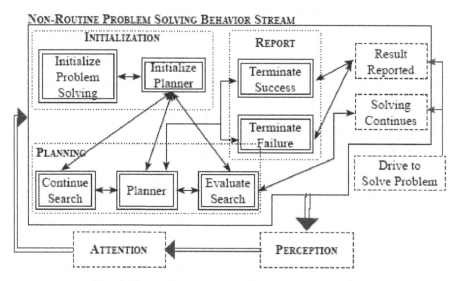

Fig. 4. Non-routine problem solving for state anticipation

that guides the solution search and generation process with a global resource filtering capability provided by selective attention (Figure 4). The NRPS behavior stream is motivated by a drive to solve encountered novel problems, and has main goals of searching for solutions and reporting search outcomes in the given problem context. When a situation is detected as a novelty, the NRPS stream is instantiated for a given problem context and then the *initialization behaviors* setup the problem solving process. Once initialized, the *solution searching behaviors* operate over multiple cycles with the help of selective attention in recruiting relevant resources (behavior codelets) in the deliberative regressive planning process – particularly certain attention codelets attempt to recruit relevant resources (building blocks to solutions) in the form of schemes from procedural memory. If the solution evaluation behavior reports an intermediate search outcome, the search continues. But, if the evaluation is final (whether a solution is found or not) one of the *report outcome* behaviors may be selected. Once an outcome is reported, the instantiated NRPS stream for the given problem context terminates.

The above description indicates that the NRPS process guides a controlled partial-order planner (Sacerdoti, 1977). While it shares similarities to dynamic planning systems, it differs from earlier approaches such as the general problem solver (Newell, Shaw, & Simon, 1958) in that selective attention is used to target relevant solutions from procedural memory, thus pruning the search space on the basis of the current world model. We conclude that the NRPS mechanism is a type of animat learning system that makes state anticipations, i.e., planning action decisions are biased towards selecting a plan operator that satisfies a required goal/sub-goal state. Please see Negatu (2006) for more details on the algorithm.

3.3 Sensorial Anticipatory Mechanism

Rather than directly influence the selection of behaviors, sensorial anticipatory mechanisms influence sensorial processing (Butz, Sigaud, & Gerard, 2002). The LIDA system recognizes two forms of sensorial anticipation, the biasing of the senses similar to a preafferent signal (Kay, Freeman, & Lancaster, 1996) and preparatory attention (LaBerge, 1995; Pashler, 1998).

3.3.1 Preafferent Signal for Sensorial Anticipation
Sensorial anticipation via a preafferent signal was observed by Kay et al. (1996) while monitoring electroencephalogram (EEG) signals over the olfactory bulb and associated processing centers (entorhinal cortex, hippocampus) in rats. They discovered that intentional action realized by motor commands was accompanied by a signal to the sensory processing areas (i.e. olfactory bulb). They called this a preafferent signal and hypothesized that it was used to bias the sensory processing areas to selectively respond to anticipated outcomes of actions.

The preafferent signal described above can be easily integrated into the LIDA model as a form of sensorial anticipation. As described above, nodes of the agent's perceptual associative memory, the slipnet, constitute the agent's perceptual symbols, representing individuals, categories and simple relations. Additionally, schemes in the agent's procedural memory represent uninstantiated actions and action sequences. The context and results of the schemes are represented by the same nodes for objects, categories, and relations in perceptual associative memory. A behavior in the behavior net can be considered an instantiated scheme, thereby sharing its context (as preconditions) and results (as postconditions). Once a behavior is selected in the behavior net, the nodes of the slipnet that compose the postconditions of the behavior have their activations increased, thus biasing them towards selection in the next cycle.

We conducted a simple experiment to highlight the performance benefits that can be obtained by the use of this type of sensory biasing. Consider a set of behaviors (N = 10) that operate in some predetermined sequence, i.e., B1 executes before B2, etc. In such a sequence each behavior has one precondition and one postcondition. The postcondition of behavior B_{N-1} is the precondition of behavior B_N. Therefore, in order for behavior B_N to execute behaviors $B_{1..N-1}$ must have previously been executed. Since we are primarily concerned with sensorial anticipation, let us assume that goal oriented motivation is set to 0, i.e. the network is purely opportunistic. Therefore, at any given time, only a single behavior (B_C) may qualify for execution if and only if its one precondition (P_C) is satisfied.

Now suppose the environment is highly uncertain, such that precondition P_C has a probability p_e of being present. For example, if $p_e = .5$ then a flip of a fair coin can determine whether this precondition is present in the environment. Furthermore, a precondition being present in the environment does not guarantee that it is perceived by the agent. The agent only perceives preconditions with probability p_s. Let us also set an execution threshold t which is the threshold required for a behavior whose precondition is currently satisfied to be executed. Therefore, for a control condition, with no sensorial bias, a behavior may be selected if and only if it exists in the environment (with probability p_e), is perceived (with probability p_s), and its activation is greater than threshold t.

Results of experimental simulations averaged over 100 runs are presented in Figure 5. For all runs we set $p_e = 0.5$, $p_s = 0.5$, and $t = 0.5$. We note that in the absence of any sensorial bias (control condition) it is expected to take 40 cycles to complete execution. For the sensorial bias conditions, the probability of a precondition being perceived (p_s) increases with respect to the amount of bias applied. Figure 5(a) reveals that performance linearly improves when a sensorial bias ranging from 0.1 - 0.5 is applied. A critical performance improvement occurs when the bias = $p_e = p_s = 0.5$. It is important to note that once the amount of sensorial bias applied exceeds the

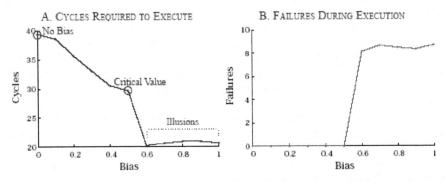

Fig. 5. Experimental simulations for sensorial anticipatory mechanism

probability of perceiving a precondition (>0.5) we witness a sharp improvement in performance. However, this should be interpreted with caution. An increase in bias beyond the probability of an object existing in the environment implies that the agent is experiencing a perceptual illusion in which it imagines sensing an object that does not exist. Therefore, even though a behavior is selected for execution, it will unduly fail as illustrated in Figure 5(b). The fact that failures do not cause a drop in performance is an artifact of our test environment.

3.3.2 Preparatory Attention for Sensorial Anticipation

Another form of sensorial biasing that would be applied at a later stage in processing is similar to preparatory attention. Preparatory attention can be differentiated from brief attention and maintenance attention in that an organism attends to a target before it is displayed (LaBerge, 1995). In LIDA, preparatory attention can be considered to be another form of sensorial biasing that occurs at a later stage, i.e. after perception. The mechanism to implement preparatory attention based sensorial anticipation is also based of the currently selected behavior. Each behavior is equipped with one or more expectation codelets, a special type of attention codelet that attempts to bring the results of the selected action to attention. Once a behavior is selected for execution, its expectation codelets attempt to bring the results of the behavior to attention, thereby biasing selective attention. In this manner the LIDA system incorporates a second form of action driven sensorial anticipation.

4 Anticipatory Learning

In this section we explore an automatization mechanism to learn low-level implicit anticipations coupled with a deautomatization system to temporarily suspend automatized tasks when a failure is detected. We also describe a procedural learning mechanism to learn the context and results of existing actions, which in turn, are used to construct a variety of anticipatory links.

4.1 Automatization

Automatization is defined as the ability to learn a procedural task to such an extent that the amount of attention devoted to the routine steps in accomplishing the task is reduced. Automatization develops as procedural tasks are rehearsed with attentional intervention. However, once tasks have been sufficiently automatized, the cognitive processing shifts from serial, attentional, and controlled to parallel, non-attentional, and automatic (Logan, 1980). In LIDA, a procedure produces a stream of actions with the execution of each action furthering the progression from the current state towards some goal state. For an unrehearsed procedure the transitions between consecutive actions are usually controlled by selective attention. The automatization process forms associations between unattended, coactive, low-level processes (behavior and attention codelets). As the associations between these processes strengthen over time (rehearsal) the need for selective attention gradually fades.

A complete description of the automatization mechanism is beyond the scope of this paper (see Negatu, 2006; Negatu, McCauley and Franklin, in review). However, Figure 6(a) illustrates the basic automatization process by depicting an execution

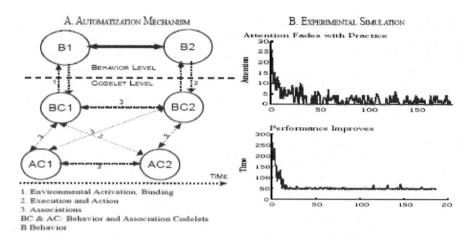

Fig. 6. Automatization mechanism with experimental results

sequence of a task at a high level (the behavioral level, B1 and B2) before automatization, and at a low level (the codelet level, BC1, BC2, AC1, AC2). In the absence of automatization selective attention is needed to activate the successor link between the high level behaviors B1 and B2. However, over time a number of associations between the low-level codelets are formed. When the associative link between BC1 and BC2 is sufficiently strong the link between behavior B1 and B2 can be implicitly activated without any attentional control, which is, without B2 being selected by the action selection mechanism.

Details of the implementation and experiments of the automatization mechanism can be found in Negatu (2006). In order to describe the effects of the automatization process we setup a simple experiment to perform a walking task that required a sequence of actions to execute. Figure 6(b) correspondingly shows that attention fades (upper right) and performance improves (lower right) as the degree of automatization increases with each iteration until it saturates.

The automatization mechanism implicitly causes a controlled task execution process to transition into a highly coordinated skill, thus improving performance and reserving attention, a limited resource, for more novel tasks. It is a type of implicit anticipatory learning mechanism since the encoding of the experiences of performing tasks is integrated in, and arises from, the payoff anticipatory process of LIDA's action selection dynamics.

4.2 Deautomatization

Automatization provides a performance improvement by sparing selective attention, an expensive resource in executing consistent and predictable tasks. However, of equal importance to an animat is a mechanism to gracefully handle failures or unexpected outcomes that occur during the execution of automatized tasks. LIDA's *deautomatization* mechanism is another anticipatory related adaptation that enables the detection of such failures, and causes the subsequent reengagement of selective attention to handle the novelty that arises from the failures (Negatu, 2006).

A failure of an automated task occurs when one of the behaviors in the sequence does not produce the anticipated outcome. The deautomatization process occurs in 3 steps as indicated in Figure 7(a). First, failure detection is achieved by a special type of attention codelet, called an expectation codelet, that calibrates the deviance between the actual and expected outcome of an action. When no failure is detected, the normal outcome is reported to the behavior that was just executed and normal automatization continues (Step 1). If a failure is detected, the associated expectation codelets then compete for attention and, if successful, provide feedback to the behavior codelet(s) that underlie the behavior (Step 2). The feedback process may require multiple cognitive cycles – the expectation codelet tries to bring the failure as a novelty to selective attention. At this point, the associations between the behavior codelets that contributed to the automatized behavior are temporarily discontinued, thus resulting in deautomatization (Step 3). Once deautomatized, selective attention is utilized in the execution of all actions in the sequence that precedes the point of failure. At a later time, the suppressed automatization may be reengaged.

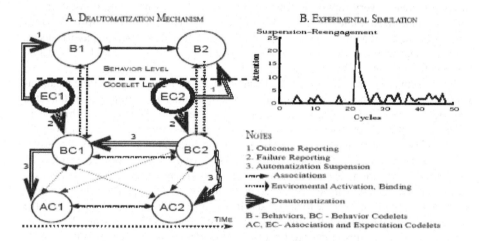

Fig. 7. Deautomatization mechanism with experimental results

The deautomatization process was tested within the context of the walking task that required a sequence of actions to execute. The task was automatized as indicated in Figure 6(b). Figure 7(b) correspondingly shows the detection of failure in cycle 22, followed the prompt suspension of automatization. Selective attention is then utilized until the automatization mechanism reengages (cycles 25 and above).

In conclusion, automatization has been shown to have some level of dependency on attention (Kahneman & Chajczyk, 1983) and expectation (Logan 1980). In our mechanism, attention and expectation codelets correspond to these roles and perform the necessary functions dictated. Also, our mechanism agrees with the view that automatization is a multistep algorithmic process as described by Schneider and Shiffrin (1977), rather than a single step memory instance retrieval of past experience as suggested by Logan (1998). Our mechanism postulates that this multistep algorithmic process underlies associative weights among low-level processes, and the implicit adaptation takes place as a result of changes in the associative weights. The deautomatization process breaks the automatic skills into individual components by disabling the effects of automatization, which mostly occurs at the behavioral level. The presented deautomatization algorithm suggests that automatization cannot happen as single step instance retrieval from memory.

4.3 Procedural Learning

The discussion so far has assumed that procedural memory in LIDA is built in and does not learn. In such situations, the use of anticipations is restricted to what was engineered into the system, thus greatly restricting its scope. In order to alleviate this problem we briefly describe a procedural learning mechanism in which the context and results of action schemes are learnt.

The procedural learning mechanism is a variant of Drescher's (1991) schema mechanism, and is based on selective attention and reinforcement learning. Reinforcement is provided via an asymmetric sigmoid function such that reinforcement initially increases very rapidly but tends to saturate. By negating and translating the same asymmetric sigmoid curve by +1 we obtain the decay curve. Therefore, schemes with low base-level activation (measure of reliability, used to determine payoff) decay rapidly, while schemes with high (saturated) base level activation values tend to decay at a much lower rate.

For learning to proceed initially, the behavior network must first select the instantiation of an empty scheme for execution. Before executing its action, the instantiated scheme spawns a new expectation codelet. After the action is executed, this newly created expectation codelet focuses on changes in the environment that result from the action being executed, and attempts to bring this information to attention. If successful, a new scheme is created, if needed. If one already exists, it is appropriately reinforced. Perceptual information selected by attention just before and after the action was executed form the context and result of the new scheme respectively. The scheme is provided with some base-level activation, and it is connected to its parent empty scheme with a link. More details on this mechanism can be found in D'Mello et al. (2006).

Of importance to this paper are the effects that the learning of a new scheme have on the anticipatory processes. The creation of a new scheme leads to a number of new anticipatory links being formed. The result of the scheme can be used to learn new expectation codelets to monitor future execution. These expectation codelets can be used to assess the reliability of this scheme, thus influencing payoff anticipations. They also serve as sensorial anticipations by biasing perceptual associative memory and selective attention.

5   Related Work

In an introductory paper, Butz et al., (2002) provide examples of a number of artificial systems that incorporate anticipations at some level. The major difference between LIDA's anticipatory mechanisms and those of other systems is that LIDA adapts the view of developmental robotics (Blank et al., 2005; Weng, 2004), which postulates that an animat is a self-organizing and adaptive agent that bootstraps from innate facilities for sensing, primitive actions, motivations and learning, with the innate architecture enabling the self-organization of knowledge on the basis of sensorimotor experiences in the world. Blank et al (2005) emphasize the merits of developmental robotics and connectionist anticipatory systems with low-level representations and the limitations of discrete and symbolic anticipatory systems such as Witkowski's system (2002). Incorporation of connectionist flavored mechanisms (perceptual associative memory and action selection) and low-level representation to encode anticipatory learning (e.g. automatization mechanism) give added merit to LIDA as an animat system.

Several control mechanisms have been proposed for payoff anticipatory animats (Butz et. al., 2002). These essentially involve variations of Learning Classifier Systems (LCS, YACS-Gerard, Stolzmann, & Sigaud, 2002; ACS-Stolzmann, 1998). These are rule-based systems that use evolutionary computing and heuristics to search the space of possible rules as well as reinforcement learning techniques to assign utility value to existing rules (i.e. the learned reinforcement values predict the action-payoff). In general, the major differences between LIDA's payoff mechanism and that of other LCS are: (a) the payoff is obtained from built in motivators (evolution) and experience (environment) rather than explicit reinforcement; (b) selective attention plays a role in both the gathering of experience and in directing awareness to what is important; and (c) LIDA uses the concepts of drive importance and discrimination among action plans to control priority among competing tasks.

LIDA's state anticipatory mechanism uses its non-routine problem solving (NRPS) behavior stream as a regressive partial-order plan generator (see section 3.2). LIDA's integrated modules, encompassing selective attention, procedural learning, and the behavior net are involved in this plan generation. There are other examples of state anticipatory animats – extended Dyna-PI model (Baldassarre, 2002), Dynamic Expectancy Model (Witkowski, 2002), and XACS (Butz & Goldberg, 2002) and all use explicit predictive model represented by a pool of schemes or rules. In our view, the major difference of LIDA's mechanism from the above three is that LIDA, as state anticipatory animat, uses explicit selective attention in the explicit deliberative process of action-plan formulation.

6 Discussion

This paper was motivated by the desire to develop and incorporate computational mechanisms for anticipatory processes in the LIDA cognitive agent, with the hope that any insights gained may generalize to other cognitive agents and animat systems. In order to develop cognitively plausible mechanisms, two constraints were enforced throughout. First, all mechanisms should be psychologically plausible and be consistent with (i.e. not contradict) known empirical evidence. Second, rather than offer microtheories of selected aspects of anticipatory adaptation, the mechanisms should be broad and diverse in scope, so as to encompass the wide gamut of anticipatory behavior documented in humans and higher animals. A crucial decision made at the time of writing involved the choice to provide a broad view of a large set of mechanisms as opposed to a more detailed exploration of a couple of mechanisms. This decision was motivated by the fact that the vast majority of the research in this area has focused on selective aspects related to anticipatory adaptation. This is in no means a criticism of previous research. However, we see the need to cover more ground by describing a larger set of mechanisms. Consequently the algorithms are presented at a conceptual level at the expense of eliminating details. However, Negatu (2006) provides elaborations on the algorithms associated with most of these mechanisms.

In this paper we have demonstrated that a wide variety of anticipatory mechanisms can be quite naturally integrated into a broad, comprehensive, autonomous agent architecture modeled after human cognition. The anticipatory mechanisms described for the LIDA model may well be suitable for incorporation into other cognitive based computational architectures designed to control autonomous software agents and/or mobile robots. In the experimental simulations described above, we have included control conditions in order to compare the behavior of the system with and without the anticipatory elements in place. In particular, we note that the enhanced behavior net with the payoff anticipatory processes provides more efficient priority (figure 3b) and discrimination control (figure 4b) than is afforded by the original behavior net without these anticipatory elements in place (figures 3a and 4a). Similarly, figure 5 demonstrates that sensorial biasing as a form off sensorial anticipation yields quicker execution. Finally, the automatization mechanism reduces the need for selective attention, a scarce resource, and speeds up execution (figure 6b), at least when compared to control conditions without this mechanism. Additionally, the deautomatization mechanism provides a degree of seamless tolerance to partial failure by detecting execution errors and invoking selective attention to scaffold the execution of an automatized task. Some may object to the characterization of deautomatization as a "learning" mechanism. In our view, deautomatization is more in line with temporary forgetting of automatized tasks and by correcting learning errors effectively scaffolds the automatization process.

Some of the mechanisms described in this paper are conceptual, i.e. algorithms exists but are not implemented. However, we can speculate on some of the unique computational benefits that these may afford. The preparatory attention system (sensorial anticipation) serves as a high-level perceptual filter that focuses on relevant content at a more abstract level. The state anticipatory mechanism involved the use of selective attention to construct a regressive action plan to organize behaviors to solve nonroutine problems. Finally, the procedural learning mechanism helps construct a low-level predictive model from sensorimotor experience caused by an animat interacting with its environment. The learned predictive model can be used to modulate sensorial and payoff anticipations.

As we explore the links between anticipation and cognition in humans and artificial systems we anticipate the need to revise, refine, and perhaps reconceptualize our theoretical perspective and our computational mechanisms. We are currently devising computational algorithms for anticipatory behavior and learning based on functional psychological frameworks. More complex models based on recent advances in biology and neuroscience may also be tested and integrated into the LIDA model. However, we suspect that the ultimate test of our theories and models of anticipatory adaptation will occur in more realistic, ecologically valid, environments. Cognitive robotics offers the ultimate challenge for all such functional and computational theories, because in essence the robot must adapt in order to survive in complex, dynamic, and sometimes unpredictable environments.

References

1. Baars, B.: A Cognitive Theory of Consciousness. Cambridge University Press, New York (1988)
2. Barsalou, L.W.: Perceptual symbol systems. Behavioral and Brain Sciences 22, 577–609 (1999)
3. Baddeley, A.D., Hitch, G.J.: Working memory. In: Bower, G.A. (ed.) The Psychology of Learning and Motivation, Academic Press, New York (1974)
4. Baldassarre, G.: A biologically plausible model of human planning based on neural networks and Dyna-PI models. In: Butz M., Sigaud O., Gérard P., Workshop on Adaptive Behaviour in Anticipatory Learning Systems, pp. 40–60 (2002)
5. Baluja, S., Pomerleau, D.A.: Using the Representation in a Neural Network's Hidden Layer for Task-Specific Focus of Attention. In: Mellish, C. (ed.) (IJCAI-95) IJCAI. International Joint Conference on Artificial Intelligence 1995, pp. 133–139. Morgan Kaufmann, San Mateo, CA (1995)
6. Blank, D.S., Lewis, J.M., Marshall, J.B.: The Multiple Roles of Anticipation in Developmental Robotics. In: AAAI Fall Symposium Workshop Notes, From Reactive to Anticipatory Cognitive Embodied Systems, AAAI Press, Stanford, California, USA (2005)
7. Brooks, R.A.: A robust layered control system for a mobile robot. IEEE Journal of Robotics and Automation RA-2, 14–23 (1986)
8. Butz, M.V., Sigaud, O., Gerard, P.: Internal models and anticipations in adaptive learning systems. In: Proceedings of the Workshop on Adaptive Behavior in Anticipatory Learning Systems, pp. 1–23 (2002)
9. Butz, M.V., Goldberg, D.E.: Generalized state values in an anticipatory learning classifier system. In: Proceedings of the Workshop on Adaptive Behavior in Anticipatory Learning Systems, pp. 78–96 (2002)
10. Cañamero, D.: Modeling Motivations and Emotions as a Basis for Intelligent Behavior. In: AA'97. Proceedings of the First International Symposium on Autonomous Agents, Marina del Rey, CA, February 5-8, The ACM Press, New York (1997)
11. Card, S., Moran, T., Newell, A.: The psychology of human-computer interaction. Hillsdale, NJ: Lawrence Erlbaum Associates (1983)
12. D'Mello, S.K., Franklin, S., Ramamurthy, U., Baars, B.J.: A Cognitive Science Based Machine Learning Architecture (Technical Report SS-06-02). In: AAAI Spring Symposia Technical Series, Stanford CA, USA, pp. 40–45. AAAI Press, Stanford, California, USA (2006)
13. Drescher, G.: Made Up Minds: A Constructivist Approach to Artificial Intelligence. MIT Press, Cambridge, MA (1991)
14. Franklin, S.: Automating Human Information Agents. In: Chen, Z., Jain, L.C. (eds.) Practical Applications of Intelligent Agents, Springer-Verlag, Berlin (2001)
15. Franklin, S., Baars, B.J., Ramamurthy, U., Ventura, M.: The Role of Consciousness in Memory. Brains, Minds and Media, vol.1, bmm150 (urn:nbn:de:0009-3-1505) (2005)
16. Gerard, P., Stolzmann, W., Sigaud, O.: YACS: a new Learning Classifier System using Anticipation.Soft. Computing 6(3-4), 216–228 (2002)
17. Glenberg, A.: What memory is for Behavioral and Brain Sciences, vol. 20, pp. 1–55 (1997)
18. Hartland, C., Bredeche, N.: Evolutionary Robotics: From Simulation to the Real World using Anticipation. In: Third Workshop on Anticipatory Behavior in Adaptive Learning Systems (2006)

19. Hofstadter, D.R., Mitchell, M.: he Copycat Project: A model of mental fluidity and analogy-making. In: Holyoak, K.J., Barnden, J.A. (eds.) Advances in connectionist and neural computation theory, logical connections, vol. 2, Ablex, Norwood N.J (1994)

20. Kahneman, D., Chajczyk, D.: Test of automaticity of reading: Dilution of the stroop effect by color-irrelevant stimuli. Journal of Experimental Psychology: Human Perception and Performance, 9, 947–509 (1983)

21. Kay, L.M, Freeman, W.J, Lancaster, L.R.: Simultaneous EEG recordings from olfactory and limbic brain structures: Limbic markers during olfactory perception. In: Gath, I., Inbar, G. (eds.) Advances in Processing and Pattern Analysis of Biological Signals, pp. 71–84. Plenum, New York (1996)

22. Kunde, W.: Response–effect compatibility in manual choice reaction tasks. Journal of Experimental Psychology: Human Perception & Performance 27, 387–394 (2001)

23. LaBerge, D.: Attentional processing: the brain's art of mindfulness. Harvard University Press, Cambridge (1995)

24. Laird, J.E., Newell, A., Rosenbloom, P.S.: SOAR: An architecture for general intelligence. Artificial Intelligence 33, 1–64 (1987)

25. Lebiere, C., Anderson, J.R.: A connectionist implementation of the ACT-R production system. In: Proceedings of the Fifteenth Annual Conference of the Cognitive Science Society. Hillsdalle NJ: Erlbaum (1993)

26. Logan, G.D.: Attention and automaticity in Stroop and priming tasks: Theory and data. Cognitive Psychology 12, 523–553 (1980)

27. Logan, G.D.: Toward an instance theory of automatization. Psychological Review 95, 583–598 (1998)

28. Maes, P.: How to do the right thing. Connection Science 1, 291–323 (1989)

29. Negatu, Aregahegn.:Cognitively Inspired Decision Making for Software Agents: Integrated Mechanisms for Action Selection, Expectation, Automatization and Non-Routine Problem Solving, Ph.D. Dissertation, The University of Memphis, USA (2006)

30. Negatu, A., McCauley, T. L., Franklin, S.: in review. Automatization for Software Agents.

31. Newell, A.: Unified theories of cognition. Harvard University Press, Cambridge MA (1990)

32. Newell, A., Shaw, J.C., Simon, H.A.: Elements of a theory of human problem solving. Psychological Review 65, 151–166 (1958)

33. Pashler, H.E.: The psychology of attention. MIT Press, Cambridge, MA (1998)

34. Piaget, J.: The Origins of Intelligence in Children. International Universities Press, New York (1952)

35. Post, E.L.: Absolutely unsolvable problems and relatively undecidable propositions account of an anticipation. In: Davis, M. (ed.) The undecidable: Basic papers on undecidable propositions, un-solvable problems and computable functions, Raven Press, New York (1965)

36. Rosen, R.: Anticipatory Systems. Pergamon Press, London (1985)

37. Sacerdoti, E.: A structure for plans and behavior. American Elsevier, New York (1977)

38. Schubotz, R.I., von Cramon, D.Y.: Functional organization of the lateral premotor cortex. fMRI reveals different regions activated by anticipation of object properties, location and speed. Cognitive Brain Research 11, 97–112 (2001)

39. Schneider, W., Shiffrin, R.M.: Controlled and automatic human information processing: I. Detection, search, and attention. Psychological Review 84, 1–66 (1977)

40. Sloman, A.: What Sort of Architecture is Required for a Human-like Agent? In: Wooldridge, M., Rao, A.S. (eds.) Foundations of Rational Agency, Kluwer Academic Publishers, Dordrecht, Netherlands (1999)

41. Sutton, R.: Planning by incremental dynamic programming. In: Eight International Workshop on Machine Learning, pp. 353–357. Morgan Kaufmann, San Francisco (1991)
42. Stolzmann, W.: Anticipatory Classifier Systems. In: Genetic Programming, University of Wisconsin, Madison, Wisconsin, Morgan Kaufmann, Seattle, Washington, USA (1998)
43. Sun, R.: An agent architecture for on-line learning of procedural and declarative knowledge. In: (ICONIP'97). Proceedings of the International Conference on Neural Information Processing, Singapore, Springer Verlag, Heidelberg (1997)
44. Tani, J.: Model-based learning for mobile robot navigation from the dynamical systems perspective. IEEE Transaction on Systems, Man. and Cybernetics 26B, 421–436 (1996)
45. Varela, F.J., Thompson, E., Rosch, E.: The Embodied Mind. MIT Press, Cambridge, MA (1991)
46. Watkins, C.: Learning with delayed rewards. Doctoral dissertation, Psychology Department, University of Cambridge, England (1989)
47. Witkowski, C.M.: Anticipatory learning: The animat as discovery engine In: Butz, M. V., Gerard, P., Siguad, O. (eds.) Adaptive Behavior in Anticipatory Learning Systems (ABiALS'02) (2002)
48. Weng, J.: Developmental Robotics: Theory and Experiments. International Journal of Humanoid Robotics 1(2), 199–235 (2004)

Schema-Based Design and the AKIRA Schema Language: An Overview

Giovanni Pezzulo[1] and Gianguglielmo Calvi[2]

[1] ISTC-CNR, Via S. Martino della Battaglia, 44 - 00185 Rome, Italy
giovanni.pezzulo@istc.cnr.it
[2] Noze s.r.l., Via Giuntini, 25 int.29 56023 Navacchio, Cascina (PI), Italy
gianguglielmo.calvi@noze.it

Abstract. We present a theoretical analysis of *schema-based design (SBD)*, a methodology for designing autonomous agent architectures. We also provide an overview of the *AKIRA Schema Language (AKSL)*, which permits to design schema-based architectures for anticipatory behavior experiments and simulations. Several simulations using AKSL are reviewed, highlighting the relations between pragmatic and epistemic aspects of behavior. Anticipation is crucial in realizing several functionalities with AKSL, such as selecting actions, orienting attention, categorizing and grounding declarative knowledge.

1 Introduction

In the last two decades several theoretical and computational models have been inspired, directly or indirectly, by theories of sensorimotor and cognitive development [9,11,53] that describe schematic structures, or 'schemas', as crucial in behavior and cognition. The term 'schema' was first introduced by Bartlett [8] to mean a map or structure of knowledge stored in long-term memory. Successively Piaget [53] described schemas in a more operational sense—roughly as mental representations of some physical or mental actions that can be performed on an object or event. He considered schemas as the building blocks of thinking, and the basic structure underlying behavior and cognition (in a process that he described as 'assimilation and accommodation').

One tenet of schema theory is that schemas are specialized subsystems realizing a tight coupling between perception and action. A schema can be used for recognizing a specific entity (say a dog) or a class of entities (say journalists or swimmers), or for controlling a specific action (say opening a door or skiing). Some of these schemas may require parameters to be filled in. There can also be more complex schemas for planning sequences of actions, as well as for more complex cognitive operations such as doing inferences. Central to schema theory is not only what schemas can individually do, but also how they are organized and what they can collectively do.

This view has inspired several other researchers in cognitive science, artificial intelligence and cognitive robotics. In these fields several schema-like structures have been proposed, including *frames* [43], *scripts* [57], *schemas* [3,4,21,44,46,60],

M.V. Butz et al. (Eds.): ABiALS 2006, LNAI 4520, pp. 128–152, 2007.

neural schemas [39], *semiotic schemas* [55], and *behaviors* [12,38]. Architectures including distributed and competitive functional units are often referred to as 'behavior-based' or 'schema-based'. Several integrated frameworks have been proposed for designing them; among the most popular ones, we can mention the behavior-based approach proposed in [6], the *NSL/ASL* in [65] and the *Robot Schema (RS)*, a formal language for designing robot controllers proposed in [37] which includes perceptual and motor schemas.

Since the term 'schema' has been used in several contexts, it has assumed several senses, too. For example, Piaget referred mainly to *sensorimotor schemas*, highlighting their action-oriented nature in contrast with other data structures that only include conceptual knowledge. Schemas for processing stimuli or controlling the perceptual apparatus are often referred to as *perceptual schemas*, while those for controlling locomotion, reaching or grasping are often referred to as *motor schemas*. Another important distinction is between *anticipatory schemas*, that include predictive components, and *reactive schemas*, that do not; and consequently between anticipatory and reactive schema-based architectures.

Reactive vs. Anticipatory Schema-Based Architectures. One important distinction among schema-based architectures is their *reactive* or *anticipatory* nature. Originally, the label 'behavior-based' has been used as a synonym of 'reactive' [5,12]. Reactive schema-based architectures, that respond quickly to dynamic environments, have challenged traditional AI models which rely on slow and costly deliberation. They are now de facto a standard in autonomous robotic systems [56]. However, recently several schema-based architectures have been proposed which include anticipatory mechanisms, such as inverse and forward internal models, which generate and exploit expectations about the next sensory stimuli [13,18,21,47,67]. These anticipatory aspects are inspired by psychological theories of action control [30,34], indicating that anticipated effects of (possible) actions play a fundamental role in regulating the agent's behavior. Several neurobiological evidences also suggest internal models and in particular forward models as plausible candidate mechanisms [40,66]. See also [22] for a comprehensive review of neural correlates of anticipation in the brain.

Aims and structure of the paper. The main contribution of the paper is twofold: illustrating the schema-based design methodology and its peculiarities, and presenting a comprehensive framework and an implementation environment for anticipatory schema-based architectures. Accordingly, in the rest of the paper we firstly introduce *schema-based design (SBD)* as a methodology for building autonomous agents architectures. We then present the *AKIRA Schema Language (AKSL)*, a framework for designing and implementing anticipatory schema-based architectures, and we review simulations realized with it.

2 Schema-Based Design (SBD)

Schema-Based Design is a methodology for designing artificial systems. It is inspired by ethological and neuroscientific empiric evidence [3,4]; and many schema-based architectures are directly inspired by ethological models, such as the praying mantis in [5], the computational frog in [2], and the computational cockroach in [10]. In SBD the functional aspects are more stressed than the actual realization and localization of schemas in the brain: several researchers find schema useful exactly because they provide an intermediate level of representation between the neural and the personal level [3]. In SBD cognition and behavior are explained in terms of schemas and their dynamics: behavior is not controlled by an unique process, but emerges from the dynamic competition and cooperation of several active schemas. More complex cognitive functionalities can emerge both by sophisticating the schemas and by permitting them to interact in more complex ways.

2.1   What's in a Schema?

Schemas are coarse-grained functional units, being approximately at the same level of description of ethological and neurobiological units of automatic action control such as *detect prey* or *escape* [5,47]. They do not only contain conceptual knowledge, but are strongly action-oriented and include perceptual and motor elements. Schemas consist of actions and sensory information organized around, and serving to realize, a *goal* or a set of related *goals*. In their simplest form, schemas can be described as sets of rules having the form *condition → action* or *condition → action → expectation* (respectively in the case of reactive and anticipatory schemas), that can act in parallel or in series, and whose success corresponds to the achievement of a goal.

Fig. 1. A sample schema: *chase prey*

As an example, Fig. 1 illustrates the main functional components of a sample motor schema, which is named after its goal: *chase prey*. It includes two sample triggering conditions: *hungry*, that indicates the value of a drive, and *prey in sight*, that indicates the presence of specific stimuli in the visual field. It also includes three actions[1]: *approach prey, grab prey, eat prey*. They can be implemented as rules, or set of rules, which receive perceptual input and send motor commands such as 'go left' or 'go right' to the motor apparatus of the agent.

Schemas have four main properties: goal-orientedness, flexibility, selectivity, and excitability.

---

[1] In this example each action is represented as a localist sub-unity of the schema. However by using a distributed representation scheme, multiple actions can be embedded implicitly, say in a single neural network; see for example [62,63].

Goal-orientedness. The goal-centered behavioral organization of a schema is its first property. This aspect also distinguishes schema-based systems from production systems and classifier systems [13,21,31,45], which also use rules and rulesets. Related views are the *ideomotor principle* [34] in psychology, the *TOTE* [42] model in cybernetics, and the definition of goal-orientedness provided by Gallese and Metzinger [24]: "Action control actually equates to the definition of the action goal: the goal is represented as a goal-state, namely, as a successfully terminated action pattern".

Flexibility. Goal-orientedness does not imply, however, that schemas have only one way to realize their goals: they can flexibly realize their goals under variable contingent conditions and by exploiting a (limited) repertoire of actions. This property can be called *flexibility*.

Fig. 2. A sample sequence of actions realized by the *chase prey* schema

When *chase prey* is active, its actions can be triggered in a different timing and order, or be skipped, depending on the context. In the example illustrated in Figure 2, the three actions are simply concatenated. *Approach prey* is imme diately triggered, since the two preconditions, *hunger* and *prey in sight*, still hold if the schema is active. If *approach prey* succeeds, it produces as a consequence *prey close*, which in turn triggers *grab prey*, and so on. The success of the last action, *eat prey*, also entails the success of the whole schema. The failure of one of the actions can instead trigger another action, or produce a context in which no actions are suitable, and thus lead to the failure of the whole schema.

Selectivity. An important consequence of schemas' goal-orientedness is their *selectivity*: in order to realize their goal, schemas do not need (and can not process) all the possible information from the environment. On the contrary, they select, attend to, and use only stimuli that are relevant for their specific goal. This implies that when a schema is operating the action-perception loop of the agent has both pragmatic effects (realizing the goal via triggering actions) and epistemic ones (gathering relevant stimuli).

In several schema-based frameworks epistemic and pragmatic aspects are implemented in different schemas: perceptual and motor. However, these schemas are either embedded one in another, [3,4], or they can pass sensory information [47]. The dotted lines in Figure 3 indicate the functional relations between one perceptual and one motor schema (*detect prey* and *chase prey*) which can pass sensory information. *Detect prey* includes as triggering conditions specific stimuli such as *red* and *moving*. It includes as actions three specialized strategies for

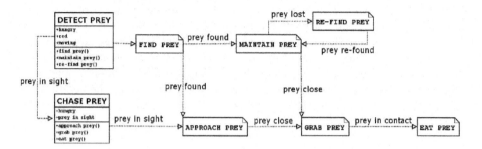

Fig. 3. Possible sequences of actions realized by a perceptual and a motor schema

finding prey the first time (*find prey*), maintain it in the visual field (*maintain prey*), and find it again if it is temporarily lost (*re-find prey*). The edge from *detect prey* to *chase prey* indicates that the former schema can trigger the latter (this is the meaning of the triggering condition *prey in sight*).

In this example the actions of the *chase prey* motor schema can be triggered by (the success of) other actions both in the same schema or in the related perceptual one, *detect prey*. Moreover, the perceptual schemas can convey to the motor schemas sensory information, for example the position of the prey to approach (not shown in the picture). Several courses of actions can emerge by the interactions of the two schemas in different contexts, depending on which functional relations are actually exploited. For example, a prey can be lost during tracking: the failure to *maintain prey* triggers *re-find prey*. Or, the prey can be captured without being lost, and in that case *re-find prey* is never activated. Notice that not all the functional relations are shown; in particular, those resulting from failure are not (for example, if a prey is lost both *maintain prey* and *approach prey* fail).

Excitability. The last important property of schemas is their *excitability*: they have a variable *activity level*. As we will discuss in detail, the activity level represents its relevance and desirability in the current situation. It can depend on motivational factors, such as active drives, and on contextual factors, such as the presence of appropriate stimuli or of other active schemas. Each schema gains resources, such as access to sensors and effectors, and as a consequence the possibility to influence the overall behavior of the agent, in a measure dependent on its relative activity level. This leads us to the next topic, which is the organization and functioning of a whole schema-based architecture.

2.2   Schema-Based Architectures and Cooperative Competition

A schema-based architecture is distributed; it includes several schemas specialized for different goals, such as *detect prey* and *detect predator*, or for realizing the same goal under different contexts, such as several instances of *detect prey* specialized for different kinds of prey. All schemas cooperate. They trigger one another, exchange sensory information, and compete for gaining priority over sensors and effectors: only some of them can be (partially) active at once. Such *cooperative competition* from which behavior emerges is considered a fundamental brain principle [3,4,43]. Competitive cooperation can be implemented in multiple ways, but it exploits three principles of self-organizing systems: (1) local excitation (e.g. active drives and schemas can excite other schemas); (2) global inhibition (e.g. schemas and drives inhibit concurrent ones); (3) emergence: behavior emerges from the influences of several schemas that can be active at once.

Since commands from several schemas can be fused, a schema-based architecture can realize complex patterns of actions, most beyond the possibilities of single schemas. An important aspect of schema-based design is the possibility to realize systems which can fulfill more than one functionality, selecting the most appropriate one on the basis of contextual factors such as current drives/goals, stimuli and expectations. We can say that the most challenging aspect of SBD is not implementing one specific functionality, but understanding how all them coordinate and realize a complex system having habits and fulfilling its drives/goals while remaining responsive to opportunities and affordances in the environment. In several architectures schemas can also be arranged hierarchically [18,27,51,54]; in this case, top-down influences channel behavior in accordance with expectations generated at the higher level, while the system remains responsive to stimuli-driven bottom-up dynamics.

2.3 Pragmatic and Epistemic Aspects of SBD

Organisms evolve schemas to successfully interact with entities in their environment, realizing their own goals. In producing behavior, schemas are thus selected for their expected success in action; if there are several schemas for realizing the same goal in different contexts, the most fit is selected. Since schemas are selected for action according to their activation level, a high activation level of a schema encodes a high confidence that it is succeeding or it will succeed in realizing its goal state (e.g. in *detect prey*: *the prey can be detected*). In summary:

> The activity level of a schema encodes a degree of confidence that it will succeed.

Although schemas are evolved for pragmatic reasons, their functioning also entails several factors which can be considered epistemic, in the sense that they are directed to acquire or process information. In traditional AI architectures these operations are traditionally dealt with by manipulating explicit representations, for example comparing an observation with an expectation, assigning a confidence level to an assumption, or matching an expectation. These operations can instead be dealt with *implicitly* and *procedurally* in SBD.

In the literature of dynamical systems it is often assumed that embodiment and structural coupling permits using information that is not explicitly represented in the system. The environment can be used as an external memory, provided that the agent's sensors can access it often enough. For example, a prey may continue to serve as a trigger of some schemas as long as it remains visible. In a similar way, certain pragmatic states of the schemas, such as their successes or failures, implicitly encode epistemic content. For example, it is possible to interpret the success of a *follow prey* schema as an indication that there is a prey, without any need to explicitly represent such state of affairs.

Therefore we discuss two functional equivalences: the degree of activation of a schema corresponds to (1) the truth of its assumptions and (2) to the desirability of its consequences, without any need to explicitly represent them (but, as we will see, we can derive explicit knowledge from them).

The First Functional Equivalence. There is a functional equivalence between the success of a schema and the assumption that the state of affairs that it is permitted to deal with is true. This is due to two reasons. First, action success or failure depends on epistemic assumptions and conditions that are verified or falsified by acting. Second, since the activity level of the schema depends on its success rate, it also indicates a confidence level that the behavior is appropriate and thus the entity to deal with is indeed there. As an example consider again the *detect prey* schema; in order to successfully track a kind of entity, the prey, the schema must be specialized to deal with prey-relevant features. Since in order to gain activation the schema has to succeed in actually matching these features, success of action is also a confirmation of such assumptions and expectations: *a prey is here, or will be here in the near future*. The functional equivalence between the success of the action and the truth of its assumptions and predictions is the *first functional equivalence*:

> The success of a schema indicates that specific (actual or expected) states of affairs, encoded in its assumptions and expectations, are true.

We can now come back to the example in Figure 3. When the preconditions of *detect prey* are verified (e.g. *red* is verified by the compliance of a visual routine), the schema triggers its actions. In turn, the compliance of its actions continuously verifies its preconditions (the success of *find prey* verifies *red* and *moving*) and produces new conditions in the same or in other schemas (the success of *find prey* produces *prey found*, which is a precondition of *approach prey*). This means that actions in two schemas are triggered by epistemic assumptions, which in turn are verified by the compliance of other actions.

As discussed above, several pieces of information can be implicitly dealt with. All the conditions shown in the picture, such as *prey found* or *prey close*, do not need to be explicitly represented, nor is it needed that symbolic information is passed among the schemas or the actions (all the labels are only for the designer's sake). Their functional meaning is implicitly encoded in the functioning of the schema mechanism, and in particular in the schema's activity level. The activity level of a schema, in fact, implicitly encodes the degree of confidence in its implications. If a schema can access the activity level of other ones, it can use this information 'as if' it was an explicitly reported condition. For example, a high activity level of *detect prey* implicitly encodes conditions (e.g. *prey found*), that are informative for *chase prey*. Further contextual elements, such as the state of the schema or the presence of specific stimuli, help *chase prey* disambiguating the information and triggering different actions such as *approach prey* or *grab prey*.

Very often epistemic information is graded and not crisp. For example I can be more or less sure that a prey is in front of me. As a corollary of the first functional equivalence, the degree of certainty, or confidence, in an assumption can be formulated according to the degree of success of the schema or action:

> The activity level of a schema is a measure of confidence in its assumptions.

Categories and Beliefs. Thanks to the first functional equivalence, schemas can not only be used for acting: their success also entails an implicit categorization of the entities to deal with, and implicitly represents beliefs such as 'there is a prey now'. Bickhard's interactivism [11] suggests a similar perspective: if an active interaction fails, then the 'indication for action' (and the content of the representation) is false. This fact has two main implications. Firstly, contrary to the typical pipeline information-processing scheme *perception → categorization → action*, a prey is categorized as a prey because of the compliance of the *detect prey* schema, and not vice versa (actually, there is a loop between all these factors). Secondly, categories and beliefs do not need to be explicitly represented anywhere, since the current activity level of schemas already indicates them. As an example, in Section 4 we present a simulation in which such dynamical, action-related categorization is realized.

As proposed by Piaget [53], in humans there is a progressive conceptualization of information which is initially only procedural[2]. However, only part of information implicitly used by the schemas becomes explicitly available, for example for categorization, and internally manipulable when coupling is broken; some information remains instead procedural. For a discussion of accessibility and awareness of procedural information used in the control of action, see [23].

Epistemic Actions. We have discussed how actions can also have, as a side effect, an epistemic value for the system. For example, knowing that the ball is there, is round, is soft, etc., is a form of implicit knowledge that in some cases can be internalized. But there is another, more sophisticated way for a system to obtain information by exploiting the first functional equivalence: performing an action with the aim to know something about the world (e.g. 'control if' or 'look whether'). That is, I can turn on the light in order to know whether or not the circuit works well, and not because I need light. In this case, we can distinguish between the *pragmatic action* (action in the most common sense)

---

[2] Some assumptions which are much more 'profound' and invariant are conceptualized very late, when they are. Consider the assumption 'under normal circumstances, the world is quite stable'. In order to remain successful for a while, several (if not all) schemas implicitly rely on this assumption, which is not however conceptualized. We could say that also the functioning of the whole schema-based system encodes several important assumptions about the world.

and the *epistemic action*, that is aimed at gathering information: the pragmatic action (turning on the light) is only a vehicle of the epistemic one (knowing if the circuit works well). This example illustrates that an action can have both pragmatic and epistemic value, and it can be executed for the former or the latter necessity. It is also worth noting that in order to know something about the world we need to act on it (either actually or 'in simulation', see [26,28]).

In Section 4 we will present a simulation showing the two main consequences of the first pragmatic principle: (1) on the basis of pragmatic actions, either real or simulated, epistemic states such as beliefs can be formulated; (2) epistemic actions are possible, too: some pragmatic actions, actually performed or simulated, can be triggered for knowing something, and not for their pragmatic effects. In both cases we interpret a belief as the result of an epistemic action, which can be either implemented through a pragmatic action, or explicitly executed by means of a pragmatic action. This view has an important implication: all cognitive operations involving beliefs, such as reasoning, refer to (and have their meaning thanks to) actual or possible pragmatic actions.

The Second Functional Equivalence. Organisms have motivations, and their actions are determined by their needs. In schema-based design this is modeled through motivational units, such as drives, causing schema activations: in this way there is no need to manipulate and reason explicitly on utility and values of entities in the world. As a consequence, typically a high activity level of a schema also encodes the fact that it has been learned to, and is expected to be effective for satisfying the organism's needs: a primitive, implicit form of means-ends reasoning. Again, success of action strengthens and tends to confirm the relationship between a schema and the satisfaction of an organism's need. Since the organism can have competing motivations, typically the schemas for realizing the most important or urgent ones are assigned the highest activity level and are thus selected (but of course, as far as they do not penalize one another, more than one schema can be selected). This implies that:

> The activity level of a schema is a measure of desirability of (the consequences of) its success.

In some cases the organism faces challenges that have to be dealt with very quickly. As an example, consider an organism successfully following a prey. If an unanticipated danger occurs, such as a predator, or the organism is near a cliff, it has to quickly change its behavior and activate another schema, whichever the activity level of the *follow prey* schema is. This means that a sudden change in allocation of resources among schemas has to occur, depending on how promptly the new situation has to be dealt with:

> The rapidity of increase in activation of a schema is a measure of its urgency.

The two factors, one epistemic and one motivational, which correspond to a high activity level in a schema seem to be at odds. For example, a high activity level of the schema *detect prey*, which depends on the organism being hungry, also corresponds to the belief or expectation that there is a prey somewhere, which may not be the case. However, in organisms there is a relation between the epistemic criterion, that is maintaining true assumptions, and the motivational criterion, that is pursuing its needs. In order to succeed (and consequently to satisfy its goals) an organism needs to maintain its epistemic states, and this is why schemas are designed for implicitly checking their conditions. For this reason an organism motivated by hunger can 'bet' on the success of the schema for *catching prey*, maintaining artificially (against evidence) a high activity level even if it is unsuccessful at the moment. Since activating a schema also means inhibiting other ones, this strategy is only good if the schema will indeed succeed in the near future, otherwise the organism will die. This means that the organism is selected by evolution to have 'good guesses' and to fuel schemas that will succeed and, in that way, verify their assumptions: the first functional equivalence is not broken, only postponed. The correspondence between the desirability of a behavior and the truth of its assumptions is the *second functional equivalence*:

> The activity level of a schema is a measure of confidence that it will be successful, and consequently that its assumptions and expectations will become true.

Notice that this principle works due to the fact that, at the end, each schema predicts its own success too. The teleonomic structure of a schema can be thus represented as *condition* → *action* → *expectation* → . . . → *expectation* → . . . → *action* → . . . → *success*. A schema is selected by evolution because of the adaptive advantage of (the implications of) its success. Its structure guarantees the desirability of its intermediate actions as well as the meaningfulness of the assumptions and expectations it produces during its execution. Schemas are learned in order to deal with the environment successfully: maintaining correct representations and betting that they will be useful are two sides of the same coin.

3 The AKIRA Schema Language (AKSL)

The AKIRA Schema Language (AKSL) has four main components: schemas, drives, routines, and actuators. It is build on the top of the AKIRA simulation framework [1] and integrated with the Irrlicht 3D engine [33] and the Ikaros simulation framework [32][3].

3.1 Schemas

Schemas can be described as tuples $(det,inv,for,urg,rel,app,con,act,thr)$. The first three parameters represent their components:

---

[3] The sourcecode of AKSL is available in the AKIRA website [1].

- *det* is a detector[4], i.e. the selector of a certain kind of stimuli (each schema only processes some information which has been learned to be significant)
- *inv* is an inverse model, deciding on a motor command to send to an effector
- *for* is a forward model, calculating the expected next stimuli

Basically each schema has a cycle in which: (1) the detector collects sensory information (received by perceptual routines) and the sensory expectation (received by the forward model), compares them (dotted circle in Fig. 4: the degree of mismatch is used for calculating *rel*, see later), and sends a sensory input to the inverse model; (2) the inverse model, on the basis of the input received, calculates a motor command and sends it to the effector (camera or wheel motors); (3) the forward model receives an efference copy of the final motor command (sent to the camera or wheel motors), generates a sensory expectation and sends it to the detector.

The cycle of each schema is run asynchronously and in parallel with each other, with an amount of computational resources (speed and memory) that depends on its *activity level*. Five other parameters are used for calculating the activity level at the beginning of each schema's cycle:

- *urg* is the urgency value, representing how promptly the schema has to be executed when its contextual conditions are met. This parameter is very high in schemas which have to deal with risky situations, in which an immediate action is needed.
- *rel* is the reliability value, representing how much that schema is (expected to be) successful in the current situation. The reliability value is set according to the degree of match of the expectations generated by the forward model *for* with respect to the actual stimuli.
- *app* represents the appropriateness with respect to currently active drives and goals. It is a learned parameter.
- *con* is a learned contextual parameter that depends on the activity level of other schemas. Schemas can in fact evolve links with an Hebbian-like mechanism explained in [49], which permit the transfer of activation. This associative mechanism permits mutual priming of schemas that are often active in the same situations.
- *act* represents the total activity level of the schema, which sums up the epistemic and motivational factors. It is calculated as $urg + rel + app + con$. Thus, the final activity level of a schema represents how much the schema is

---

[4] Optionally more sophisticated operations can be realized inside the detector. For example, as in Kalman filtering [36], a reliability value can be assigned to stimuli and expectations, and the final input be calculated as their weighed sum. In this case, if the stimulus is lacking or inaccurate, the expectation can be used for (partially) replacing it. Another possibility is to erase from the stimulus the self-generated part, predicted by the forward model. This functionality is useful e.g. for avoiding tracking our own hand when it is in the visual field. Moreover, as in Smith predictors [61], the prediction of the forward model can be fed at different time intervals for compensating long loop delays.

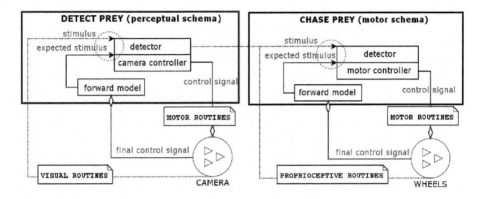

Fig. 4. Example of the coupled perceptual and motor schemas

expected to be both effective (successful) and desirable in the given context
(according to current drives/goals and other contingent factors).
- *thr* is a threshold for sending motor commands. Under the threshold the
  schema functions normally but its motor commands to the actuators are
  inhibited.

Coupled Perceptual and Motor Schemas. Perceptual and motor schemas
can be *coupled*. The functioning of two sample coupled schemas, *detct prey* and
chase prey, is illustrated in Figure 4. The perceptual schema receives as input
perceptual information from the camera (data are preprocessed by perceptual
routines). As indicated by the dotted circle, sensed stimuli are compared with
sensory information that is predicted by the forward model, and the error is used
for setting the reliability value *rel* of the schema. The detector thus sends sensory stimuli to the controller (inverse model), which in turn generates a motor
command and sends it to the camera motor (via motor routines), and (optionally) sensory information (e.g. the position of the detected prey) to the coupled
motor schema. The motor schema receives as input the activity level of the
coupled perceptual schema, proprioceptive information about the current state
of the wheels' motor, and optionally additional sensory information from the
perceptual schema. Like in the perceptual schema, sensed and predicted stimuli
are compared and reliability values are assigned. Sensory information is con-
veyed to the controller, which sends motor commands to the wheels' motor (via
motor routines). Notice that in both schemas, in order to generate predictions,
the forward models receive an efference copy of the (final) motor commands
received by the camera or wheel motors, and learn to predict their sensory
effects.

Three Modes of Operation. Schemas can operate in three distinct modes:
(1) generation; (2) simulation; (3) imitation.

The *generation* mode is the default one and serves for generating behavior appropriate to the context. The functioning is the one previously explained.

The *simulation* mode is used for predicting the long-term effects of schemas. When a schema runs in simulation, the motor commands generated by its inverse model are inhibited and not sent to the actuators; however, efference copies are sent as usual to the forward model, which generates expectations and sends them to the detector. Inside the detector, expectations are not matched against stimuli (and their accuracy can not be calculated), but directly fed to the inverse model. The simulation mode thus produces a loop between the forward and inverse model, generating a simulation of the possible courses of events, which extends over several steps. Schemas in simulation mode can also be run faster than real time, thus actually simulating several steps ahead[5]. The simulation mode can be used for producing, testing and selecting in advance multiple alternative courses of events 'proposed' by different schemas. For example, if an agent has alternative schemas for navigating, by running them in simulation it can explore 'virtually' (and not by trial and error) its environment. This permits to foresee possible dangers that can arise during navigation, to anticipate if by following a path it is actually possible to reach a target location, or to calculate which is the shortest path to a target location by comparing the time spent for simulating the alternative ones. In principle all schemas can be run in simulation, regardless of their actual reliability and activity level (but notice that simulating is a costly operation). However, only schemas whose reliability value is significantly high are able to generate predictions which are adapt to the current context.

The *imitation* mode serves for understanding (and possibly reproducing) behavior observed in a demonstrator (see [17]). When schemas run in imitation mode the perceptual state (observed in the demonstrator) is fed to the inverse models which generate the motor command that would have been produced in that situation. The motor commands to the actuators are inhibited, but the efference copies are fed to the forward models which generate the next predicted perceptual state, which is thus compared with the next perceptual state (observed in the demonstrator). This process roughly corresponds to the question: "which of the schemas could have generated the perceptual states I observe?", the answer being the schema(s) which are accurate in predicting. By knowing that, the agent is now able both to understand the demonstrator's actions, and to imitate them. Of course depending on the differences between the agent's and demonstrator's behavior repertoires, imitation can be more or less accurate.

---

[5] A more complex possibility is running in simulation the whole schema-based system. In this case, predictions generated by one schema are also fed to other schemas. The effects of the motor commands on actuators and sensors are simulated too, and then predictions replace sensory information in the whole system. This mechanism permits to test in advance not only the long-term effects of single schemas, but also their combinations. Notice that it is impossible to use the schema-based system in generation and simulation at the same time.

3.2   Determining the Activity Level of Schemas

Schemas in AKSL couple perception and action via anticipation: they learn to associate stimuli with behavior that produces appropriate expected results[6]. As already discussed, thanks to their dual nature, schemas are well suited for both pragmatic activity such as guiding behavior, and for epistemic activity such as categorization. This is due to the fact that their activity level depends both on their reliability, and on their appropriateness with respect to current drives/goals.

Schemas are assigned a reliability level depending on how well they predict the sensorimotor flow. The reliability of a perceptual schema is a confidence level that a certain entity, encoded in the schema, is expected to be present. For example, if the schema *detect prey* has a high activity level, not only it can be assumed that it is successfully tracking the prey (with the camera), but also that there is, or there will be, a prey in the visual field. For this reason, the most important aspect of perceptual schemas is their epistemic side: if an architecture has several perceptual schemas, they can be seen as competing hypotheses for representing/categorizing the current perceptual situation, such as *detect prey* vs. *detect predator*. The reliability of a motor schema is instead a confidence level that the behavior encoded in the schema is applicable in the context.

As an example, consider the *catch prey* schema. The most important aspect of motor schemas is pragmatic: they can be seen as competing behaviors, or as different means to realize the same behavior. However, even perceptual schemas have relevant pragmatic aspects (they orient attention by moving the camera) and motor schemas have epistemic aspects (they contribute to categorization). Each schema has thus both aspects, since its success implies both achievement (of action) and categorization (of object/event).

The current motivational state of a schema-based agent also influences the activity level of schemas. This means that when a schema-based agent has *hunger* or is *fearful*, its schemas for detecting and catching prey in the former case, or for detecting and escaping from predators in the latter case, will gain activation. In this way the agent's pragmatic activity is oriented toward desired goal states, such as prey or hiding places. Attention is channelized toward relevant stimuli too: the agent spends many resources in deciding whether or not there is a prey than, say, deciding whether or not there is a hiding place. This is obtained by drives (such as *hunger*) providing activation to relevant perceptual schemas (such as *detect prey*). This happens even when there are not prey in the visual field; and since a high activity level of *detect prey* can in turn be interpreted as an evidence that there is a prey, this process causes visual imagery: the agent 'imagines' what it is searching for. However, this is only temporary: although *hunger* can artificially maintain a high activity level for some time, if there are

---

[6] Several machine learning methodologies have been used in literature for learning the inverse and forward models, and for evaluating the degree of mismatch between stimuli and expectations; for example, in [67] responsibility signals are used. AKSL currently permits the usage of both fuzzy logic and feed-forward or recurrent neural networks libraries; see [49].

no prey the schemas for detecting and catching them will still not be relevant and thus will lose activation.

There is also a general constrain on how much activation can be assigned to schemas. Activation, to be divided among schemas, is in fact limited. Schemas compete for acquiring resources: they can not access them while they are used by other schemas, but have to wait until they are released (the mechanism is based on the AKIRA Energetic Model, explained in [49]). This means that active schemas inhibit one another via the allocated resources but without lateral inhibitions. By modifying the total amount of activation available it is also possible to channelize behavior in different ways, since few or many schemas can be active at once.

3.3   Motivations and Routines

AKSL also permits the design of simple motivational systems: as in several ethological studies, drives such as *hunger* and *fear* can be implemented. We have also included in AKSL several routines such as *detect red* or *move left* for pre-processing information (e.g. sensory data).

Although they have very different roles, drives and routines are implemented in a similar way. Each drive and routine embeds a simple operation. In the case of drives, this may consist of a 'biological clock' (e.g. raising the activity level of *hunger*). In the case of routines, this may consist of reading the value of a sensor, or sending a motor command to an actuator. Schemas do not receive input from sensors and they do not send output to actuators: three kinds of routines (perceptual, motor and proprioceptive) have the role to mediate between them.

Drives and routines have an activity level *act* and can exchange activation with other components. For example, drives typically fuel appropriate schemas, and schemas fuel routines (and vice versa). The more a drive is active, the more it can fuel the schemas which satisfy it; thus drives introduce a motivational influence on the agent's behavior. The more a routine (e.g. *detect red*) is active, the more it reliably finds a pattern in the sensor (e.g. *red* is detected). Energetic links between schemas, drives and routines can be learned via a Hebbian-like system. The rationale is that schemas whose success reliably satisfy drives become associated with them, while schemas become associated with routines that provide useful input or output facilities. See [47] for the details.

3.4   Sensors and Actuators

Perceptual and motor routines receive input from sensors such as a camera. The actuators (camera or wheel motors) receive asynchronous commands from the motor routines and perform command fusion (libraries based on fuzzy logic and a mixture of Gaussians are available; see [49]). In many systems in the literature (see [14] for a review) several schemas can be partially active at once but only one is selected for commanding the actuators. In our model each active schema sends concurrently its motor commands to the actuators via the motor routines. Since more active schemas receive more resources and can perform their operations faster, they also have a higher firing rate when sending motor commands. The actuators fuse all the motor commands, which were received asynchronously. This means that a schema with an higher activity level (and as a consequence sending commands with higher firing rate) also has more influence on the actuators. Notice that this has not necessarily only impact on the agent's movements. Since one of the actuators is the camera motor, that actuator also determines which sensory information the agent pays attention to.

3.5   Comparison with Related Literature

AKSL shares resemblances with other schema-based architectures such as [3,37]. With respect to them, the two main differences are the presence of internal models and the fact that perceptual and motor schemas are separate units, which can, however, be coupled. AKSL shares resemblances with MOSAIC [67] and HAMMER [18] too. Differently from them, AKSL uses a parallel architecture in which the activity level influences directly the amount of computational resources (speed and memory) assigned to each schema, without explicitly calculating responsibility values. The second difference is that schemas compete for limited resources and active schemas inhibit other ones. The third relevant difference is that commands are received and fused asynchronously by the effectors; a high activity level permits schemas to have a higher firing rate and thus to send more commands. Lastly, AKSL also permits Hebbian-like learning and spreading activation between the schemas, which can thus provide activation to one another, being in the same or in different hierarchical layers. See [49] for the details of the architecture.

4   Exemplar Capabilities of AKSL

We have used AKSL for addressing several research fields. Here we review our simulations in the fields of (1) action selection and attention, (2) category formation, (3) simulation of future behavior, (4) grounding, and (5) hierarchical control of action.

4.1   Action Selection and Attention

Recently several anticipatory schema-based systems have been proposed for action control in robotics which base action selection on predictive success, both in distributed approaches [62,63] and in localist ones, such as MOSAIC and HAMMER (but others exist, [64]). They use a combination of forward and inverse models for generating competing motor plans for the same or for different targets, and the models with better predictions are selected for the control of action. This responds to two related questions: which action is preferable given the sensory and goal context? Which schema can successfully actuate the action? These questions become related if success of prediction is used for action selection: a *successful* schema performs an action (and satisfies a drive/goal), thus a schema predicting well also predicts its own success. On the contrary, not only schemas which fail, but also schemas which are expected to fail, can be assigned less activity. Differently from several schema-based models, in MOSAIC and HAMMER there is not a one-to-one correspondence between a schema and a behavior, but each behavior (e.g. 'grasp teapot') is realized by the cooperation and competition of several schemas, specialized for different contexts (e.g. 'light' or 'heavy' cup). Thus, competing models generate alternative motor plans, such as *grasp full teapot* vs. *grasp empty teapot*, which are selected according to the basis of how accurately the models predict the right sensorimotor flow. When a motor command is generated for grasping a teapot, an efference copy is used by the forward models in the two modules for generating the sensory consequences under two different contexts. These predictions are thus compared with actual sensory feedback, and the most appropriate one is selected for action control. Commands of the two modules can be combined linearly, providing generalization.

In [47] we have described a schema-based architecture (see Fig. 5), inspired by an ethological model of the praying mantis, which shares resemblances with MOSAIC and HAMMER but uses AKSL. In that architecture predictions generated by the forward models are not only used for determining schemas' reliability, but also used for orienting the motor apparatus (camera and wheels). *Perceptual schemas* gather information relevant for the current task, and orient the camera toward relevant inputs (e.g., relevant colors and trajectories), also determining part of the next stimuli, as in *active sensing. Motor schemas* select the most appropriate motor action (e.g., specialized for following or escaping from quick or slow, big or small entities). The first novelty of this architecture is the possibility to deal with multiple concurrent drives, whose urgency changes over time (depending on internal regulatory mechanisms or by external stimuli). The challenge is generating the appropriate behavior for satisfying the currently active motivation. Depending on the current motivational state, affordances offered by the environment are selected: for example, a hungry agent tries to catch prey, while a fearful agent that is escaping from a predator avoids prey. The second novelty is that perceptual and motor processes are integrated in the same framework and coupled, and the agent is able to orient attention for gathering information necessary for satisfying its current needs. The two aspects, determining behavior and orienting attention, are closely related in AKSL thanks to the coupling of perceptual and motor schemas.

4.2 Category Formation

Schema activity has (real or anticipated) epistemic implications; it can be exploited for categorizing and distinguishing objects from background. In an anticipatory framework, since the activity level of schemas depends on their predictions, objects and categories are defined by the typical, coherent patterns of (expected) transformations under a given set of the agent's actions. For example, in [21] it is investigated how to build *action → effect* sensorimotor schemas through interaction with a simple environment. Moreover, the agent interactively enlarges its ontology by learning new objects (called *synthetic items*), which in turn permits the learning of incrementally new abilities and competences.

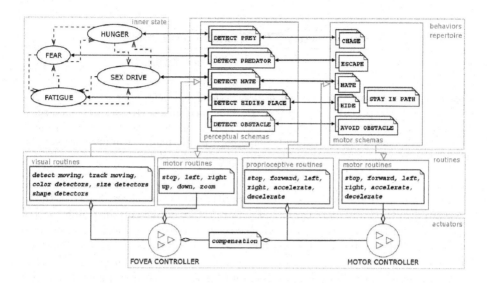

Fig. 5. The schemas-based architecture of the praying mantis in [47]

AKSL has been used for categorization, too. In [48] we have demonstrated how to develop *perceptual categories* (such as different types of insects) and *abstract categories* (such as two roles played by those insects, predators and prey), on the basis of the theory of *perceptual symbol systems* [7]. Shortly, the agent learns to activate the most relevant schemas, specialized for dealing with features of the entities (such as color and size) and is able to track, follow or escape the entities by means of the dynamics of collaboration and competition among the schemas. The specific novelty of AKSL comes from the possibility to evolve energetic links among schemas in an Hebbian-like way. In such a way 'clusters' of active schemas emerge during interaction with a given entity (say insect_one or predator) behave as *simulators* in the sense of [7]. They can then generate a simulation of categories of objects or events by rehearsing and priming the associated schemas, even in absence of environmental cues. Thanks to simulative capabilities, specific runs of a simulator reenact the multimodal experience of a category, while adapting to the current situation.

4.3   Simulation of Future Behavior

Simulative theories of cognition [7,26,28] suggest that by rehearsing the motor programs for interacting with an object an agent can anticipate the sensorial stimuli it will receive and simulate the consequences of its motor commands one or many steps beyond current time. Recently several neuroscientists [20,41] proposed that the cerebellum and the basal ganglia could create a loop permitting the simulation and selection of multiple alternative courses of actions, providing neural support for these theories.

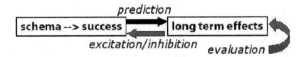

Fig. 6. Simulation and Long Term Effects. The long term effects of schemas are predicted, and the expected states evaluated. This mechanism can be used for exciting/inhibiting the schemas depending on their long-term effects. If the long-term effects of several schemas are generated and compared, this mechanism can also be used for selecting among alternative courses of actions (simulative planning).

Simulation permits the realization of multiple functionalities. Some of them are related to the immediate control of action: for example, actual stimuli can be replaced when the sensors are unavailable or unreliable [19]. More complex and future-oriented capabilities are possible, too. The alternative courses of events can be evaluated before acting, for example with a *somatic marker* mechanism [15]: if the sensory predictions have already been experienced and had been categorized as negative, the schemas generating them can be stopped by a 'command from the future' (see Fig. 6). A simulated exploration of the environment can also be exploited for selecting a plan: the effects of several competing plans can be produced off-line, and their expected sensory consequences evaluated against actual or expected drives/goals. Neuroscientific evidence indicates that reward prediction is used for selecting, for example, a path in a maze [58].

Simulative capabilities have been used in cognitive robotics too. For example, in [25,68] internal simulation of the sensory consequences of multiple possible motor actions is used to perform robust planning in the presence of noise. Similarly, in [16] a simulative process is used for checking in advance if the selected behavior will cause problems in the future. This method permits to avoid the costs of performing complete planning, since only the usual behavior path is checked in anticipation. Simulative capabilities have also been used in social tasks, such as imitation, joint attention, plan recognition, perspective taking, prediction of intent, etc.; for example, the HAMMER architecture has been used for modeling all these aspects [18,17].

Schemas in AKSL, if used in the *generation mode*, already simulate few steps in the future for the sake of predicting the next sensory stimuli. If a prediction of the long-term effects of the agent's actions is instead needed, they can be used in *simulation mode*. Running schemas in simulation permits the realization of several novel functionalities, such as realizing a 'somatic marker' mechanism or simulative planning. At the moment we have only conducted preliminary experiments on these topics; in our future work we plan to continue investigating them.

4.4   Grounding

Since schemas can be run in simulation, the agent is able to self-generate sensory information that would be provided by the environment as a consequence of its actions. This means that not only actually performed actions provide an epistemic access to the environment, but also simulated ones; as a consequence, not only experienced objects and events can be grounded, but also these 'virtually' experienced via simulated interaction. In [29] internal simulation of possible trajectories is used for grounding concepts related to navigation; for example, distance from obstacles is grounded and estimated by running simulations until they encounter the obstacle. Dead-ends are recognized through simulated obstacle avoidance, while passages are grounded in successfully terminated simulations of navigation.

In the schema-based framework proposed by [55] expectations about the sensorimotor flow are used instead for grounding the meaning of words and sentences in natural language. Words for perceptual features are grounded into sensory information; for example, 'red' is grounded in some (expected) values of the robot's sensors. More complex attributes are grounded thanks to (actual and potential) actions. Concepts for objects which are for example reachable or graspable are grounded by schemas which regulate actual behavior and at the same time encode predictions of the consequences of expected interaction. For example, the meaning of 'sponge' is the set of expected consequences of own actions with a sponge (e.g. the anticipated softness), which constitutes the grounding of the word.

In [50] AKSL has been used for designing a 2-layered architecture, corresponding to the two systems for automatic and willed control of action in [46]. The lower layer, which is called *sensorimotor*, is very similar to the already presented architecture of the praying mantis, but includes schemas for navigating in a simulated house. The higher layer, which is called *deliberative*, includes instead declarative knowledge (beliefs and goals states) and pre-compiled sequences of schemas (plans) for navigating the house scenario; this layer is used for reasoning. One novelty of this system is that beliefs are dynamically added to the deliberative layer on the basis of how schemas perform in the sensorimotor layer. For example, the belief *the door is open* is added when the schema(s) for traversing the door is being successful, or is expected to be successful. Not only actual actions, but also simulated ones have been used for forming beliefs: in this case schemas have been used in the *simulation mode*. The rationale is that 'I can believe that the door is open since I expect that, if I try to traverse it, my attempt will succeed' (I also have to assume that the context for acting will be appropriate; for example, I have to start the action in front of the door). The system can also perform explicit epistemic actions (for example in order to check if a belief is true). By exploiting the same machinery that serves for building up the belief x, I can know under which conditions I will come to believe x by using counterfactual reasoning (e.g., what do I have to do in order to know whether or not the door is open?). We have argued that beliefs which are built in this way are grounded, and their verofunctional value can be verified or falsified on the basis of success or failure of schemas in the sensorimotor layer[7].

---

[7] Some beliefs are about other beliefs and not about stimuli. In all cases, however, a relation (direct or indirect) can be identified with the sensorimotor layer.

4.5   Hierarchical Control of Action

Several hierarchical architectures exist in the literature for action control and the orientation of attention [18,27] in which schemas that include representations at different level of abstraction are used. We have used AKSL for hierarchical control is the above mentioned 2-layered architecture. Plans in the deliberative layer are simply pre-compiled sequences of schemas. If a plan is selected by reasoning, it influences the dynamics of the sensorimotor layer by triggering schemas in sequence, much in the way drives do. Depending on the amount of resources assigned to the deliberative layer (and to plans), this influence can constrain more or less the behavior of the agent.

Another example of the use of AKSL for hierarchical control is the architecture for visual search in [51]. Similarly to 'pandemonium' models [59], schemas at the higher layers encode increasingly abstract representations and expectations. They learn to predict the activity level of those at the lower layers, and expectations produced at the high level canalize in a top-down way search at the lower level, while bottom-up error signals serve mainly to confirm or disconfirm concurrent running hypotheses. Empirical evidence exist for a hierarchical organization of the visual apparatus; a comprehensive theoretical framework and implementation is *predictive coding* [54]. This approach is also consistent with simulative theories of cognition. According to Grush [26], simulations can nest to produce increasingly abstract levels of description, in which the criteria for 'matching' are increasingly distant from perceptual matching, although they remain grounded on (actual or simulated) sensorimotor interaction.

5   Conclusions

AKSL permits the design and implementation of *anticipatory* schema-based architectures, and anticipation plays a crucial role in realizing several functionalities. For example, action selection is influenced by the anticipatory capabilities of the schemas, and in particular of their forward models. One of the elements for assigning activity level to a schema is its accuracy in predicting the next sensory input in case one pattern of actions is selected. Attention has an anticipatory component too: predictions generated by the forward models permit the orientation of attention toward the expected position of entities such as a prey to detect, or toward parts of the environment in which the agent expects to find information relevant for its current task. Category formation depends on anticipation, too, and in particular by expectations generated by several schemas at once. Schemas specialized for features of the same entity are likely to have coordinated patterns of prediction and for this reason can evolve energetic links in an Hebbian-like way. Categories thus emerge as clusters of schemas which are expected to be more active during interaction with the same entities. Simulative capabilities, that are based on the substitution of actual stimuli with self-generated expectations, permit a number of other functionalities, such as generating and comparing alternative courses of events on the basis of their long-term effects, or grounding concepts on the basis of self-generating sensory stimuli. Lastly, in designing hierarchical architectures, an interplay of top-down and bottom-up signals are used which convey sensory expectations and prediction errors.

Anticipation is thus crucially involved in several functions, from the simplest to more complex ones. But is it really advantageous to anticipate in all circumstances? Running the forward models, and especially using schemas in simulation mode, is very costly in terms of time and computational resources, and the adaptive advantage of predicting the future could be lost if this means less responsiveness to the contingencies of the environment. We have conducted several preliminary experiments (reported in [47,48]) and compared schema-based models having anticipatory and reactive strategies (i.e. without the forward models). We have found a significant adaptive advantage of anticipatory strategies when the agent has to deal with complex and dynamical environments offering multiple possibilities for action, while this may be not the case if only simpler tasks are required; see also the experiments reported in [35].

These experiments seem to indicate that a selective pressure for developing anticipatory capabilities, and consequently using anticipation as a 'lever' to develop increasingly complex functionalities, could be the increased level of complexity and dynamicity of the environment. More anticipation also permits an agent to deal successfully with more drives and motivations, since it can allocate attentive resources and orient its behavior by taking into account present, but also future needs. In turn, more motivations and more anticipation make the environment of the agent increasingly complex, and thus demand for even more anticipatory capabilities; for a discussion, see [52]. Of course, it is an open challenge to understand which is the level of complexity and dynamicity of the environment which makes anticipation advantageous, and for this reason we plan to continue using AKSL in the future in a number of simulations comparing reactive and anticipatory strategies.

Acknowledgments. This work is supported by the EU project **MindRACES**, FP6-511931. We wish to thank Martin V. Butz for his many helpful comments and discussions.

References

1. akira, 2003. http://www.akira-project.org/
2. Arbib, M.: Levels of modelling of mechanisms of visually guided behavior. Behavioral and Brain Science 10, 407–465 (1987)
3. Arbib, M.: Schema theory. In: Shapiro, S. (ed.) Encyclopedia of Artificial Intelligence, 2nd edn. vol. 2, pp. 1427–1443. Wiley, Chichester, UK (1992)
4. Arbib, M.A.: The metaphorical brain 2: Neural networks and beyond. Wiley, New York (1989)
5. Arkin, R., Ali, K., Weitzenfeld, A., Cervantes-Prez, F.: Behavioral models of the praying mantis as a basis for robotic behavior. Robotics and Autonomous Systems 32(1), 39–60 (2000)
6. Arkin, R.C.: Behavior-Based Robotics. MIT Press, Cambridge (1998)

7. Barsalou, L.W.: Perceptual symbol systems. Behavioral and Brain Sciences 22, 577–600 (1999)
8. Bartlett, F.C.: Remembering. Cambridge University Press, Cambridge (1932)
9. Bates, E.: The Emergence of Symbols. Academic Press, London (1979)
10. Beer, R.D.: Intelligence as Adaptive Behavior: An Eperiment in Computational Neuroethology. Academic Press, San Diego (1990)
11. Bickhard, M.H.: Levels of representationality. Journal of Experimental and Theoretical Artificial Intelligence 10(2), 179–215 (1998)
12. Brooks, R.A.: Intelligence without representation. Artificial Intelligence 47(47), 139–159 (1991)
13. Butz, M.V.: Anticipatory learning classifier systems. Kluwer Academic Publishers, Boston, MA (2002)
14. Crabbe, F.L.: Optimal and non-optimal compromise strategies in action selection. In: Proceedings of the Eighth International Conference on Simulation of Adaptive Behavior (2004)
15. Damasio, A.R.: Descartes' Error: Emotion, Reason and the Human Brain. Grosset/Putnam, New York, 1994. trad. it. di Cartesio, L., Adelphi, Milano (1999)
16. Davidsson, P.: A framework for preventive state anticipation. In: Butz, M.V., Sigaud, O., Gérard, P. (eds.) Anticipatory Behavior in Adaptive Learning Systems. LNCS (LNAI), vol. 2684, pp. 151–166. Springer, Heidelberg (2003)
17. Demiris, Y.: Prediction of intent in robotics and multi-agent systems. to appear in Cognitive Processing (2007)
18. Demiris, Y., Khadhouri, B.: Hierarchical attentive multiple models for execution and recognition (hammer). Robotics and Autonomous Systems Journa 54, 361–369 (2005)
19. Desmurget, M., Grafton, S.: Forward modeling allows feedback control for fast reaching movements. Trends Cogn. Sci. 4, 423–431 (2000)
20. Doya, K.: Complementary roles of basal ganglia and cerebellum in learning and motor control. Curr. Opin. Neurobiol 10(6), 732–739 (December 2000)
21. Drescher, G.L.: Made-Up Minds: A Constructivist Approach to Artificial Intelligence. MIT Press, Cambridge, MA (1991)
22. Fleischer, J.G.: Neural correlates of anticipation in cerebellum, basal ganglia, and hippocampus. In: this volume
23. Frith, C.D., Blakemore, S.J., Wolpert, D.M.: Abnormalities in the awareness and control of action. Philos Trans R Soc. Lond. B Biol. Sci. 355(1404), 1771–1788 (December 2000)
24. Gallese, V., Metzinger, T.: Motor ontology: The representational reality of goals, actions, and selves. Philosophical Psychology 13(3), 365–388 (2003)
25. Gross, H.-M., Volker, S., Torsten, S.: A neural architecture for sensorimotor anticipation. Neural Networks 12, 1101–1129 (1999)
26. Grush, R.: The emulation theory of representation: motor control, imagery, and perception. Behav Brain Sci. 27(3), 377–396 (June 2004)
27. Haruno, M., Wolpert, D., Kawato, M.: Hierarchical mosaic for movement generation. In: Ono, T., Matsumoto, G., Llinas, R., Berthoz, A., Norgren, H., Tamura, R. (eds.) Excepta Medica International Congress Series, Elsevier Science, Amsterdam (2003)
28. Hesslow, G.: Conscious thought as simulation of behaviour and perception. Trends in Cognitive Sciences 6, 242–247 (2002)
29. Hoffmann, H.: Perception through visuomotor anticipation in a mobile robot. Neural Networks 20, 22–33 (2007)

30. Hoffmann, J.: Anticipatory behavioral control. In: Butz, M.V., Sigaud, O., Gerard, P. (eds.) Anticipatory Behavior in Adaptive Learning Systems: Foundations, Theories, and Systems, pp. 44–65. Springer-Verlag, Berlin Heidelberg (2003)
31. Holland, J., Holyoak, K., Nisbett, R., Thagard, P.: Induction: Processes of Inference, Learning, and Discovery. MIT Press, Cambridge, Massachussets (1986)
32. ikaros,(2002) http://www.lucs.lu.se/IKAROS.
33. irrlicht, (2003) http://irrlicht.sourceforge.net/.
34. James, W.: The Principles of Psychology. Dover Publications, New York (1890)
35. Johansson, B., Balkenius, C.: An experimental study of anticipation in robot navigation. In: this volume
36. Kalman, R.E.: A new approach to linear filtering and prediction problems. Journal of Basic Engineering 82(1), 35–45 (1960)
37. Lyons, D.M., Arbib, M.A.: A formal model of computation for sensory-based robotics. IEEE Journal of Robotics and Automation 5(3), 280–293 (1989)
38. Maes, P.: Situated agents can have goals. In: Maes, P. (ed.) Designing Autonomous Agents, pp. 49–70. MIT Press, Cambridge (1990)
39. Mccauley, L.: Neural schemas: A mechanism for autonomous action selection and dynamic motivation. In: the 3rd WSES Neural Networks and Applications Conference (2002)
40. Miall, R.C., Wolpert, D.M.: Forward models for physiological motor control. Neural Networks 9(8), 1265–1279 (1996)
41. Middleton, F.A., Strick, P.L.: Basal ganglia output and cognition: evidence from anatomical, behavioral, and clinical studies. Brain Cogn. 42(2), 183–200 (2000)
42. Miller, G.A., Galanter, E., Pribram, K.H.: Plans and the Structure of Behavior. Holt, Rinehart and Winston, New York (1960)
43. Minsky, M.: The Society of Mind. Simon & Schuster (1988)
44. Neisser, U.: Cognition and reality. Freeman, San Francisco, CA (1976)
45. Newell, A., Simon, H.A.: Human problem solving. Prentice-Hall, Englewood Cliffs, NJ (1972)
46. Norman, D.A., Shallice, T.: Attention to action: Willed and automatic control of behaviour. In: Davidson, R.J., Schwartz, G.E., Shapiro, D. (eds.) Consciousness and Self-Regulation: Advances in Research and Theory, Plenum Press, New York (1986)
47. Pezzulo, G., Calvi, G.: A schema based model of the praying mantis. In: Nolfi, S., Baldassarre, G., Calabretta, R., Hallam, J.C.T., Marocco, D., Meyer, J.-A., Miglino, O., Parisi, D. (eds.) SAB 2006. LNCS (LNAI), vol. 4095, pp. 211–223. Springer Verlag, Heidelberg (2006)
48. Pezzulo, G., Calvi, G.: Toward a perceptual symbol system. In: Proceedings of the Sixth International Conference on Epigenetic Robotics: Modeling Cognitive Development in Robotic Systems. Lund University Cognitive Science Studies, 118 (2006)
49. Pezzulo, G., Calvi, G.: Designing modular architectures in the framework akira. To appear in Multiagent and Grid Systems, 3(1) (2007)
50. Pezzulo, G., Calvi, G., Castelfranchi, C.: Dipra: Distributed practical reasoning architecture. In: Proceedings of the Twentieth International Joint Conference on Artificial Intelligence, pp. 1458–1464 (2007)
51. Pezzulo, G., Calvi, G., Ognibene, D., Lalia, D.: Fuzzy-based schema mechanisms in akira. In: CIMCA '05. Proceedings of the International Conference on Computational Intelligence for Modelling, Control and Automation and International Conference on Intelligent Agents, Web Technologies and Internet Commerce, Washington, DC, USA. Vol-2, pp. 146–152. IEEE Computer Society, Los Alamitos, CA, USA (2005)

52. Pezzulo, G., Castelfranchi, C.: The symbol detachment problem. Cognitive Processing (to appear, 2007)
53. Piaget, J.: The Construction of Reality in the Child. Ballentine (1954)
54. Rao, R.P., Ballard, D.H.: Predictive coding in the visual cortex: a functional interpretation of some extra-classical receptive-field effects. Nat. Neurosci 2(1), 79–87 (January 1999)
55. Roy, D.: Semiotic schemas: a framework for grounding language in action and perception. Artificial Intelligence 167(1-2), 170–205 (2005)
56. Saffiotti, A.: Handling uncertainty in control of autonomous robots. In: Artificial Intelligence Today, pp. 381–407 (1999)
57. Schank, R.C., Abelson, R.P.: Scripts, Plans, Goals and Understanding: an Inquiry into Human Knowledge Structures. Erlbaum, L., Hillsdale, NJ (1977)
58. Schultz, W.: Predictive reward signal of dopamine neurons. Journal of Neurophysiology 80, 1–27 (1998)
59. Selfridge, O.: The Mechanisation of Thought Processes, volume 10, chapter Pandemonium: A paradigm for learning, pp. 511–529. National Physical Laboratory Symposia. Her Majesty's Stationary Office, London (1959)
60. Shapiro, D., Schmidt, R.: The schema theory: Recent evidence & developmental implications. In: Clark, J.K.J. (ed.) The development of movement control and co-ordination, Wiley, New York (1982)
61. Smith, O.J.M.: A controller to overcome dead time. ISA Journal 6(2), 28–33 (1959)
62. Tani, J.: Learning to generate articulated behavior through the bottom-up and the top down interaction processes. Neural Netw. 16(1), 11–23 (January 2003)
63. Tani, J., Ito, M., Sugita, Y.: Self-organization of distributedly represented multiple behavior schemata in a mirror system: reviews of robot experiments using rnnpb. Neural Netw. 17(8-9), 1273–1289 (2004)
64. Tani, J., Nolfi, S.: Learning to perceive the world as articulated: an approach for hierarchical learning in sensory-motor systems. Neural Netw. 12(7-8), 1131–1141 (October 1999)
65. Weitzenfeld, A., Peguero, O., Gutiérrez, S.: NSL/ASL: Distributed simulation of modular neural networks. In: MICAI, pp. 326–337 (2000)
66. Wolpert, D.M., Gharamani, Z., Jordan, M.: An internal model for sensorimotor integration. Science 269, 1179–1182 (1995)
67. Wolpert, D.M., Kawato, M.: Multiple paired forward and inverse models for motor control. Neural Networks 11(7-8), 1317–1329 (1998)
68. Ziemke, T., Jirenhed, D.-A., Hesslow, G.: Internal simulation of perception: a minimal neuro-robotic model. Neurocomputing 68, 85–104 (2005)

Training and Application
of a Visual Forward Model
for a Robot Camera Head

Wolfram Schenck and Ralf Möller

Computer Engineering Group, Faculty of Technology, Bielefeld University,
Bielefeld, Germany
wschenck@ti.uni-bielefeld.de

Abstract. Visual forward models predict future visual data from the
previous visual sensory state and a motor command. The adaptive acquisition of visual forward models in robotic applications is plagued by
the high dimensionality of visual data which is not handled well by most
machine learning and neural network algorithms. Moreover, the forward
model has to learn which parts of the visual output are really predictable
and which are not because they lack any corresponding part in the visual input. In the present study, a learning algorithm is proposed which
solves both problems. It relies on predicting the mapping between pixel
positions in the visual input and output instead of directly forecasting
visual data. The mapping is learned by matching corresponding regions
in the visual input and output while exploring different visual surroundings. Unpredictable regions are detected by the lack of any clear correspondence. The proposed algorithm is applied successfully to a robot
camera head under additional distortion of the camera images by a retinal mapping. Two future applications of the final visual forward model
are proposed, saccade learning and a task from the domain of eye-hand
coordination.

1 Visuomotor Prediction

Sensorimotor control is an important research topic in many disciplines, among
them cognitive science and robotics. These fields tackle the questions of how
complex motor skills can be acquired by biological organisms or robots, and how
sensory and motor processing are interrelated to each other. So-called "internal
models" help to clarify ideas of sensorimotor processing on a functional level
[14,22]. "Inverse models" or controllers generate motor commands based on the
current sensory state and the desired one; "forward models" (FWM) predict
future sensory states as outcome of motor commands applied in the current
sensory state. The present study focuses on the anticipation of visual data by
FWMs.

The anticipation of sensory consequences in the nervous system of biological
organisms is supposed to be involved in several sensorimotor processes: First,
many motor actions rely on feedback control, but sensory feedback is generally

M.V. Butz et al. (Eds.): ABiALS 2006, LNAI 4520, pp. 153–169, 2007.
© Springer-Verlag Berlin Heidelberg 2007

too slow. Here, the output of FWMs can replace sensory feedback [15]. Second, FWMs may be used in the planning process for complex motor actions [21]. Third, FWMs are part of a controller learning scheme called "distal supervised learning" [13]. Fourth, FWMs can help to distinguish self-induced sensory effects (which are predicted) from externally induced sensory effects (which stand out from the predicted background) [1]. Fifth, it has been suggested that perception might rely on the anticipation of the consequences of motor actions that could be applied in the current situation. The anticipation would be accomplished by FWMs [16].

Regarding the fourth function mentioned above, a classical example is the reafference principle suggested by Holst and Mittelstaedt [12]. It explains why (self-induced) eye movements do not evoke the impression that the world around us is moving. As long as the predicted movement of the retinal image (caused by the eye movement) coincides with the actual movement, the effect of this movement is canceled out in the visual perception. Additional evidence for the fourth function of FWMs stems from a study of Blakemore et al. [1]: In their experiments, subjects had to tickle themselves via a robotics interface. The closer the actual movement of the robot corresponded to the tickling movements of the subjects, the less the subjects rated their tactile sensation as tickly, pleasant, and intense. This effect has been ascribed to FWMs whose output cancels out the sensory effect of self-executed tickling.

In fields like robotics or artificial life, studies using FWMs for motor control focus mainly on navigation or obstacle avoidance tasks with mobile robots. The sensory input to the FWMs are rather low-dimensional data from distance sensors or laser range finders (e.g.: [21,23]), optical flow fields [7], or preprocessed visual data with only a few remaining dimensions [9]. Only in a recent study by Hoffmann, a visual FWM is implemented which predicts images with a size of 40×40 pixels. It is used for distance estimation and dead-end recognition to demonstrate that perception by anticipation actually works (the fifth function of FWMs mentioned above) [11].

We are especially interested in the learning of FWMs in the visual domain, and its application to robot models. In our understanding, visual FWMs predict representations of entire visual scenes. In the nervous system, this could be the relatively unprocessed representation in the primary visual cortex or more complex representations generated in higher visual areas. Regarding robot models, the high-dimensional sensory input and output space of visual FWMs poses a tough challenge to any machine learning or neural network algorithm. Moreover, there might be unpredictable regions in the FWM output (because parts of the visual surrounding only become visible after execution of the motor command). In the present study, we suggest a learning algorithm which solves both problems in the context of robot "eye" movements. In doing so, our main goal is to demonstrate a new efficient learning algorithm for image prediction. Furthermore, we suggest two different future applications for the resulting visual FWM in the domain of motor learning and behavioral control.

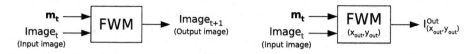

Fig. 1. Left: Visual forward model (FWM). Right: Single component of a visual forward model predicting the intensity of a single pixel $(x_{\text{Out}}, y_{\text{Out}})$ of the output image.

2   Visual Forward Model for Camera Movements

In our robot model, we attempt to predict the visual consequences of eye movements. In the model, the eye is replaced by a camera which is mounted on a pan-tilt unit. Prediction of visual data is carried out on the level of camera images. In analogy to the sensor distribution on the human retina, a retinal mapping is applied which decreases the resolution of the camera images from center to border. Although this mapping copies an important aspect of visual processing in humans, we do not intend to develop, implement, or test a model of the human visual pathway. Instead, we explore a learning mechanism for visual FWMs and suggest a possible architecture for this purpose. The retinal mapping is part of this architecture, but it is not required for the proposed learning mechanism. Instead, its only purpose is to make the prediction task more difficult and to prove the robustness of the learning mechanism. The input of the visual FWM is a "retinal image" at time step t (called "input image" in the following) and a motor command $\mathbf{m}_t$. The output is a prediction of the retinal image at the next time step $t + 1$ (called "output image" in the following; see left part of Figure 1).

The question is how such an adaptive visual FWM can be implemented and trained by exploration of the environment. A straight-forward approach is the use of function approximators which predict the intensity of single pixels. For every pixel $(x_{\text{Out}}, y_{\text{Out}})$ of the output image, a specific forward model $\text{FWM}_{(x_{\text{Out}}, y_{\text{Out}})}$ is acquired which forecasts the intensity of this pixel (see right part of Figure 1). Together, the predictions of these single FWMs form the output image as in Figure 1 (left). Unfortunately, this simple approach suffers from the high dimensionality of the input space (the retinal image at time step t is part of the input), and does not produce satisfactory learning results [8]. In the work of Hoffmann, where images with a size of 40×40 pixels are directly predicted, an additional denoising model is required. This model has to be trained for the specific environment of the mobile robot [11].

Hence, in this study we pursue a different approach. Instead of forecasting pixel intensities directly, our solution is based on a "back" prediction where a pixel of the output image was in the input image before the camera's movement. The necessary mapping model (MM) is depicted in Figure 2: as input, it receives the motor command $\mathbf{m}_t$ and the location of a single pixel $(x_{\text{Out}}, y_{\text{Out}})$ of the output image; as output it estimates the previous location $(\widehat{x}_{\text{In}}, \widehat{y}_{\text{In}})$ of the

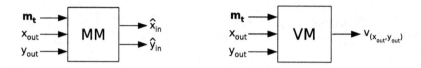

Fig. 2. Left: Mapping model (MM). Right: Validator model (VM) (for details see text).

corresponding pixel (or region) in the input image. The overall output image is constructed by iterating through all of its pixels and computing each pixel intensity as $\widehat{I}^{Out}_{(x_{Out},y_{Out})} = I^{In}_{(\widehat{x}_{In},\widehat{y}_{In})}$ (using bilinear interpolation).[1] Moreover, an additional validator model (VM) generates a signal $v_{(x_{Out},y_{Out})}$ indicating whether it is possible at all for the MM to generate a valid output for the current input. It predicts which pixels of the output image are in a position that does not correspond to any pixel of the input image. This is necessary because even for small camera movements parts of the output image are not present in the input image. In this way, the overall FWM (Figure 1, left) is implemented by the combined application of a mapping and a validator model.

The basic idea of the learning algorithm for the MM can be outlined as follows for a specific $\mathbf{m}_t$ and (x_{Out}, y_{Out}): during learning, the motor command is carried out in different environmental settings. Each time, both the actual input and output image are known afterwards, thus the intensity $I^{Out}_{(x_{Out},y_{Out})}$ is known as well. It is possible to determine which of the pixels of the input image show a similar intensity. These pixels are candidates for the original position (x_{In}, y_{In}) of the pixel (x_{Out}, y_{Out}) before the movement. Over many trials, the pixel in the input image which matches most often is the most likely candidate for (x_{In}, y_{In}) and therefore chosen as MM output $(\widehat{x}_{In}, \widehat{y}_{In})$. When none of the pixels match often enough, the MM output is marked as non-valid (output of VM).

3   Method

To acquire such a MM and VM as in Figure 2, the following steps are executed. First, a grid of points is defined in the input space of the MM and VM (composed of $\mathbf{m}_t$, x_{Out}, and y_{Out}), ranging from the minimum to the maximum value in each input dimension. For each grid point, the most likely estimate $(\widehat{x}_{In}, \widehat{y}_{In})$ is determined by collecting candidate pixels in many different visual surroundings. Along the way, the VM output $v_{(x_{Out},y_{Out})}$ is determined as well. Thereafter, one radial basis function network (RBFN) [17] is trained to interpolate the MM output between the grid points, and another RBFN to interpolate the VM output. The resulting networks can be applied to image prediction afterwards. In the following, the methods are outlined in more detail.

---

[1] In this study, pixel intensities of the retinal input and output images are three-dimensional vectors in RGB color space.

3.1   Setup

The robot setup is shown in Figure 3 (left). Only the right camera is used. A central quadratic region of the original camera image (captured in RGB color) with a resolution of 240×240 pixels is used for further processing (and called "camera image" in the following for simplicity). The horizontal and the vertical angle of view of this region amount to 48.5 degrees. The camera is mounted on a pan-tilt unit with two degrees of freedom. In this study, the valid range for the pan angle is between -60.4 and 23.8 degrees, for the tilt angle between -42.9 and 21.4 degrees. In this range, the camera image always captures at least a small part of the white table shown in Figure 3 (left) below the cameras.

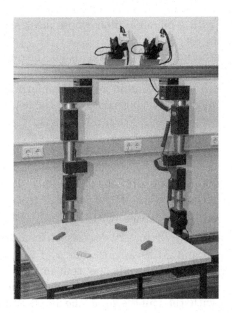

Fig. 3. Left: Setup used as basis for the visual prediction task. Right: Retinal mapping. Upper right image: Original image. Lower right image: Retinal image (enlarged by factor two).

The pan and tilt axes cross in close vicinity to the nodal point of the camera-lens system. For this reason, the effect of changing the pan and tilt position by a certain amount (Δpan, Δtilt) is almost independent of the current camera position. Accordingly, the motor input $\mathbf{m}_t$ of the FWM just consists of Δpan and Δtilt. Both values can vary between -29 and $+29$ degrees. For the same reason, object displacements in the camera images during camera movements are virtually independent from the object distance to the camera. Thus, depth information is irrelevant for our learning task.

3.2   Retinal Mapping

As mentioned before, the input and output images of the FWM are "retinal" images with decreasing resolution from image center to border. Camera images are converted to such retinal images by a "retinal mapping". The effect of this conversion is depicted in Figure 3 (right). The basic idea of this mapping is best outlined in polar coordinates. The origins of the coordinate systems are located at the image centers. They are scaled in a way so that in both images the maximum radius (along the horizontal/vertical direction) amounts to 1.0. r_R is the radius of a point in the retinal image, r_C is the radius of the corresponding point in the camera image, the angle of the polar representation is kept constant. r_C is computed by $r_C = \lambda r_R^\gamma + (1 - \lambda) r_R$, $\gamma > 1$, $0 \leq \lambda \leq 1$. Here we use $\gamma = 2.5$ and $\lambda = 0.7$. The resolution of the final retinal image is 69×69 pixels. To avoid aliasing artifacts in the heavily subsampled outer regions of the original image, adaptive smoothing is applied (with a binomial filter whose mask size is proportional to the local subsampling factor).

While the input image of the FWM is an unmodified retinal image, the output image is a center crop with a size of 53×53 pixels. It is necessary to clip the white corners of the retinal image, which do not contain any valid information (see Figure 3, right). This is just a technical artifact but could spoil the learning algorithm.

3.3   Grid of Cumulator Units

The input space of the MM and VM consists of four dimensions: Δpan, Δtilt, x_{Out}, and y_{Out}. A four-dimensional grid $\mathbf{P}$ of points $\mathbf{p}_{ijkl} = \left(\Delta\text{pan}^{(i)}, \Delta\text{tilt}^{(j)}, x_{\text{Out}}^{(k)}, y_{\text{Out}}^{(l)} \right)$ is inscribed in this space, with $i, j = 1, .., 7$ and $k, l = 1, .., 11$. $\Delta\text{pan}^{(i)}$ and $\Delta\text{tilt}^{(j)}$ vary from -29 to $+29$ degrees with constant step size (covering the whole valid Δpan and Δtilt range), while $x_{\text{Out}}^{(k)}$ and $y_{\text{Out}}^{(l)}$ form an equally spaced rectangular grid covering the whole output image.

To each point $\mathbf{p}_{ijkl}$, a so-called "cumulator unit" C_{ijkl} is attached. Such a unit is basically a single-band image with the same size as the input image. Thus, the input image and the cumulator units have the same number of pixels in the horizontal and vertical direction. Each "pixel" of a cumulator unit can hold any positive integer value including zero. They are used to accumulate and store the number of matches between input and output image at their specific position (the position within the cumulator unit addresses the position in the input image, the position within the grid $\mathbf{P}$ addresses the position in the output image and in the motor space). This is explained in further detail in the next section.

3.4   Learning Process

The goal of the learning process is to accumulate activations in the cumulator units. At the beginning, all pixels of these units are set to zero. In each learning trial, the pan-tilt unit is first moved into a random (pan, tilt) position. The input

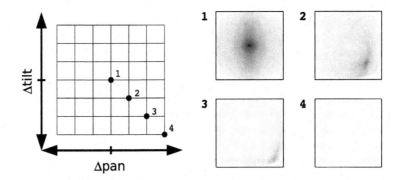

Fig. 4. Cumulator units for the center pixel for four different (Δpan, Δtilt) positions. All depicted cumulator units were normalized by the same scaling factor so that a pixel value of zero corresponds to white and the overall maximum pixel value to black.

image for the FWM is recorded and processed. Afterwards, the algorithm iterates through all points of the grid $\mathbf{P}$, the corresponding motor command is executed (relative to the initial random position), and the output image is generated from the camera image after the movement. For each point $\mathbf{p}_{ijkl}$, the intensity of the output image at the coordinates $\left(x_{\text{Out}}^{(k)}, y_{\text{Out}}^{(l)}\right)$ is compared to the intensities of all pixels $(x_{\text{In}}, y_{\text{In}})$ in the current input image. Whenever the intensity difference is below a certain threshold α, the value of pixel $(x_{\text{In}}, y_{\text{In}})$ in cumulator unit C_{ijkl} is increased by one. The intensity difference is computed as Euclidean distance in RGB color space. The threshold α is set to 3.5% of the overall intensity range in a single color channel.

In the present study, 100 trials were carried out, each with $7 \times 7 \times 11 \times 11 = 5929$ iteration steps (size of the grid $\mathbf{P}$). Each trial took place in a slightly different visual environment because the initial camera position varied. 42 colored wooden blocks were placed on the table to enhance the visual richness of the environment (see Figure 3, right).

Figure 4 illustrates four final cumulator units C_{ijkl} in the grid $\mathbf{P}$. Their positions along the Δpan and Δtilt dimensions are marked on the two-dimensional grid on the left (camera movements to the lower right of increasing length, starting at position 1 with zero movement). Their position $\left(x_{\text{Out}}^{(k)}, y_{\text{Out}}^{(l)}\right)$ in output image coordinates is the center pixel. The pixel color in the cumulator units reflects the size of the accumulated sum from white (zero) to black (maximum sum). Unit 1 with zero camera movement shows a clear maximum exactly in the center. Thus, the most likely origin of the center pixel in the output image is the center pixel in the input image. This is exactly what is expected when no camera movement takes place. Unit 2 is associated with a small camera movement to the lower right. The intensity maximum is no longer in the center of the unit, but in the lower right corner: when the camera moves in a certain direction, the new image center has its origin in the direction of the movement. Because of the retinal mapping, the intensity maximum moves a large distance towards

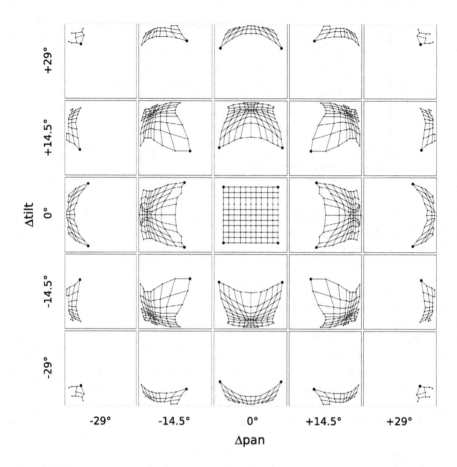

Fig. 5. Mapping from pixel coordinates $\left(x_{\text{Out}}^{(k)}, y_{\text{Out}}^{(l)}\right)$ (grid points) in the output image to pixel coordinates $(\widehat{x}_{\text{In}}, \widehat{y}_{\text{In}})$ in the input image for 5×5 different $(\Delta\text{pan}, \Delta\text{tilt})$ positions

the border of the cumulator unit although the corresponding camera movement is rather small. Unit 3 with a larger camera movement shows a similar effect. Moreover, its maximum intensity is obviously weaker than in unit 1. This is mainly caused by the retinal mapping with its heavy subsampling in the outer image regions (causing fewer matches with the correct candidate pixel). Finally, unit 4 shows no visible maximum in print at all. Actually, the corresponding camera movement is so large that the center pixel of the output image has no valid counterpart in the input image, therefore it is unpredictable.

3.5   Generating a Raw Version of the MM and VM

After the cumulator units have been acquired in the learning process, raw versions of the MM and VM can be created whose output is defined at the grid positions $\mathbf{p}_{ijkl}$ in input space. The output $(\widehat{x}_{\text{In}}, \widehat{y}_{\text{In}})$ of the MM at grid point

$\mathbf{p}_{ijkl}$ is the coordinate of the pixel with maximum intensity in the cumulator unit C_{ijkl}. The output $v_{(x_{\mathrm{Out}},y_{\mathrm{Out}})}$ of the VM at point $\mathbf{p}_{ijkl}$ is set to 1 (signaling valid output of the MM at this point) whenever the maximum pixel intensity in unit C_{ijkl} is above a certain threshold. Otherwise, $v_{(x_{\mathrm{Out}},y_{\mathrm{Out}})}$ is set to 0. The threshold is computed as the product of the maximum pixel intensity of all cumulator units and a factor $\beta = 0.45$. This proved to be the value resulting in the most correct separation.

Figure 5 shows the output of the MM and VM for 25 different motor commands $\left(\Delta\mathrm{pan}^{(i)}, \Delta\mathrm{tilt}^{(j)}\right)$. For each motor command, the pixel coordinate space of the input image is shown in a single panel. The two-dimensional grid in each panel connects points along the $x_{\mathrm{Out}}^{(k)}$ and $y_{\mathrm{Out}}^{(l)}$ directions of $\mathbf{P}$. The position of each grid point corresponds to the output $(\widehat{x}_{\mathrm{In}}, \widehat{y}_{\mathrm{In}})$ of the MM at this point. Only points with valid output are shown (determined by the VM). The central panel with no movement shows an identity mapping between $(x_{\mathrm{Out}}^{(k)}, y_{\mathrm{Out}}^{(l)})$ and $(\widehat{x}_{\mathrm{In}}, \widehat{y}_{\mathrm{In}})$ (as expected). The other panels reflect the relationship between the camera movement and the pixel shift between input and output image. The strong curvature of the grid is mainly caused by the retinal mapping.

3.6 Network Training

The output of the raw versions of the MM and the VM is only defined at the grid points $\mathbf{p}_{ijkl}$. To get the output between grid points, function interpolation is necessary. For this purpose, the raw versions of the MM and the VM were replaced by radial basis function networks (RBFN) in the final step of the learning algorithm. These networks have the same input/output structure as the MM and the VM, respectively (see Figure 2). The training data for both networks was generated from the output of the raw versions of the MM and the VM at the grid points $\mathbf{p}_{ijkl}$ (overall, there are $7 \times 7 \times 11 \times 11 = 5929$ grid points). For the MM network, training data was restricted to the 2935 grid points with valid output (signaled by the raw version of the VM).

For both networks, the hyperbolic tangent was used as activation function for the output units. Both the MM and the VM network were initialized with the k-means algorithm, afterwards they were trained for 1000 epochs with gradient descent. Input and output values were scaled to the range $[-0.6; 0.6]$.

The MM network is a RBFN with 200 Gaussians for each output unit (x_{Out} and y_{Out}). The training set consisted of the 2935 valid input-output pairs of the raw MM. The mean squared error per pattern per output unit amounted to $2.3 \cdot 10^{-4}$ after the last epoch.

The VM network has 250 Gaussians in the hidden layer for its single output unit. It basically had to learn a classification task with a training set covering all 5929 grid points. While the mean squared error per pattern per output unit still amounted to $5.3 \cdot 10^{-2}$ after the last epoch, only 1.3% of the grid points were misclassified.

It is possible to use alternative methods for function interpolation, e.g., to construct the RBFNs directly from the grid points without learning (even during

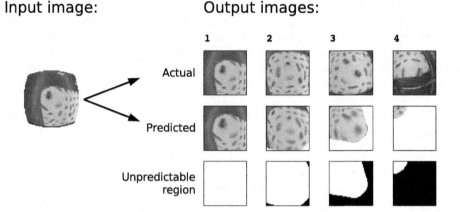

Fig. 6. Comparison of actual and predicted output images at four different (Δpan, Δtilt) positions (the same as in Figure 4)

the acquisition of the cumulator units as a kind of "online" method), or to use other non-linear regression methods.

4   Results

The MM and VM network are used to implement the overall visual FWM for predicting the output image as explained in Sect. 2. Especially, non-predictable regions of the output image are marked by the VM network. The prediction works rather precisely, as shown exemplary in Figure 6. The actual and the predicted output images are compared for four different motor commands (Δpan, Δtilt) (camera movements to the lower right of increasing length as in Figure 4). Moreover, the region of each output image that is marked as non-predictable by the VM is shown in black in the third row of images. The input image (the same for all four movements) is displayed as well. Movement 1 is a zero movement. The actual and the predicted output images are very similar and show the center crop from the input image. Movements 2 and 3 are of increasing size. The non-predictable regions mask parts of the output images which have no correspondence in the input image. The center of the predicted images is slightly blurred and distorted because the mapping generated by the MM network has to enlarge a region of a few pixels in the input image to a much larger area (especially for movement 3). Movement 4 is so large that the center of the output image is non-predictable. Nevertheless, the small upper left part of the output image which is predicted corresponds closely to the actual output.

This visual inspection of a few exemplary camera movements demonstrates the learning success. At the current stage of development, the additional application of quantitative evaluations is not useful because of the lack of competing learning algorithms for visual FWMs. Furthermore, quantitative measures like the Euclidean distance in pixel space are difficult to interpret because the FWM has to enlarge parts of the input image while the actual output maintains the optimum resolution in the image center.

We pointed out in Sect. 3.1 that depth information is irrelevant for our learning task because of the camera geometry. Therefore, it is possible to rearrange objects in the field of view of the camera without any harm to the prediction performance of the visual FWM.

5   Future Applications

As outlined in the introduction, there is strong evidence that FWMs play an important role for the sensorimotor coordination in biological organisms. Here, we outline two modeling approaches in which the acquired visual FWM is used for motor learning and behavioral control. It is planned to implement both approaches on a robot setup. Beyond the robotics application, they put forward specific hypotheses for which purposes visual FWMs might be used in biological organisms.

5.1   Saccade Learning

In a previous study, we presented an adaptive saccade controller for the robot camera head shown in Figure 3 (left) [19]. The task of the controller is to fixate target objects with both cameras so that the target object is projected onto the center of both camera images. Overall, the camera head has four degrees of freedom: a conjoint pan-tilt direction (pan, tilt), and a horizontal and vertical vergence value ($\mathrm{verg}_{\mathrm{hor}}$, $\mathrm{verg}_{\mathrm{vert}}$).

As input, the saccade controller receives the current sensory state $\mathbf{s}_t$, composed of a kinesthetic and a visual part (see Figure 7). The kinesthetic input consists of the current position of the cameras. The visual part represents the position of the target object in the left and right camera image: x_{left}, y_{left}, x_{right}, y_{right}. The motor output $\mathbf{m}_t$ of the saccade controller is defined as a change of the motor position. It consists of four values: $\Delta\mathrm{pan}$, $\Delta\mathrm{tilt}$, $\Delta\mathrm{verg}_{\mathrm{hor}}$, and $\Delta\mathrm{verg}_{\mathrm{vert}}$.

After the movement, the position of the target object in the camera images has changed. Its deviation from the image center is the *sensory error*. In the beginning of the learning process for adaptive saccade control, this error is large. Several motor learning schemes exist which make use of the sensory error to generate a *motor error* signal which is necessary to adjust the motor output of the saccade controller. In [19], their performance is compared.

This comparison study used a simulated geometrical model of the robot camera head and its environment instead of the real-world setup [19]. In this simulated setting, the position of the target object is always clearly defined, whereas for the real-world setup, this is not the case. First, a suitable target object for fixation has to be determined in either the left or right camera image. Afterwards, the object has to be identified in the other camera image as well. Most

Kinesthetic input:
pan, tilt, verg$_{hor}$, verg$_{vert}$

Visual input:
x$_{left}$, y$_{left}$, x$_{right}$, y$_{right}$

Saccade controller

Motor output:
Δpan, Δtilt, Δverg$_{hor}$, Δverg$_{vert}$

Fig. 7. Input and output of the saccade controller (for details see text)

importantly, the target object has to be re-identified by both camera images after the saccade for the computation of the sensory error. This re-identification is computationally very expensive since it involves extensive matching of image regions. Moreover, when working with retinal images like the ones used for the visual FWM, simple matching algorithms do not work.

A straightforward solution for the re-identification of target objects is the application of the visual FWM developed in this study. For each camera, it can predict from the executed motor command to which image location the target object has moved, or if it has been lost. Afterwards, the exact position of the target object can be determined by a search process which is restricted to the very close neighborhood of the predicted position. No extensive matching or search process is necessary. Used in the way, the visual FWM offers improved sensory processing and faster behavioral learning. We have not implemented this kind of re-identification yet. However, we don't expect any considerable difficulties.

Several studies on human saccades show that small displacements of target objects during saccades go unnoticed (e.g., [2] and [4]). This finding implies that the mechanisms which implement the reafference principle are not precise enough to detect such small target shifts. Deubel [5] proposes that visual stability between saccades is maintained by matching visual landmarks before and after the saccade. Based on our robot model, we suggest that landmark re-identification is based on a visual FWM which predicts approximately the position of the landmarks after the saccade. Such a FWM is not precise enough for the detection of small displacements, but it suffices to point to a search region. If the target is not found within this region, the mismatch is detected (in the study of Bridgeman [2], target shifts of 4 degrees are detected by the subjects in at least 40% of the trials). Furthermore, we state the hypothesis that this target re-identification mechanism based on visual FWMs is also used during saccade adaptation (as in our saccade learning model).

5.2   Grasping of Extrafoveal Targets

The second potential application of the visual FWM belongs to the domain of eye-hand coordination. It is based on the premotor theory of attention [18] which states that spatial attention is a consequence of the preparation of goal-directed, spatially coded movements. Because of the strong development of the neural mechanisms for foveal vision in primates and humans, oculomotor maps coding

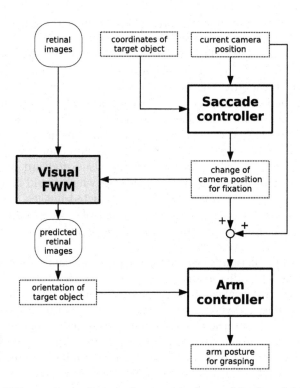

Fig. 8. Flow of information and processing steps in the combined model for the grasping of extrafoveal targets (for details see text). The visual FWM (shaded) is described in-depth in this paper.

space for eye movements play a central role in selective attention. Experimental evidence for the close coupling of saccade preparation and visual attention has been found in several studies, for example in [3]. The link between manual response preparation and shifts of spatial attention has been less convincing, but recent studies provide support for the claim that covert preparation of manual responses is linked to shifts of spatial attention as well [6].

We suggest a computational model of grasping to extrafoveal targets, which is implemented on the robot setup shown in Figure 3 (left), consisting of two cameras with four degrees of freedom (a conjoint pan-tilt direction (pan, tilt), and a horizontal and vertical vergence value ($\mathrm{verg}_{\mathrm{hor}}$, $\mathrm{verg}_{\mathrm{vert}}$)), and a robot arm with six rotatory degrees of freedom and a two-finger gripper. This model is based on the premotor theory of attention and adds one specific hypothesis: Attention shifts caused by saccade programming imply a prediction of the retinal foveal images after the saccade. For this purpose, the visual FWM is used. The predicted foveal images are required to determine movement parameters for the manual interaction with objects at the target location of the attention shift.

Our model consists of three parts (see Figure 8). First, a saccade controller as described in Sect. 5.1; second, the visual FWM of this study; and third, an

arm controller for grasping movements, which receives the output of the saccade controller and the orientation of the target object as inputs (similar to the
controller presented in [10]). The orientation of the target object is determined
from the foveal region of the retinal images. A single grasping trial starts with
the presentation of the grasping target, a colored wooden block, at a random
location within the working space on a table surface. The cameras are in a random posture. The saccade controller generates the necessary motor command
for proper fixation with the cameras, but this movement is not carried out,
only the suggested motor command is recorded as input for the visual forward
model and the arm controller. Afterwards, the visual forward model predicts
the retinal images after the (hypothetical) saccade which would move the object
into the foveae of both cameras. From these predicted images, the orientation
of the block is determined. Finally, the arm controller uses both the saccadic
motor command and the block orientation in the predicted images as inputs to
generate the grasping movement.

Without visual prediction, grasping towards extrafoveal target objects is difficult because of the heavy distortions found in retinal images. Whenever an object
is depicted at an extrafoveal position, its picture is significantly different from its
foveal representation. To overcome this problem, the object can be fixated first
so that it is projected onto the fovea. Actually, grasping movements in primates
and humans are often targeted towards already fixated objects [20]. For grasping
towards extrafoveal target objects, visual prediction offers an elegant solution.
Having predicted the foveal representation of the object after the (purely mental) saccade, the same sensorimotor model linking the foveal representation and
the grasping movement can be used.

In this application, the visual FWM is linked to attentional mechanisms.
Furthermore, by substituting the real visual feedback after a saccade with the
predicted visual feedback (without the need to actually carry out the saccade),
it allows for more efficient behavioral control.

6   Discussion and Conclusions

The proposed learning algorithm for visual FWMs overcomes the problem of
these models having a high-dimensional input and output space due to the size
of visual data. Forecasting pixel intensities is replaced by forecasting a mapping
between output and input pixel locations. The only restriction regarding image
size is imposed by the size of the computer memory because it has to hold the
cumulator units during the learning process.

The learning process relies on matching pixels between the output and input
images. By imposing a retinal mapping, it is demonstrated that this learning
principle even works when strong image distortions are involved (including color
changes caused by smoothing and subsampling in the outer areas of the camera
images). Future research will reveal to which extent the performance of the
learning algorithm deteriorates in response to even more ambiguous visual data
(e.g., by using monochrome images).

The distinction between cumulator units with a large and a small maximum pixel intensity offers a natural solution for the detection of unpredictable image regions. A small maximum indicates that no correct pixel match exists, while an existing correct match accumulates to a large maximum during the learning process.

At the current stage of development, the application of a grid of cumulator units spanned in the input space of the MM and VM only allows for low-dimensional motor commands $\mathbf{m}_t$ because of the storage requirements of these units. To overcome this problem, the next step of development is an online learning scheme to adapt to the maximum (the modal value) of the intensity distribution in each cumulator unit without the need to store the units. This would allow us to extend the scheme towards more dimensions in motor space. Even further, the goal is to replace the fixed grid structure in motor space by random movements (while maintaining the grid in $(x_{\mathrm{Out}}, y_{\mathrm{Out}})$ space with the appropriate spacing for the distortions caused by the imaging system).

The visual FWM of this study belongs to the class of anticipatory mechanisms which generate sensory anticipations reflecting the state of the outer space (and of the own body, but only from an outside perspective). The time horizon of the prediction is of the order of magnitude of tenths of seconds. The FWM works at the lowest level of abstraction by predicting direct sensor input (that is, continuous pixel intensities in RGB color space of a retinal image in this model). It remains an open question at which level visual FWMs work in biological organisms. At the lowest plausible level, it could be the activation of the optic nerve, or it could be a more abstract representation of the visual surroundings generated in one of the visual cortical areas.

However, the basic ideas of the proposed learning algorithm might even offer an explanation for the acquisition of visual FWMs in biological organisms: first, learning the input-output relationship by matching low-level visual features, and second, identifying predictable regions by detecting that a good match emerges during the learning process.

In robot models of sensorimotor processing, visual FWMs can be used to explore the various functions of FWMs stated in the introduction. We suggested two different models, one for saccade learning, the other for grasping to extrafoveal targets, where visual FWMs are used for behavioral learning and for the replacement of sensory feedback. These models demonstrate that visual FWMs are actually very useful from a biological modeling perspective. Moreover, for robotics applications, visual FWMs may become an important building block of truly autonomous systems, both for motor control and for perceptual competences.

References

1. Blakemore, S.J., Wolpert, D., Frith, C.: Why can't you tickle yourself? NeuroReport 11(11), R11–R16 (2000)
2. Bridgeman, B.: Failure to detect displacement of the visual world during saccadic eye movements. Vision Research 15(1), 719–722 (1975)

3. Deubel, H., Schneider, W.X.: Saccade target selection and object recognition: Evidence for a common attentional mechanism. Vision Research 36(12), 1827–1837 (1996)
4. Deubel, H., Schneider, W.X., Bridgeman, B.: Postsaccadic target blanking prevents saccadic suppression of image displacement. Vision Research 36(7), 985–996 (1996)
5. Deubel, H.: Localization of targets across saccades: Role of landmark objects. Visual Cognition 11(2-3), 173–202 (2004)
6. Eimer, M., Van Velzen, J., Gherri, E., Press, C.: Manual response preparation and saccade programming are linked to attention shifts: ERPevidence for covert attentional orienting and spatially specific modulations of visual processing. Brain Research 1105(1), 7–19 (2006)
7. Gross, H.M., Heinze, A., Seiler, T., Stephan, V.: Generative character of perception: A neural architecture for sensorimotor anticipation. Neural Networks 12(7-8), 1101–1129 (1999)
8. Große, S.: Visuelle Vorwärtsmodelle für einen Roboter-Kamera-Kopf, Diploma Thesis. Computer Engineering Group, Faculty of Technology, Bielefeld University (2005)
9. Hoffmann, H., Möller, R.: Action selection and mental transformation based on a chain of forward models. In: Schaal, S., Ijspeert, A., Billard, A., Vijayakumar, S., Hallam, J., Meyer, J.A. (eds.) From Animals to Animats 8. Proceedings of the Eighth International Conference on the Simulation of Adaptive Behavior, Los Angeles, CA, pp. 213–222. MIT Press, Cambridge (2004)
10. Hoffmann, H., Schenck, W., Möller, R.: Learning visuomotor transformations for gaze-control and grasping. Biological Cybernetics 93(2), 119–130 (2005)
11. Hoffmann, H.: Perception through visuomotor anticipation in a mobile robot. Neural Networks 20(1), 22–33 (2007)
12. von Holst, E., Mittelstaedt, H.: Das Reafferenzprinzip. Die Naturwissenschaften 37(20), 464–476 (1950)
13. Jordan, M.I., Rumelhart, D.E.: Forward models: Supervised learning with a distal teacher. Cognitive Science 16(3), 307–354 (1992)
14. Kawato, M.: Internal models for motor control and trajectory planning. Current Opinion in Neurobiology 9(6), 718–727 (1999)
15. Miall, R.C., Weir, D.J., Wolpert, D.M., Stein, J.F.: Is the cerebellum a smith predictor? Journal of Motor Behavior 25(3), 203–216 (1993)
16. Möller, R.: Perception through anticipation—a behavior-based approach to visual perception. In: Riegler, A., Peschl, M., von Stein, A. (eds.) Understanding Representation in the Cognitive Sciences, pp. 169–176 Plenum Academic / Kluwer Publishers, New York (1999)
17. Moody, J., Darken, C.J.: Fast learning in networks of locally-tuned processing units. Neural Computation 1, 281–294 (1989)
18. Rizzolatti, G., Riggio, L., Sheliga, B.M.: Space and selective attention. In: Umiltà, C., Moscovitch, M. (eds.) Attention and Performance VI: Conscious and Nonconscious Information Processing, pp. 231–265. MIT Press, Cambridge, MA (1994)
19. Schenck, W., Möller, R.: Learning strategies for saccade control. Künstliche Intelligenz Iss. 3/06, 19–22 (2006)
20. Snyder, L.H., Batista, A.P., Andersen, R.A.: Saccade-related activity in the parietal reach region. Journal of Neurophysiology 83(2), 1099–1102 (2000)
21. Tani, J.: Model-based learning for mobile robot navigation from the dynamical systems perspective. IEEE Transactions on Systems, Man, and Cybernetics—Part.B 26(3), 421–436 (1996)

22. Wolpert, D.M., Kawato, M.: Multiple paired forward and inverse models for motor control. Neural Networks 11(7-8), 1317–1329 (1998)
23. Ziemke, T., Jirenhed, D.A., Hesslow, G.: Internal simulation of perception: A minimal neuro-robotic model. Neurocomputing 68, 85–104 (2005)

A Distributed Computational Model of Spatial Memory Anticipation During a Visual Search Task

Jérémy Fix, Julien Vitay, and Nicolas P. Rougier

Loria, Campus Scientifique, BP239
54506 Vandoeuvre-les-Nancy, France

Abstract. Some visual search tasks require the memorization of the location of stimuli that have been previously focused. Considerations about the eye movements raise the question of how we are able to maintain a coherent memory, despite the frequent drastic changes in the perception. In this article, we present a computational model that is able to anticipate the consequences of eye movements on visual perception in order to update a spatial working memory.

1 Introduction

In the most general framework of behavior, the notion of anticipation is intimately linked with the possibility to predict the consequences and the outcomes of a given action. If we consider that any action is goal-motivated, then an action is carried out in the first place because it is anticipated that this action will lead to a situation in which the goal can be reached more directly. In this framework, anticipation can be viewed as a prediction of the future and is tightly linked to the notion of goal-directed behavior. However, there also exists more structural reasons why anticipation is necessary.

For example, when dealing with both accurate and very fast movements like catching a ball or scanning a visual scene, brain representations should be updated very quickly (even in advance in some cases) in accordance with the task that is carried out. The problem in this context is that the time scale required for carrying out such tasks may be dramatically smaller than the time scale of a single neuron. Moreover, those neurons are also in interaction with other neurons in the network and the resulting dynamic may be even slower. One solution to cope with this problem is to use a forward predictive model that is able to anticipate the consequences and outcomes of a motor action. The resulting dynamic at the level of the model is then faster than the dynamic of its components.

Let us consider the ability to anticipate changes in the visual information resulting from an eye saccade. This anticipation is known to be largely based on unconscious mechanisms that provide us with a feeling of stability while the whole retina is submerged by different information at each saccade; producing a saccade results in a complete change in the visual perception of the outer world. If a system is unable to anticipate its own saccadic movements, it cannot pretend

M.V. Butz et al. (Eds.): ABiALS 2006, LNAI 4520, pp. 170–188, 2007.

to obtain a coherent view of the world, because each image would be totally uncorrelated from the others. One stimulus being at one retinal location before a saccade could not be easily identified as being the same stimulus at another retinal location after the saccade. Consequently, the saccadic eye movements should be anticipated in order to keep the coherence of the scene and to be able to track down interesting targets. A number of works have already addressed the specific problem of visual search of a target among a set of distractors. However, most of the resulting models do not deal with the problem of saccadic eye movements that produce drastic changes in the available visual information.

Using neural fields introduced by Amari [1] for the one dimensional case and later extended to higher dimensions by Taylor [34], we would like to address in this paper the specific problem of anticipation during visual search using a purely distributed and numerical neural substrate. After briefly reviewing literature related to visual search in the first section, we introduce a very simple visual experiment that helps to illustrate the underlying mechanisms of the model that is detailed in that same section.

2   Visual Search

Visual search is a cognitive task that most generally involves an active scan of a visual scene to find one or several given targets among distractors. It is deeply anchored in most animal behaviors, from a predator looking for a prey in the environment, to the prey looking for a safe place to avoid being seen by the predator. Psychological experiments may be less ecological and may propose, for example, to find a given letter among an array of other letters, measuring the efficiency of the visual search in terms of reaction time (the average time to find the target given the experimental paradigm). In the early eighties, [35] suggested that the brain actually extracts some basic features from the visual field in order to perform the search. Among these basic features, which have been recently reviewed by [40], one can find features such as color, shape, motion, or curvature. Finding a target is then equivalent to finding the conjunction of features, which may be unique, that best describes the target. In this sense, [35] distinguished two main paradigms (a more tempered point of view can be found in [6]).

Feature search refers to a search where the target differs from all distractors by exactly one feature.
Conjunction search refers to a search where the target differs from distractors by at least one of two or more features.

What characterizes feature search best is a constant search time that does not depend on the number of distractors. The target is sufficiently different from the distractors to pop out. However, in the case of conjunction search, the mean time needed to find the target is roughly proportional to the number of distractors that share at least one feature with the target (cf. Figure 1). These observations lead to the question of how a visual stimulus could be represented in the

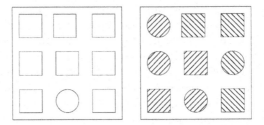

Fig. 1. Feature search can be performed very quickly as illustrated on the left part of the figure; the disc shape pops out from the scene. However, as illustrated in the right figure, if the stimuli share at least two features, the pop out effect is suppressed. Hence, finding the disc shape with the stripes going from up-left to down-right requires an active scan of the visual scene.

brain. The explanation given by Treisman and Gelade [35], the so-called *Feature-Integration Theory*, proposes that elementary features are processed in separated feature maps. Competition inside one map would lead to feature search, based on the idea that the item differing the most from its background would win the competition and be represented. For targets differing from distractors by more than two features, there cannot be any global competition. This would mean that finding the target requires successively scanning every potential candidate until the correct target is found. This explains the dependence of the search time on the number of similar distractors in conjunction search tasks.

The main prediction of this theory is that processing visual inputs is not a global feed-forward processing, but more an iterative and sequential process on sensory representations. We describe below the strategies used by the brain to achieve this sequential search, by putting emphasis on saccadic eye movements and visual attention. The scope of this article is therefore to model the cognitive structures involved in the sequential processing of visual objects, and not the visual processing of the features alone.

2.1   Saccadic Eye Movements

The eye movements may have different behavioral goals, leading to five different categories of movements: saccades, vestibulo-ocular reflex, optokinetic reflex, smooth-pursuit and vergence. However, in this article we will only focus on saccades (for a detailed study of eye movements, see [17], [3]).

Saccades are fast and frequent eye movements that move the eye from the current point of gaze to a new location in order to center a visual stimulus on the fovea, a small area on the retina where the resolution is at its highest. The velocity of the eyes depends on the amplitude of the movement and can reach up to 700 degrees per second at a frequency of 3 Hz. The question we would like to address is how the brain may give the illusion of a stable visual space while the visual perception is drastically modified every 300 ms.

While the debate to decide whether or not the brain is blind during a saccade has not been settled (see [18,2,14,29] for the notion of saccadic suppression and [24] for a discussion about the necessity of a saccadic suppression mechanism), the coherence between the perception before and after a saccade cannot be established accurately solely based on perception. One solution to consider is that the brain may use an efferent copy of the voluntary eye movement to remap the representation it has built of the visual world. Several studies shed light on presaccadic activities in areas such as V4 and LIP where the locations of relevant stimuli are supposed to be represented. In [22], the authors suggest that "the pre-saccadic enhancement exhibited by V4 neurons [...] provides a mechanism by which a clear perception of the saccade goal can be maintained during the execution of the saccade, perhaps for the purpose of establishing continuity across eye movements." In [20], the authors review evidence that LIP neurons, whose receptive field will land on a previously stimulated screen location after a saccade, are excited even if the stimulus disappears during the saccade. In a recent study, Sommer and Wurtz [33] showed neurons in FEF that receive projections from the superior colliculus that could explain the origin of a corollary discharge signal responsible for the pre-saccadic activity exhibited by these neurons.

2.2 Visual Attention

Focusing on a given stimulus of the visual scene is a particular aspect of the more general concept of attention that has been defined as the capacity to concentrate cognitive resources on a restricted subset of sensory information ([12]). In this context of visual attention, only a small subset of the retinal information is available at any given time to elaborate motor plans or cognitive reasoning (cf. *change blindness* experiments presented in [24], [32]). A visual scene is not processed as a whole but rather processed by successively focusing on interesting parts of it, possibly involving eye movements, but this is not necessary. The selection of a target for an eye movement is then closely related to the notion of spatial attention ([21]) that is classically divided into two types: **overt attention**, which involves a saccade to center a stimulus on the fovea, and **covert attention**, in which no eye movement is triggered. These two types of spatial attention were first supposed to be independent ([26]) but recent studies such as the premotor theory of attention proposed in [28] (see also [4], [16], [5]) consider that covert and overt attention rely on the same neural structures but the movement is inhibited in covert attention. A more general discussion about the covert and overt stages of action can be found in [13].

The deployment of attention on a specific part of the visual information can be the consequence of two phenomena. Firstly it can rely on the saliency of a stimulus, compared to its surrounding (for example a sudden strong flash light); this is known as bottom-up attention. Secondly, it can also depend on the task in which the subject is involved, which may need to enhance some parts of the perception (for example, imagine that you have to find an orange among apples and bananas, the color information could be a good criteria to find the target rapidly).

In [23], the authors shed light on the neural correlates of attention on the response of neurons in the visual and temporal cortices. If we consider a specific neuron tuned to a given orientation in its receptive field, one can distinguish several cases:

- the response of the neuron is high when an oriented bar with the preferred orientation (called good stimulus) is presented in its receptive field
- the response of the neuron is low when an oriented bar with an orientation different from the preferred one (called bad stimulus) is presented in its receptive field
- the response is between the two preceding ones when both a good and bad stimulus are presented

When a monkey is involved in a task that requires to select one of the two stimuli, for example the good one, the response of the neuron is enhanced. The study of this suppressive interaction phenomena was extended by further authors ([19], [27], [36]).

As we will see in section 3.2, we do not deal with how the salience of the visual stimuli is computed, whether or not it is a bottom-up or top-down processing. The main points are that for each location in the visual space, we are able to compute its behavioral relevance, and that considering eye movements necessarily implies dealing with overt attention.

2.3 Computational Models

Over the past few years, several attempts at modeling visual attention have been engaged ([15], [37], [41], [11], [10]). The basic idea behind most of these models is to find a way to select interesting locations in the visual space given their behavioral relevance and whether or not they have already been focused. The two central notions in this context have been proposed by [15] and [25]:

- saliency map
- inhibition of return (IOR).

The saliency map is a single spatial map, in retinotopic coordinates, where all the available visual information converge in order to obtain a unified representation of stimuli, according to their behavioral relevances. A winner-take-all algorithm can be easily used to find which stimulus is the most salient within the visual scene, and thus identify its location as the locus of attention. However, in order to be able to go to the next stimuli, it is important to bias the winner-take-all algorithm in such a way that it prevents going backward to an already focused stimulus. The goal of the inhibition of return mechanism is precisely to feed the saliency map with such a bias. The idea is to have another neural map that records focused stimuli and inhibits the corresponding locations in the saliency map. Since an already focused stimulus is actively inhibited by this map, it cannot pretend to win the winner-take-all competition, even if it is the most salient.

The existence of a single saliency map is still not proved. In [10] the author proposes a more distributed representation of these relevances, making a clear anatomical distinction between the processing of the visual attributes of an object and its spatial position (according to the What and Where pathways hypothesized by [38], see also [9]). In this model, spatial competition occurs in a motor map instead of a perceptive one. It exhibits good performances regarding visual search task in natural scene, but is restricted to covert attention. In most of the previously proposed models, the authors do not take into account eye movements and the visual scene is supposed to remain stable: scanning is done without any saccade. During the rest of this article, we will keep the saliency map hypothesis, even if controversial, in order to illustrate the anticipatory mechanism.

3 A Model of Visual Search with Overt Attention

The goal of our model is to show the basic mechanisms necessary to achieve sequential search in a visual scene using both overt and covert attention. Using a saliency map, we need to compute the location of the most interesting stimulus that will be processed to achieve recognition. This focus of attention on a stimulus has to be displaced in two situations. First, in covert attention this focus has to be dynamically inhibited to represent another stimulus. There is therefore a need for an inhibition-of-return mechanism than can inhibit the current focus of attention. Moreover, we have to memorize the locations of previously attended stimuli, by the means of a dynamic spatial working memory.

The second situation to consider is when eye movements can center the stimulus that is being attended to. The spatial working memory has to be updated by the eye movement so that its state corresponds to the post-saccadic locations of memorized stimuli. This is where an anticipatory mechanism is mandatory.

To describe these mechanisms, we first present an experimental setup for which previous computational models would fail to achieve efficient sequential search. We then present the architecture of our model and report simulated results.

3.1 Experiment

In order to accurately evaluate the model, we setup a simple experimental framework in which some identical stimuli are drawn on a blackboard and are observed by a camera. The task is to successively focus (i.e. center) each one of the stimuli without focusing twice on any of them. We estimate the performance of the model in terms of how many times a stimulus has been focused. Hence, the point is not to analyze the strategy of deciding which stimulus has to be focused next (see [7,8] for details on this matter). In the context of the proposed model, the strategy is simply to go from the most salient stimulus to the least salient one, and to randomly pick one stimulus if the remaining ones are equally salient.

Figure 2 illustrates an experiment composed of four identical stimuli where the visual scan path has been materialized. The effect of making a saccade from one

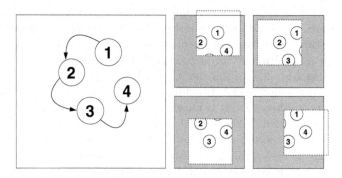

Fig. 2. When scanning a visual scene, going for example from stimulus 1 to stimulus 4, as illustrated in the left figure, the image received on the retina is radically changed when each stimulus is centered on the retina, as illustrated in the right figures. The difficulty in this situation is to be able to remember which stimuli have already been centered in order to center another one. The figures on the stimuli are shown only for explanation purpose and do not appear on the screen; all the stimuli are identical.

stimulus to another is shown and underlines the difficulty (for a computational model) of identifying a stimulus before and after a saccade. Each one of the stimuli being identical to the others, it is impossible to perform an identification based solely on features. The only criteria that can be used is the spatial location of the stimuli.

3.2   Model

The model is based on three distinct mechanisms (cf. Figure 3 for a schematic view of the model). The first one is a competition mechanism that involves potential targets represented in a saliency map that were previously computed according to visual input. Second, to be able to focus only once on each stimulus, the locations of the scanned targets are stored in a memory map using retino-topic coordinates. Finally, since we are considering overt attention, the model is required to produce a camera movement, centering the target on the fovea, used to update the working memory. This third mechanism works in conjunction with two inputs: current memory and parameters of the next saccade. This allows the model to compute quite accurately a prediction of the future state of the visual space, restricted to the targets that have already been memorized.

The model is based on the computational paradigm of two dimensional discrete neural fields (the mathematical basis of this paradigm can be found in [1] for the one dimensional case, extended to a two dimensional study in [34]). The model consists of five $n{\times}n$ maps of units, characterized by their position, denoted $\mathbf{x} \in [1..n]^2$ and their activity as a function of their position and time, denoted $u(\mathbf{x},t)$. The basic dynamical equation that follows the activity of a unit at position $\mathbf{x}$, depends on its input $I(\mathbf{x}, t)$. Equation (1) is the equation proposed in [1], discretized in space.

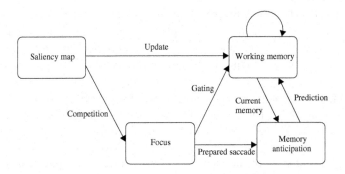

Fig. 3. Schematic view of the architecture of the model. The image captured by the camera is filtered and represented in the saliency map. This information feeds two pathways: one to the memory and one to the focus map. A competition in the focus map leads to the most salient location that is the target for the next saccade. The anticipation circuit predicts the future state of the memory with its current content and the programmed saccade.

$$\tau.\frac{\partial u(\mathbf{x},t)}{\partial t} = -u(\mathbf{x},t) + baseline + \frac{1}{\alpha}I(\mathbf{x},t) \tag{1}$$

We distinguish two kinds of units. The first are sigma units that compute their input as a weighted sum of the activity of afferent neurons, where afferent neurons are defined as neurons in other maps. We also consider lateral connections that involve units in the same map. If we denote w_{aff} the weighting function for the afferent connections and w_{lat} the weighting function for the lateral connections, the input $I(\mathbf{x},t)$ of a unit $\mathbf{x}$ at time t can be written:

$$I(\mathbf{x},t) = \sum_{aff} w_{aff}u_{aff}(t) + \sum_{lat} w_{lat}u_{lat}(t), \tag{2}$$

where equations 3 and 4 define the lateral and afferent weighting functions as a Gaussian and difference of Gaussians, respectively.

$$w_{aff}(\mathbf{x},\mathbf{y}) = A.e^{\frac{\|\mathbf{x}-\mathbf{y}\|^2}{a^2}} \text{ with } A,a \in \mathbb{R}^{*+}, \mathbf{x},\mathbf{y} \in [1..n]^2 \tag{3}$$

$$w_{lat}(\mathbf{x},\mathbf{y}) = B.e^{\frac{\|\mathbf{x}-\mathbf{y}\|^2}{b^2}} - C.e^{\frac{\|\mathbf{x}-\mathbf{y}\|^2}{c^2}} \text{ with } B,C,b,c \in \mathbb{R}^{*+}, \mathbf{x},\mathbf{y} \in [1..n]^2 \tag{4}$$

The second kind of units we consider are sigma-pi units ([31]), which compute their input as a sum of the product of the activity of afferent neurons. We also consider the lateral connection term so that the input of a unit $\mathbf{x}$ at time t can be written:

$$I(\mathbf{x},t) = \sum_{i \in I} w_{aff_i} \prod_{j \in E_i} u_{aff_j}(t) + \sum_{lat} w_{lat}u_{lat}(t). \tag{5}$$

All the parameters of the previous equations used in the simulation are summarized in the appendix.

We now describe how the different maps interact. Since the scope of this article is the anticipation mechanism, the description of the saliency map, the focus map and the working memory will not be accurate; a more detailed explanation, with the appropriate dynamical equations, can be found in [39].

Saliency Map. The saliency map, also referred to as INPUT in the following, is computed by convolving the image captured with the camera of a robot used for the simulation with Gaussian filters. The stimuli we use are easily discriminable from the background on the basis of the color information. This computation leads to a representation of the visual stimuli with Gaussian patterns of activity in a single saliency map. We do not deal with how this saliency map is computed, whether or not it is due to bottom-up or top-down attention. We only consider that we are able to compute a spatial map, in retinotopic coordinates, that represents the behavioral relevance of each location in the visual space. We point out again that this is one of our working hypothesis, detailed in section 2.3.

Focus. The units in the FOCUS map have direct excitatory feedforward inputs from the saliency map. The lateral connections are locally excitatory and widely inhibitory so that a competition between the units within the map leads to the emergence of only one stimulus in the focus map. This mechanism is not just a dynamical *winner-take-all* algorithm because the winning stimulus will still be represented in this map, even if the other stimuli in the visual scene become comparatively more salient through time, but it has to be explicitly inhibited. This focused stimulus is considered the next target to focus on and the movement to perform to center it on the fovea is decoded from this map. This map then codes the parameters of the next saccade to make.

Working Memory. Once a stimulus has appeared within the focus map and because it is also present in the saliency map at the same location, it emerges within the working memory. Both the excitations from the focus map and the saliency map (at a same location) are necessary for the emergence of a stimulus in the working memory area. If the focused stimulus changes, it will not be present anymore in the focus map such that an additional mechanism is needed to maintain it in the memory. It is not shown on the schematic illustration (3) but the memory consists of two maps, WM and THAL_WM, that share excitatory connections in two ways: the first map excites the second and the second excites the first, weighted so that the excitation is limited in space.

Memory Anticipation. The memory anticipation mechanism aims at predicting what should be the state of the working memory after an eye movement centers another stimulus in the focus map before the movement is triggered. This map, filled with sigma-pi units, has two inputs: units of the focus map and units of the working memory. If we denote $wm(\mathbf{x}, t)$ the activity of unit $\mathbf{x}$ of the working memory at time t, and $f(\mathbf{x}, t)$ the activity of unit $\mathbf{x}$ of the focus map at time t, we define the input $I(\mathbf{x}, t)$ of unit $\mathbf{x}$ in the anticipation map as:

$$I(\mathbf{x}, t) = w_{sigma-pi} \sum_{\mathbf{y} \in \mathbb{R}^2} wm(\mathbf{y}, t) f(\mathbf{y} - \mathbf{x}, t) + \sum_{lat} w_{lat} u_{lat}(t) \qquad (6)$$

The input of each unit in the anticipation map is computed as a convolution product of the working memory and the focus map, centered on its coordinates. To make (6) clearer, the condition of the sum is weaker than the one that should be used: since the input maps are discrete sets of units, the two vectors $\mathbf{y}$ and $\mathbf{y}$-$\mathbf{x}$ mustn't exceed the size of the maps. The equation (6) should also take into account that the position *eye centered* is represented by a bell-shaped pattern of activity centered in the focus map, so that an offset should be included in the first sum when determining which unit of the focus map multiplies $wm(\mathbf{y}, t)$ From (1) and (6), the activity of the units in the anticipation map, without lateral connections, satisfies (7).

$$\tau . \frac{\partial u(\mathbf{x}, t)}{\partial t} = -u(\mathbf{x}, t) + baseline + w_{sigma-pi} \sum_{\mathbf{y} \in \mathbb{R}^2} wm(\mathbf{y}, t) f(\mathbf{y} - \mathbf{x}, t) \qquad (7)$$

Then, the shape of activity in the anticipation map converges to the convolution product of the working memory and the focus map. Since the activity in the focus map has a Gaussian shape and the working memory can be written as a sum of Gaussian functions, the convolution product of the working memory and the focus map leads to an activity profile that is the profile in the working memory translated by the vector represented in the focus map. This profile is the prediction of the future state of the working memory and is then used to slightly excite the working memory. After the eye movement, and when the saliency map is updated, the previously scanned stimuli emerge in the working memory as a result of the conjunction of the visual stimuli in the saliency map and the prediction of the working memory, that is, the prediction is combined with the new perception. This is exactly the same mechanism as the one used when a stimulus emerges in the working memory owing to the conjunction of the activity in the saliency map and the focus map.

3.3   Simulation and Results

The visual environment consists of three distributed but identical stimuli that the robot is expected to scan successively exactly once. A stimulus is easily discriminable from the background, namely a green lime on a white table. A complete activation sequence of the different maps is illustrated on Figure 4. The saliency map is filled by convolving the image captured from the camera by a green filter in HSV coordinates such that it leads to three distinct stimuli[1].

At the beginning of the simulation (Figure 4a), only one of the three stimuli emerges in the focus map, thanks to the strong lateral competition that occurs within this map. This stimulus, which present in both the focus map and the saliency map, emerges in the working memory. The activation within the

---

[1] A video of the model is available at http://www.loria.fr/~fix/publications.php

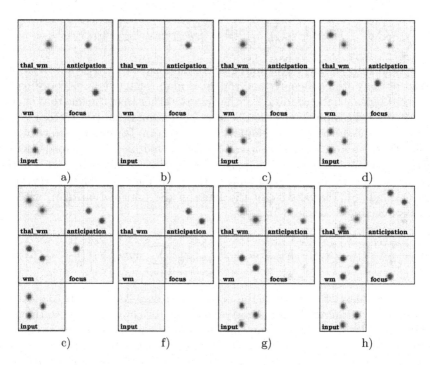

Fig. 4. A sequence of evolution of the model during an overt visual scan trial. a) One of the three stimuli emerges in the focus map and the anticipation's units predict the future state of the visual memory (the maps wm and thal_wm). b) During the execution of the saccade, only the units in the anticipation map remain active. c) The focused stimulus emerge in the memory since it is both in the saliency map and the anticipation map at the same location. d) A new target to focus is elicited. e) The future state of the memory is anticipated. f) The saccade is executed and only the prediction remains. g) The two already focused stimuli emerge in the memory. h) The attentional focus lands on the last target.

anticipation map reflects what should be the state of the saliency map, restricted to the stimuli that are in the working memory after the movement that brings the next targeted stimulus into the center of the visual field. During the eye movement (Figure 4b), no visual information is available and the parameter τ in (1) and (7) is adjusted so that only the units in the anticipation map remain active, whereas the activity of the others approach zero. After the eye movement and as soon as the saliency map is fed with the new visual input, the working memory is updated thanks to the excitation from both saliency and anticipation map at a same location: the prediction of the state of the visual memory is compared with the current visual information. A new target can now be elicited in the focus map thanks to a switch mechanism similar to that described in [39], but not detailed here. This mechanism acts like the inhibition of return

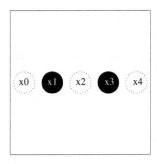

Fig. 5. The scene consists in two identical stimuli, the black blobs, initially symmetrically positioned around the center of gaze. The task is to successively focus on each target. During a trial, we measure the activity of neurons whose receptive field covers the five positions represented by the dashed circles and denoted as x0, x1, x2, x3 and x4, in several maps.

presented in section 2.3; the memorized locations in the working memory are inhibited in the focus map, therefore biasing the competition in it, so that only a stimulus that was not already focused can be the next target to focus.

In order to illustrate more explicitly the role of the anticipatory signal, we now consider a second experiment. In this experiment, the visual scene consists of only two identical stimuli (Figure 5).

The task is the same as the previous one, namely, the robot must scan each stimulus only once, but the experimental conditions are slightly different: we enforce the robot to scan these targets in a predefined order. To bias the spatial attention toward one of the two targets, we first increase the intensity of the leftmost target. Then, when the saccade to center that target is performed, we refresh the display and increase the intensity of the rightmost target. In that way, the scenario is as follows:

1. Select the leftmost target.
2. Focus on that target.
3. After the saccade, when the display is refreshed, select the rightmost target.
4. Perform the saccade to center the rightmost target.

The visual bias we add makes us able to get the same experimental conditions over the trials. During a trial, we record the activity of the neurons whose receptive field covers one of the five positions, denoted x0, x1, x2, x3, and x4 in the figure, in the four maps: visual, focus, wm and anticipation. In a typical trial, we will have a target at x1 and x3, then, after the first saccade, the targets will be at x2 and x4, to finally occupy, after the last saccade, the positions x0 and x2 (Figure 6, top).

Moreover, two conditions are considered; in the first one (Figure 6), the anticipation is enabled, whereas in the second one (Figure 7), the anticipatory signal is disabled. At the beginning of the trial, the targets are at positions x1 and x3 so that the neurons in the visual map at these positions are excited (dashed

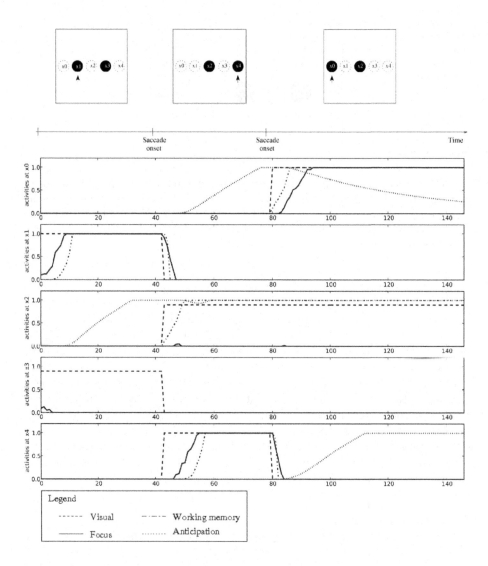

Fig. 6. Case with the anticipatory signal enabled. We record the activity of neurons whose receptive field covers one of the five positions x0, x1, x2, x3 and x4, in the four maps: visual, focus, wm and anticipation. During the trial, we add a bias toward one of the targets so that the attention directs to the biased target (that target is shown by the arrow). Each subplot represents the activity of the neurons in each map at a given position. The dashed line represents the activity of the neuron in the visual map, the solid line the activity of the neuron in the focus map, the dash-dot line the activity of the neuron in the working memory and the dotted line the activity of the neuron in the anticipation map. Please read the text for explanations on these curves.

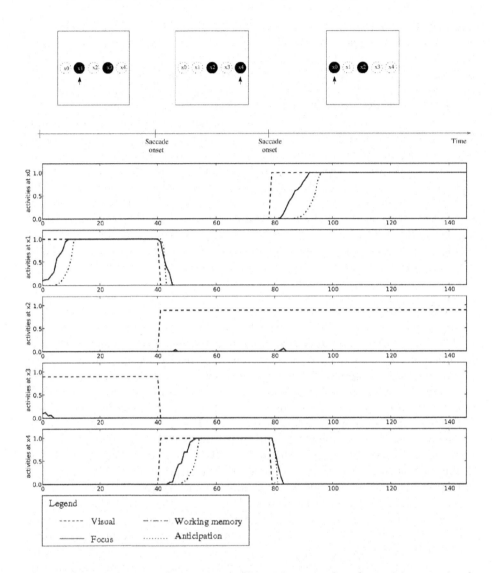

Fig. 7. The experiment is the same as in Figure 6 except that the anticipatory signal is disabled

line) whereas the neurons at the other positions remain silent. The two positions x1 and x3 compete for the spatial attention. Since we added a bias toward the target at position x1, the spatial attention is on target x1, rather than on target x3, so that the activity of the neuron at position x1 in the focus map (solid line) grows, whereas the activity of the neuron at position x3 in the same map decreases to zero. The attention on target x1 enables it to emerge in the working memory (dash-dot line). The task is now to produce an eye movement that will center that target. The anticipatory mechanism predicts that when that target is centered, it will occupy the position x2; the activity of the neuron at position x2 grows (dotted line). As soon as the saccade is performed, we refresh the display. The two targets now occupy the positions x2 and x4. The bias toward the rightmost target enforces that target to be attended. The activity of the neurons at position x4 in the focus map and the working memory grows. Whereas the target at position x4 emerges in the working by the conjunction of an activity in the visual input and the focus map, the target at position x2 emerges thanks to the visual input and the anticipatory signal. As we can see in Figure 7, in which the anticipatory signal was disabled, the position of the first attended target cannot be updated at position x2. Finally, a saccade to center the target at position x4 is performed. In the case that the anticipation is present, the new positions of the two targets are in the working memory at x0 and x2, whereas when there is no anticipation, only the last attended target is in the working memory.

4   Discussion

In this paper, we have presented a continuous attractor network model that is able to anticipate the consequences of its own movements by actually predicting the visual scene as it is supposed to be after the execution of an action. Furthermore, the model also illustrates how this information is used in the context of a serial search of a target among a set of distractors. Each already focused target is kept within a working memory area that is updated with regards to eye movements.

The model is of a completely distributed nature and does not require any central supervisor. All the units in the model satisfy a dynamical equation. When dealing with this kind of dynamic model, the integration time of the units is a critical factor as shown in [30], which shares some ideas with the present model. It means that in our case, even if we make the hypothesis that the perception is available during the saccade (ignoring also that the perception is smeared), the working memory could be updated dynamically with the perception only if the movement's speed doesn't exceed a critical limit. In the case of saccadic eye movements, it is then necessary to have an anticipatory mechanism. We are definitely speaking about anticipation since a prediction about the future perception is used to maintain a coherent memory which is mandatory to accomplish the task we designed. It is nonetheless not limited to that particular case since scanning several potential targets is one of the basic primitives we use when performing a visual search task.

The question of learning the underlying transformation of the anticipatory mechanism, namely the convolution product of the focus map and the working memory, remains open and is still under study. We did implement a learning mechanism, under restrictions and strong hypotheses that rely heavily on the difference between the pre-saccadic prediction and the post-saccadic actual perception. This self generated signal is able to measure to what extent the prediction is correct or not. Hence, it is quite easy to modify the weights accordingly. The main difficulty during learning remains the sampling distribution of examples within the input space, which is a well known problem in information and learning theory. Without any additional motivational system that could bias the examples according to a given task, it is quite unrealistic to rely on a regular distribution of examples.

Finally, the coherence of the visual world is solely based on an anticipatory mechanism that ultimately allows the identification of identical targets before and after a saccade, despite drastic changes in the visual perception. The prediction of the future state of the visual memory enriches the perception of the visual world in order, for example, to prevent focusing twice on the same stimulus. Of course, this model does not pretend to be complete nor accurate and does not tackle a number of problems that are directly related to visual perception. However, we think that the possibility to unconsciously anticipate our own actions using a dynamic working memory could be extended to other motor tasks involving other types of perception as well.

References

1. Amari, S.I.: Dynamical study of formation of cortical maps. Biological Cybernetics 27, 77–87 (1977)
2. Burr, D.: Eye movements, keeping vision stable. Current Biology 14 (2004)
3. Carpenter, R.: Movements of the Eyes, 2nd edition. Pion Ltd, London (1988)
4. Chelazzi, L., Miller, E.K., Duncan, J., Desimone, R.: A neural basis for visual search in inferior temporal cortex. Nature 363, 345–347 (1993)
5. Craighero, L., Fadiga, L., Rizzolatti, G., Umilta, C.: Action for perception: a motor-visual attentional effect. Journal of Experimental Psychology, vol. 25 (1999)
6. Duncan, J., Humphreys, G.W.: Visual search and stimulus similarity. Psychological Review 96, 433–458 (1989)
7. Findlay, J.M., Brown, V.: Eye scanning of multi-element displays: I. scanpath planning. Vision Research 46, 179–195 (2006)
8. Findlay, J.M., Brown, V.: Eye scanning of multi-element displays: II. Saccade planning. Vision Research 46, 216–227 (2006)
9. Goodale, M.A., Milner, A.D.: Seperate visual pathways for perception and action. Trends in Neurosciences 15, 20–25 (1992)
10. Hamker, F.H.: A dynamic model of how feature cues guide spatial attention. Vision Research 44, 501–521 (2004)
11. Itti, L., Koch, C.: Computational modeling of visual attention. Nature Reviews Neuroscience 2, 194–203 (2001)
12. James, W.: The principles of psychology. Holt, New York (1890)
13. Jeannerod, M.: Neural simulation of action: A unifying mechanism for motor cognition. NeuroImage 14, S103–S109 (2001)
14. Kleiser, R., Seitz, R.J., Krekelberg, B.: Neural correlates of saccadic suppression in humans. Current Biology 14, 386–390 (2004)
15. Koch, C., Ullman, S.: Shifts in Selective Visual Attention: Towards the Underlying Neural Circuitry. Human Neurobiology 4, 219–227 (1985)
16. Kowler, E., Andersen, E., Dosher, B., Blaser, E.: The role of attention in the programming of saccade. Vision Research 35, 1897–1916 (1995)

17. Leigh, R.J., Zee, D.S.: The Neurology of Eye Movements, 3rd edn. FA Davis Company, Philadelphia (1999)
18. Li, W., Martin, L.: Saccadic suppression of displacement: separate influences of saccadic size and of target retinal eccentricity. Vision Research, vol. 37 (1997)
19. Luck, S.J., Chelazzi, L., Hillyard, S.A.: Neural mechanisms of spatial attention in areas v1, v2 and v4 of macaque visual cortex. Journal of Neurophysiology 77 (1997)
20. Merriam, E.P., Colby, C.L.: Active vision in parietal and extrastriate cortex. The. Neuroscientist 11, 484–493 (2005)
21. Moore, T., Fallah, M.: Control of eye movements and spatial attention. PNAS 98, 1273–1276 (2001)
22. Moore, T., Tolias, A.S., Schiller, P.H.: Visual representations during saccadic eye movements. Neurobiology 95, 8981–8984 (1998)
23. Moran, J., Desimone, R.: Selective attention gates visual processing in the extrastriate cortex. Science 229, 782–784 (1985)
24. O'Regan, J.K., Noë, A.: A sensorimotor account of vision and visual consciouness. Behavioral and Brain Sciences 24, 939–1031 (2001)
25. Posner, M.I., Cohen, Y.: Components of visual orienting. In: Bouma, H., Bouwhuis, D. (eds.): Attention and performance X, pp. 531–556 (1984)
26. Posner, M.I., Petersen, S.E.: The attentional system of the human brain. Annual Review of Neurosciences 13, 25–42 (1990)
27. Reynolds, J.H., Desimone, R.: The role of neural mechanisms of attention in solving binding problem. Neuron 14, 19–29 (1999)
28. Rizzolatti, G., Riggio, L., Dascola, I., Ulmita, C.: Reorienting attention accross the horizontal and vertical meridians. Neuropsychologia 25, 31–40 (1987)
29. Ross, J., Morrone, C., Goldberg, M.E., Burr, D.C.: Changes in visual perception at the time of saccades. Trends in Neurosciences 24, 113–121 (2001)
30. Rougier, N.P., Vitay, J.: Emergence of Attention within a Neural Population. Neural Networks 19, 573–581 (2006)
31. Rumelhart, D.E., Hinton, G.E., McClelland, J.L.: A general framework for parallel distributed processing. In: Parallel Distributed Processing, vol. 1, MIT Press, Cambridge (1987)
32. Simons, J.S.: Current approaches to change blindness. Visual Cognition, vol. 7 (2000)
33. Sommer, M.A., Wurtz, R.H.: Influence of the thalamus on spatial visual processing in frontal cortex. Nature 444, 374–377 (2006)
34. Taylor, J.G.: Neural bubble dynamics in two dimensions. Biological Cybernetics 80, 5167–5174 (1999)
35. Treisman, A., Gelade, G.: A feature-integration theory of attention. Cognitive Psychology 12, 97–136 (1980)
36. Treue, S., Maunsell, J.H.R.: Attentional modulation of visual motion processing in cortical areas MT and MST. Nature 382, 539–541 (1996)
37. Tsotsos, J.K., Culhane, S.M., Lai, W.Y.K., Davis, N.: Modeling visual attention via selective tuning. Artificial Intelligence 78, 507–545 (1995)
38. Ungerleider, L.G., Mishkin, M.: Two cortical visual systems. In: Analysis of Visual Behavior, pp. 549–586. MIT Press, Cambridge (1982)
39. Vitay, J., Rougier, N.P.: Using neural dynamics to switch attention. In: International Joint Conference on Neural Networks, IJCNN (2005)
40. Wolfe, J.M.: Visual search. In: Attention, University College London Press, London, UK (1998)
41. Wolfe, J.M.: Visual attention. In: Seeing: Handbook of Perception and Cognition, 2nd edn. De Valois KK, pp. 335–386 (2000)

Appendix

Dynamics of the Neurons

Each sigma neuron loc in a map computes a numerical differential equation given by equation (8), which is a numerized version of that proposed in [1] and [34]:

$$act_{loc}(t+1) = \sigma(act_{loc}(t) + \frac{1}{\tau}(-(act_{loc}(t) - baseline) + \frac{1}{\alpha}\sum_{aff} w_{aff}act_{aff}(t)$$

$$+ \frac{1}{\alpha}\sum_{lat} w_{lat}act_{lat}(t))) \tag{8}$$

Each sigma-pi neuron loc in a map computes a numerical differential equation given by equation (9):

$$act_{loc}(t+1) = \sigma(act_{loc}(t) + \frac{1}{\tau}(-(act_{loc}(t) - baseline) + \frac{1}{\alpha}\sum_{lat} w_{lat}act_{lat}(t)$$

$$+ \frac{1}{\alpha}\sum_{(i,j)\in E_{loc}} w_{sigmapi}act_{aff_i}(t)act_{aff_j}(t)) \tag{9}$$

where $\sigma(x)$ is a semi-linear function assuring that $0 \leq \sigma(x) \leq 1$, τ is the time constant of the equation, α is a weighting factor for external influences, aff is a neuron from another map and lat is a neuron from the same map. To know how the set of afferent neurons E_{loc} is determined in the case of a sigma-pi map, please refer to the section 3.2 describing the model.

The size, τ, α and baseline parameters of the different maps are given in the following table:

Map	Size	Type	Baseline	τ	α
INPUT	40*40	Sigma	0.0	0.75	6.0
FOCUS	40*40	Sigma	-0.05	0.75	13.0
WM	40*40	Sigma	-0.2	0.6	13
THAL_WM	40*40	Sigma	0.0	0.6	13
ANTICIPATION	40*40	Sigma-Pi	0.0	2.0	5.0

Connections Intra-map and Inter-map

The lateral weight from neuron lat to neuron loc is:

$$w_{lat} = Ae^{-\frac{dist(loc,lat)^2}{a^2}} - Be^{-\frac{dist(loc,lat)^2}{b^2}} \text{ with } A, B, a, b \in \Re^{*+}, loc \neq lat . \tag{10}$$

where $dist(loc, lat)$ is the distance between lat and loc in terms of neuronal distance on the map (1 for the nearest neighbor). In the case of a "receptive field"-like connection between two maps, the afferent weight from neuron *aff* to neuron *loc* is:

$$w_{aff} = Ae^{-\frac{dist(loc,aff)^2}{a^2}} \text{ with } A, a \in \Re^{*+} \qquad (11)$$

In the case of the sigma-pi connections, all the weights are the same:

$$w_{sigma-pi} = A \text{ with } A \in \Re^{*+} \qquad (12)$$

The connections in the model are described the following table:

Source Map	Destination Map	Type	A	a	B	b
INPUT	FOCUS	receptive-field	0.25	2.0	-	-
FOCUS	FOCUS	lateral	1.7	4.0	0.65	17.0
INPUT	WM	receptive-field	0.25	2.0	-	-
FOCUS	WM	receptive-field	0.2	2.0	-	-
WM	WM	lateral	2.5	2.0	1.0	4.0
WM	THAL_WM	receptive-field	2.35	1.5	-	-
THAL_WM	WM	receptive-field	2.4	1.5	-	-
ANTICIPATION	ANTICIPATION	lateral	1.6	3.0	1.0	4.0
WM, FOCUS	ANTICIPATION	sigma-pi	0.05	-	-	-
ANTICIPATION	WM	receptive-field	0.2	?	-	-

A Testbed for Neural-Network Models Capable of Integrating Information in Time

Stefano Zappacosta, Stefano Nolfi, and Gianluca Baldassarre*

LARAL-ISTC-CNR Laboratory of Autonomous Robotics and Artificial Life,
Istituto di Scienze e Tecnologie della Cognizione, Consiglio Nazionale delle Ricerche
Via San Martino della Battaglia 44, I-00185 Roma, Italy
{stefano.zappacosta, stefano.nolfi, gianluca.baldassarre}@istc.cnr.it
http://laral.istc.cnr.it

Abstract. This paper presents a set of techniques that allow generating a class of testbeds that can be used to test recurrent neural networks' capabilities of integrating information in time. In particular, the testbeds allow evaluating the capability of such models, and possibly other architectures and algorithms, of (a) categorizing different time series, (b) anticipating future signal levels on the basis of past ones, and (c) functioning robustly with respect to noise and other systematic random variations of the temporal and spatial properties of the input time series. The paper also presents a number of analysis tools that can be used to understand the functioning and organization of the dynamical internal representations that recurrent neural networks develop to acquire the aforementioned capabilities, including periodicity, repetitions, spikes, and levels and rates of change of input signals. The utility of the proposed testbeds is illustrated by testing and studying the capacity of Elman neural networks to predict and categorize different signals in two exemplary tasks.

Keywords: Testbed, Time Series, Waves, Time Information Integration, Signal Processing, Recurrent Neural Networks, Passive and Active Perception, Dynamical Systems, Analysis of Internal Representations, Attractors.

1 Introduction

The capability of integrating information in time is a critical functionality, which lies at the core of the functioning of several anticipatory learning systems. For example, consider a rat sampling the profile of an object with its whiskers [14], an organism scanning the environment with its eyes [15], or a robot moving in

---

* This research has been supported by the European Integrated Project "ICEA – Integrating Cognition, Emotion and Autonomy", contract no. FP6-IST-027819-IP, and by the European Specific Targeted Research or innovation Project "MindRACES – from Reactive to Anticipatory Cognitive Embodied Systems", contract no. FP6-511931-STREP.

M.V. Butz et al. (Eds.): ABiALS 2006, LNAI 4520, pp. 189–217, 2007.
© Springer-Verlag Berlin Heidelberg 2007

an office and sampling the walls with proximity sensors [16]. Imagine they have to categorize the object or the portion of the environment they are experiencing, or to predict future sensations on the basis of past ones. In all these examples, the systems need to integrate information in time. That is, they need to capture signal regularities that manifest in time in the form of periodicity, repetitions, spikes, numbers, rates of change, levels of signals, etc.

Given the importance of integrating information in time for anticipatory systems, artificial intelligence has proposed a number of models that possess such capabilities. This paper focuses in particular on recurrent neural-network models, but the testbeds, and some of the analysis tools it proposes, are also applicable to other models. The neural networks relevant for the topic tackled here are based on recurrent architectures [4], [5], [6], and [20], as these allow systems to compare signals in time, for example, by counting signals' duration, by accumulating evidence in favor of different options, by synchronizing internal dynamics with perception dynamics, etc. Section 3 briefly reviews few important examples of these neural networks, namely *Elman neural networks* ([7]: these networks are based on a hidden unit layer with a memory of the past), *echo state networks* ([11]: these networks are provided with a layer of fixed recurrent connections that provides a "reservoir" of various dynamics), *leaky integrator networks* ([21]: these networks are based on neurons with an internal memory), and *long short-term memory networks* ([10]: these networks are based on special neurons with a self-recurrent connection and gated input and output channels).

The ways recurrent neural networks integrate information in time is particularly interesting for two reasons. First, these networks exploit internal dynamical processes such as fixed-point attractors, limit cycles, chaotic attractors, etc., to "get in resonance" and synchronize with the dynamics of stimuli and so perform the integration. Second, if one assumes that neural networks of real brains exploit similar dynamics on the basis of their omnipresent recurrent connections, one can hope to understand how real organism integrate information in time by studying artificial recurrent neural networks.

Given the interest of the aforementioned models, the ABiALS community (Adaptive Behavior in Adaptive Learning Systems) has a great need of identifying a number of specific testbeds to compare the models and, given their different features, to understand how they self-organize to solve different tasks. Indeed, such models should be compared both on the basis of their capabilities of mimicking real systems and on the basis of more general criteria such as scalability properties, computational power, and robustness against noise. The first of these two "checks" should be accomplished on the basis of the evaluation of the general biological plausibility of the models and by comparing the model's behavior and functioning with data of real systems, provided by behavior and brain sciences. The second check should be based on standard testbeds developed and circulated within the community, such as those proposed here.

The *testbed* presented in this paper, which is actually a set of techniques that can be used to produce a *class of testbeds* with particular features, allows testing two anticipatory capabilities of anticipatory systems:

(1) The capacity of categorizing different signals perceived in time.
(2) The capacity of predicting future signal values on the basis of past ones.

In this respect, this paper proposes some techniques to generate different signals with various time regularities to systematically test the models' capabilities of integrating information in time. The paper also proposes some techniques to test the *robustness* of such capabilities with respect to:

(1) Noise of the signal level and of the signal speed.
(2) Biased expansions and compressions of the signal in duration.
(3) Biased variations of the phase of periodic signals.
(4) Biased variations of the signal amplitude levels.
(5) Biased expansions and compressions of the signal levels.

The paper also presents a number of techniques that can be used to analyze the internal representations that the models develop to solve the various tasks. The understanding of such representations is rather challenging given the complex dynamical systems nature of the considered models. Notwithstanding these difficulties, we believe that these studies are necessary to understand the detailed mechanisms underlying the information integration in time that such systems exhibit.

The functioning of the testbed and the analysis techniques are illustrated through some experiments using Elman neural networks. These experiments represent the preliminary investigations of a research agenda directed to investigate how recurrent neural networks internally self-organize and form abstract dynamical representations in order to integrate information in time (see also the European Projects "ICEA" and "MindRACES", which provided funding for this research).

The rest of the paper is organized as follows. Section 2 presents the testbed, in particular, the type of time series it generates, the anticipatory capabilities it allows to test, the type of noise and signal variations it allows to create, the measures of performance it uses, and some techniques that can be used to analyze the emergent internal organization of the tested models. Section 3 presents a brief review of dynamical neural networks that might be tested and compared with the testbed presented. Section 4 gives some examples, based on Elman neural networks, that illustrate the functioning of the testbed and the analysis techniques. Finally, Section 5 draws conclusions and indicates future work.

2   Testbed Description

As mentioned in the introduction, the *testbed* presented here is actually a set of techniques that can be used to generate a class of different testbeds having certain properties. These testbeds have the following features:

(a) They involve problems where the model to be tested receives a one-variable input time series (henceforth called *signal* or *input time series*).

Table 1. Summary of the testbed's features

Possible different types of input signals

TYPE	PROPERTIES OF SIGNAL THAT CAN BE MANIPULATED
Wall profile	Levels, linear changes, sudden changes of levels (steps)
Object	Non-linear changes, derivatives, discontinuities, sudden changes of levels (steps)

Tasks that can be used to test the systems' capabilities

TASK	METRICS TO MEASURE THE CAPABILITY
Prediction	Mean square errors between predicted and actual input pattern, capacity to reproduce the signal for several steps by using the prediction as input
Categorization	Percent of correct categorizations after the pattern is perceived for a certain time

Table 2. Summary of the variations of the input time series that can be used to test the robustness of the systems' capabilities

Types of noise

SOURCE OF NOISE	DESCRIPTION
Signal noise	White noise added to signal
Step noise	White noise added to size of automaton's translation movement

Systematic random variations of signal

ELEMENT VARIED	DESCRIPTION
Phase of signal	The phase of the signal is set randomly in different wall/object presentations
Period of signal	The step size of the automaton (and hence its speed) is set randomly in different wall/object presentations so as to have compressions/expansions of the signal
Signal level	The distance of the automaton from the wall/object is set randomly in different wall/object presentations
Signal range	The distance of the automaton from the wall/object and the size of the object are multiplied by a random parameter in different wall/object presentations

(b) They allow testing the models' capacity of both categorizing different signals and predicting future signal values based on past ones.

(c) They allow testing the robustness of these capacities with respect to noise and various systematic random transformations of the signal.

The features of the testbed(s), the possible variations of the input signal that can be used to test the robustness of the models, and the analysis techniques are summarized respectively in Table 1, Table 2, and Table 3. Note that, as the implementation of the testbed is quite straightforward, the software used

Table 3. Summary of the techniques that can be used to analyze the internal dynamics of systems

TECHNIQUE	ASPECTS INVESTIGATED
Cross-correlograms	Time correlations between different variables
Phase space analysis	Identification of different types of attractors
Ad-hoc input time series	Systems' reaction to different signal features
Hinton plot	Roles of different connection weights
Targeted lesions	Roles of different connection weights

to implement it can be easily re-generated by the reader. Moreover, the implementation of the analysis techniques suggested in the paper can be found in any standard statistical analysis package, for example Matlab^{TM}, which was used to carry out the results' analysis illustrated in the paper. Next, all the features and techniques are analyzed in detail.

2.1 Input Time Series

The testbed proposes two alternative techniques for generating the input time series. These two techniques allow manipulating different aspects of the input time series with different implementation ease (see Table 1). Nevertheless, notice that there is a precise correspondence between the input time series that can be generated in two ways, as further illustrated below.

The first technique involves an automaton that travels along a "wall" that has a certain profile, formed by a sequence of segments. An example of this is given in the top-left graph of Figure 1 that shows a "punctiform automaton" that moves at a distance R from a wall having a saw-like profile. The signal samples (shown in the bottom-left graph of the figure) encode the distance between the automaton's central point and the intersection between the sensor ray and the segments composing the wall. Notice that this technique allows generating signals that correspond to rather complex "objects", as shown by the right graph of Figure 1.

The second technique of generating the input time series involves an automaton that follows a circular path around an object, and perceives its profile with a proximity sensor. An example of this is given in Figure 2 where the automaton moves around a cross-shaped object following a circle with radius R while its sensor detects the distance to the object at each step. Notice that this technique allows generating signals that correspond to rather complex "walls", as shown in Figure 2.

From an implementation point of view, the general idea behind the way of generating the input signal with the two techniques is that the walls or objects are composed by a set of segments. In this way, the signals are easily obtained on the basis of the computation of the distance between the automaton's central point and the (closest) intersection between the sensor's ray and the segments

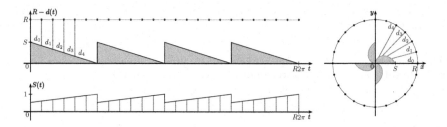

Fig. 1. Top left: profile of the wall perceived by the automaton, and positions from which the automaton senses it (dots along the straight trajectory followed by the automaton, marked by the light gray line). Right: equivalent setting showing an object, and the automaton circulating around it, that generates the same sensor reading as the wall setting. The curve of the object has been obtained with a very dense sampling of the wall. Bottom left: automaton's sensor reading, $S(t)$ normalized in $[0, 1]$, equal for the two settings.

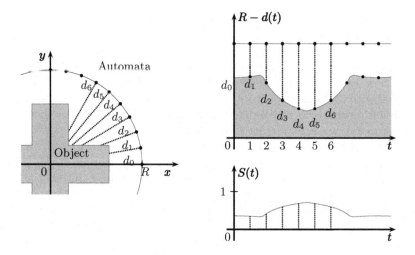

Fig. 2. Left: time series of the distances $d_0, d_1, d_2, \ldots$ (dotted lines) detected by the testbed's automaton (represented by the dots on the circumference) while moving around a cross-shaped object. Top right: the corresponding wall-setup; the curve of the wall has been obtained with a very dense sampling of the object. Bottom right: intensity of the normalized signal $S(t)$ detected by the automaton at each time step, equal for both setups.

composing the object. The two techniques give the researcher the possibility of generating a great number of signals having different time regularities, as shown in the examples reported in Section 4 and indicated in Table 1, with computationally rather easy effort. Moreover, the analysis of results can be aided by the fact that the first technique allows plotting the signal produced by walls in terms of objects, performing a dense sampling of the latter (see Figure 1, right). Vice versa, the second technique allows plotting object signals in terms of walls (see Figure 2, top right).

2.2   Metrics for Measuring Prediction and Categorization Capabilities

The testbed allows testing models' capabilities of prediction and/or categorization. The prediction task consists of producing an output at time t that matches the value of the signal that will be perceived at time $t+1$. This capability requires the integration of past signal values to capture the regularities of the signal in time.

Regarding the metrics that can be used to measure prediction performance, an immediate way of measuring the model's capacity to predict is to compute the quadratic error E of prediction with respect to the next input, averaged over the duration of the test:

$$E = \sum_{t=1}^{T-1} (P(t) - S(t+1))^2 \tag{1}$$

where T is the duration of the test, $P(t)$ is the value of the prediction, and $S(t)$ is the value of the input unit activated by the automaton's sensor.

Here "categorization" is referred to the models' capacity of distinguishing between several different signals. To this purpose, the models should have some output units whose activation can be trained in a supervised fashion and can be interpreted as the category assigned to the perceived signal. For example, in the tasks considered in Section 4, the automaton experiences two/three different objects/signals and has to categorize them with two/three units using a local code: the unit with the highest activation corresponds to the chosen category.

Performance of categorization can be measured as the percentage of time steps in which the categorization of the current perceived object produced by the automaton is correct. As the signals last more than one step, it has to be decided when to detect the categorization answer of the network. For example, in the examples shown in Section 4, this detection is done at the end of the pattern presentation. Nevertheless, dependent to the task, other points in time may be preferred.

Another way to measure the accuracy of the systems' prediction capabilities is to use prediction at time t as a new self-generated input pattern for time $t + 1$, to use the latter to produce a new prediction at $t + 1$ and again use it as input pattern for $t + 2$, and so on in a cyclical way. By monitoring how the patterns so generated diverge from those of the original input time series (i.e. by measuring the mismatch of the self-generated and real time series at a given time step in the future), it is possible to evaluate the goodness of the systems' prediction capabilities (cf. [22], where the quality of the prediction capability was measured not in terms of this mismatch, which can be used only when the input time series does not depend on the system's actions, but in terms of the capacity of the self-generated time series to produce accurate behavior).

2.3 Test of Robustness vs. Noise and Systematic Transformations of the Input Signal

Another important feature of the testbed is the possibility of determining the robustness of the model's prediction and categorization capabilities mentioned in Section 2.2. In particular, the testbed allows the evaluation of performance changes when the input time series is modified on the basis of step by step noise or on the basis of systematic random variations applied to the signal.

There are two types of step-by-step noise that can be applied to the simulated automaton (see examples in Figure 3):

1. *Translation noise.* This noise affects the size of translation of the automaton along the circular or linear trajectory it follows. This noise is set as a percent p of the automaton step size s: the noise is obtained adding a random number uniformly distributed over $[-ps, ps]$ to each step of the automaton. Notice that this noise can be cumulative, so that a certain long distance might be covered by the automaton with a different number of steps: this effect is particularly important as it can produce an overall random compression or expansion of the signal duration.
2. *Sensor noise.* This noise affects the automaton's sensor readings. The size of this noise is set as a percent of the original signal, similar to the translation noise.

The testbed allows setting four possible types of random systematic transformations to affect the signal in a biased way. We have seen in Section 2.1 that the input time series can be represented as a wave signal represented in an $x - y$ plot where the x-axis corresponds to time whereas the y-axis corresponds to the signal level. The four random systematic transformations correspond to *linear transformations* of the wave signal with respect to the two axes. As we show below, these transformations also have an interpretation in terms of specific manipulations of the relation existing between the automaton and the object (or

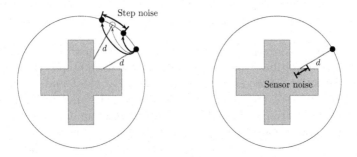

Fig. 3. Left: first type of noise affecting the step-size of the automaton, and hence the time regularity of the input time series. Right: second type of noise affecting the automaton's sensor reading.

wall) it perceives. Referring to the plot that represents the wave signal, we can formally define the transformation as follows, which are now illustrated in detail.

$$x \longmapsto a_x x + b_x \qquad (2)$$
$$y \longmapsto a_y y + b_y \qquad (3)$$

where a_x, b_x, a_y, and b_y are numerical coefficients. Each of these coefficients, when different from zero, causes one of the following four transformations:

1. *Compression/expansion of the signal time duration.* The coefficient a_x of the mapping in (2) sets the duration of the signal that corresponds to the speed of the automaton. In the testbed this parameter is set in terms of the number of steps that the automaton takes to sense the whole input time series (i.e. to complete a whole circle around the object or to complete one wall profile). The left graph of Figure 4 shows the effects of this transformation with respect to the original wave signal generated by the cross-shaped object reported in Figure 3.
2. *Phase of the signal.* The coefficient b_x of the mapping in (2) sets the start (phase or shift) of the signal and corresponds to initial position of the automaton with respect to the object or the wall. The right graph of Figure 4 shows the effects of this transformation.

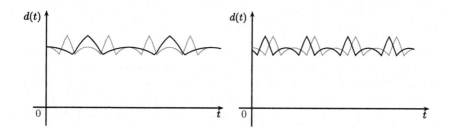

Fig. 4. Examples of random systematic transformations affecting the time variable of the input signal: $x \longmapsto \frac{1}{2}x$ (left), and $x \longmapsto x + 3$ (right)

3. *Compression/expansion of the signal level.* The coefficient a_y of the mapping in (3) sets the expansion/compression on the y-axis of the signal level. The left graph of Figure 5 shows the effects of this transformation with respect to the original signal generated by the cross-shaped object of Figure 3.
4. *Absolute signal level.* The coefficient b_y of the mapping in (3) sets the absolute position on the y-axis of the signal level. The right graph of Figure 5 shows the effects of this transformation, which corresponds to a variation of the distance of the automaton from the center of the object.

Notice that the third transformation implies a compression/expansion of both the object's size and the distance of the automaton from it, which is a rather

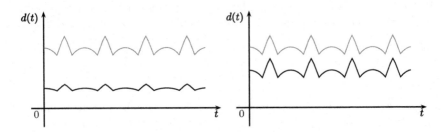

Fig. 5. Linear transformations in the space variable (distance): $y \longmapsto \frac{1}{3}y$ on the left, and $y \longmapsto y - 0.5$ on the right

uncommon situation in real experiments (e.g. with real robots). Nevertheless, notice that if it is combined with the fourth transformation it corresponds to a variation of the size of the object.

2.4   Analysis Techniques

This section proposes a number of techniques that can be used to analyze the functioning of the recurrent neural-network models tested on the testbed. The analysis tools described here are particularly important as the functioning and internal representations autonomously developed by dynamical neural networks are particularly difficult to be understood. The analysis tools are now presented and the reader is referred to Section 4 for some examples of them.

1. *Cross-correlograms.* One way to understand the functioning of the tested models is to study the *time correlations* existing between some variables of the models, such as the activation of hidden and output units, and between such variables and the input time series. Cross-correlograms and other statistical techniques directed to detect and represent correlations between time series can be used for this purpose (see Figure 11 and Figure 12 for some examples).
2. *Phase space analysis.* Given the dynamical nature of the problems tackled, and of the neural networks tested, the analysis of the *system's activity trajectory within the state space of selected variables* might shed light on the mechanisms that allow the learner to solve tasks. In particular, the identification of limit cycles, fixed point attractors, and chaotic attractors might allow understanding the properties of the solution and the properties emerging from the coupling between the dynamical process occurring within the neural controller and the dynamical process relative to the automaton/environmental interactions (see Figure 20 for an example).
3. *Special input time series.* Particular aspects of the behavior and functioning of the models might be analyzed by studying how they react to *special input time series.* These special input time series might be directly obtained by simplifying the ones used during training so as to isolate few features of interest (see Figure 14 and Figure 15 vs. Figure 10 for some examples).

4. *Hinton plot.* The study of the models' *connection weights* is of crucial importance because, after training, the models' performances depend on them (and the architecture). In this regard, a key tool to understand the role played by different weights is the *Hinton plot*, which allows representing the weights' intensities and signs in a visually comprehensive graph (see Figure 13 for an example).

5. *Targeted lesions.* The role of the models' components, either connection weights or units, might be understood by *lesioning* them and by observing how the performance accuracy of the models is disrupted. This can be done by setting connection weights to zero or by clumping the activation of the neurons of interest to zero (see Section 4.2 for some examples).

This list of analysis tools is of course non exhaustive. The use of these analysis tools, and how they can be used in a complementary fashion, is exemplified in Section 4.

3 Neural Networks for Integrating Information in Time

As mentioned in the introduction, dynamical neural networks are one of the most interesting classes of models that are capable of integrating information in time. Their integration capabilities are also supported by a wide literature that analyzes their relation with statistical algorithms for time series analysis (see [3], [4], [5], [8], and [19]). This section illustrates few important examples of these models. They have been selected for various reasons:

(1) they are widely used within the neural-network community;
(2) the testbed presented here has been originally developed as a tool to test and compare them within the research thread mentioned in the introduction;
(3) they allow the reader to envisage the models' properties that might be analyzed with the tools presented here;
(4) in the future the testbed will be used to systematically compare them.

3.1 Elman Neural Networks

The first neural architecture considered here is the *Elman neural network* (see [2], [7] and [16]). This is a feed-forward network with three layers: an input layer, a hidden layer, and an output layer (Figure 6). The core of this type of network is the presence of recurrences in the hidden layer that implies that, at each time step, the activations of hidden units depend on the activations of the same units at the previous time step.

This architecture allows the network to store information from the past so that the network is capable of detecting periodicity and regularities of the input patterns in time. In particular, if we let N_I, N_H, and N_O denote the number of units in the input, hidden and output layer, respectively, then the input to the N_H hidden units will be formed not only by the N_I units activated by the

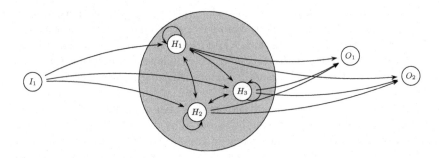

Fig. 6. Architecture of an Elman network with 1 input unit, 3 hidden units and 2 output units

current input, but also by further N_H units encoding the activation of the hidden units at the previous time step (these play the role of a *memory buffer*).

The activation function of the hidden and the output layer units is the logistic function:

$$\Phi_\sigma(x) = \frac{1}{1 + e^{\sigma x}} \, , \tag{4}$$

where σ is a "temperature" parameter (set to 1 in the experiments herein). The learning rule used is the error back-propagation algorithm (see [17] or [20]).

3.2   Echo State Neural Networks

The second neural architectures considered, called *echo state neural networks* (Figure 7), are only briefly reviewed here (see [11] and [13] for a detailed description and literature review). The important aspect of these networks is the presence of a hidden layer, called *dynamical reservoir*, formed by linear or sigmoid units which have hard-wired recurrent connections (connections in solid black in Figure 7). These connections, that form a W matrix, are initially set randomly, and then are normalized with the highest eigenvalue of the matrix so that W has a spectral radius slightly smaller than 1. This setting of the weights implies that the hidden neurons do not produce a chaotic behavior, do not explode or do not saturate on maximum or minimum values. Moreover, it implies that the variation in the activation state of the hidden neurons produced by transient inputs tends to slowly decay after the end of the stimulation.

If the set of hidden units is large enough, the dynamical reservoir is capable of producing a large number of dynamics. The units of the reservoir are connected to the output units in a simple linear fashion (the output units also feedback to the reservoir units with random connections – hence the term "echo" in the name: the signals vehiculated by these connections contribute to modulate the reservoir's dynamics). During training, the weights that connect the reservoir units with the output units are updated with a supervised algorithm to reproduce

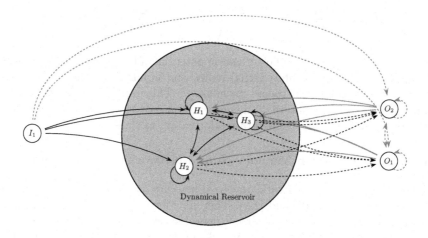

Fig. 7. Architecture of an Echo State Network with 1 input unit, 3 hidden units, and 2 output units

a target output signal (e.g., a periodic signal). This training leads the hidden-output weights to "select" few relevant dynamics from the internal reservoir among the possible ones, and allows the output units to learn to reproduce, in principle, any desired output signal having some correlation with the input signal.

3.3   Leaky Integrator Neural Networks

The third type of neural network considered here is formed by *leaky integrator neurons*. The core property of these neurons is that their activation potential depends not only on the input from other internal and external neurons, but also on own previous activation potential (see [1], [18] and [21]). Let $u_i(t)$ denote the i-th unit's potential at time t, $I_i(t)$ its external input, $h_i(t)$ its resting level, Φ_σ is the sigmoid activation function in (4), $u_j(t)$ the activation potential of another j-th unit of the network, and w_{ij} the weight from the unit j to the unit i; the dynamics of $u_i(t)$ is governed by the following dynamic equation:

$$\tau \dot{u}_i(t) = -u_i(t) + I_i(t) + h_i + \sum_j w_{ij}\, \Phi_\sigma\left(u_j(t)\right) \ . \tag{5}$$

Equation (5) implies that in absence of external inputs I_i, and with zero connection weights w_{ij}, the neuron exponentially relaxes to the resting state h_i with a rate equal to $-\frac{1}{\tau}$ (hence the name "leaky"). If Δt denotes the integration time step, then (5) has a discrete version with the following form (that can also be used to numerically integrate equation (5) in the simulations):

$$u_i(t + \Delta t) = \left(1 - \frac{\Delta t}{\tau}\right) u_i(t) + \frac{\Delta t}{\tau} \left(I_i(t) + h_i + \sum_j w_{ij} \, \Phi_\sigma \left(u_j(t)\right)\right) . \quad (6)$$

This form highlights that the activation potential of leaky neurons approaches the sum of the resting level, external input, and input from other neurons, on the basis of a "partial adjustment mechanism". This implies that leaky neurons have a ready available "internal" memory of the past that can be exploited by the whole neural network to integrate information in time.

3.4   Long Short-Term Memory Neural Networks

The last types of neural networks reviewed here are the *Long Short-Term Memory Networks* (see [9], [10] and [12]). They have been introduced to extend the memory capacity of standard recurrent neural networks, in particular, they have been shown to efficiently solve many tasks involving integration of information in time that are unlearnable for other neural networks (e.g. the recognition of temporally very long extended patterns in noisy input sequences, the recognition of the temporal order of widely separated events in noisy input streams, or the stable generation of precisely timed rhythms).

The key feature of these neural networks resides in the special type of neurons that form them, which are characterized by a self-recurrent connection and gates that exert multiplicative effects on the input and output channels (Figure 8). The functioning of one neuron of this type can be described as follows:

$$y_i(t) = \Phi_\sigma \left(\sum_j w_{ij}^{go} u_j(t)\right) \Phi_\sigma \left(u_i(t)\right) , \quad (7)$$

$$u_i(t) = u_i(t-1) + \Phi_\sigma \left(\sum_j w_{ij}^{gi} u_j(t)\right) \Phi_\sigma \left(\sum_j w_{ij} u_j(t)\right) , \quad (8)$$

where $y_i(t)$ and $u_i(t)$ are the i-th unit's activation and action potential at time t, respectively, w_{ij} is the weight from the unit j to the unit i, w_{ij}^{gi} and w_{ij}^{go} are the weights from the units j to the input and output gates, respectively, and Φ_σ is the sigmoid activation function of Equation (4).

These features allow neural networks formed by several of these special neurons to produce highly complex dynamics. The networks so formed can be trained on the basis of supervised learning algorithms.

4   Examples of Applications

This section illustrates the potential of the testbed by testing an Elman neural network with two specific tasks. In the first task, the automaton perceives two walls with different profiles, while following a linear trajectory, while in the second task the automaton senses three different objects while following a circular

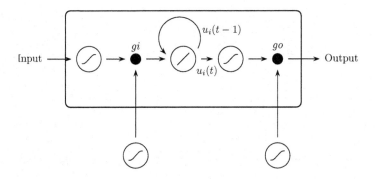

Fig. 8. Structure of a single neuron of a long short-term memory neural network

trajectory. In both cases the automaton's goal is to categorize the perceived signal and to predict, at each step, the signal level at the next step on the basis of the previously experienced signal levels. Note that the experiments reported here also represent the initial work of a research agenda directed to understand the internal dynamical mechanisms exploited by recurrent neural networks to capture regularities in time.

4.1   Wall Task: Experimental Setup

In this task the automaton moves along a straight trajectory along a wall which can have one of two different profiles, shown in Figure 9. The various settings of the experimental setup can be summarized as follows:

1. *Wall profiles.* The two possible wall profiles (Figure 9) had "hollows" with same depth (this caused the normalized automaton's sensor reading return 1, so this portion of the two walls was ambiguous for the automaton), and "humps" with different heights (these caused a normalized sensor reading equal to 0.36 and 0.68 respectively for the first and second profile).
2. *Model.* The tested model was an Elman neural network with $N_I = 1$ input units, $N_H = 2$ hidden units, and $N_O = 3$ output units. The input unit was activated by the sensor reading normalized in $[0, 1]$. The first output unit was devoted to predict the next input pattern while the other two output units were devoted to encode the categories of the two wall profiles. Such categories were locally encoded as $\{1, 0\}$ and $\{0, 1\}$ respectively for the two profiles.
3. *Training.* During training, the walls were repeatedly presented one by one to the automaton. The wall used in each presentation was randomly chosen, and at each presentation the automaton performed a whole circle around it. Training lasted $1,000,000$ presentations, and used a $\lambda = 0.005$ learning rate. For each time step, the teaching input was formed by the next input pattern (i.e. the value of the signal at time $t + 1$) and by the binary value that encoded the category of the current wall.

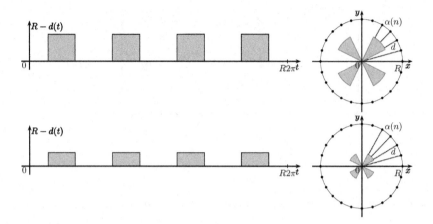

Fig. 9. The two wall profiles used in the wall task (left) and the two equivalent objects in the corresponding object task (right)

4. *Steps (Angular Speed).* The automaton covered a single lap around the patter in a number of steps denoted by #*Steps*. In the experiment this parameter was randomly assigned one of the following values in every object presentations (and kept constant during each presentation):

$$\#Steps \in \{16, 24, 32, \ldots, 128\}, \tag{9}$$

The angle of the trajectory covered by one step of the automaton, that is, its angular speed, depended on the total number of steps of a lap, and was equal to $2\pi/\#Steps$.

5. *Starting Point.* This was the angle of the circular trajectory where the automaton started to perceive the wall. Let $\alpha(n)$ denote the angle at the n-th step. For any $n = 0, \ldots, \#Steps$:

$$\alpha(n) \in \left\{ 0, \frac{2\pi}{\#Steps}, \frac{4\pi}{\#Steps}, \ldots, (\#Steps - 1)\frac{2\pi}{\#Steps} \right\}. \tag{10}$$

The starting point $\alpha(0)$ was set randomly at each presentation of the input time series.

6. *Compression/Expansion of the Signal Level (Height of Profiles).* The size of the maximum height of the two walls was kept constant: 1.6 and 0.8 for profiles 1 and 2, respectively.

7. *Absolute Signal Level (Radius/Distance from Walls).* The parameter of the distance from the walls hollows, denoted by R, was set to 2.5.

8. *Noise.* The two sources of noise illustrated in Section 2.3 were both set to 5%.

4.2   Wall Task: Results

The training of the system was rather successful. At the end of training, the neural network shows a rather good categorization ability. More precisely, the

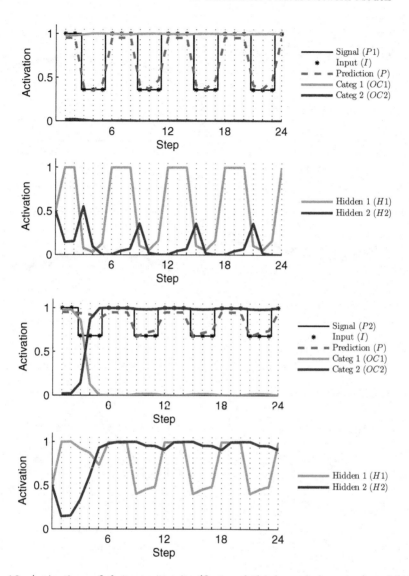

Fig. 10. Activations of the output units (first and third graph from top) and hidden units (second and fourth graph from top) when the automaton perceives the signal from the wall profile 1 and 2 (respectively first/second and /third/foruth graphs from top). Stars in the first and third graph indicate the actual sensor's readings (noise has been switched off to ease the analysis of results), whereas the continuous black lines show the wall's profiles obtained with a very dense sensor reading.

network produces the right categorization output after few steps, and after that, keeps producing the same categorization output even during the phase in which the signal is ambiguous, that is, when scanning a hollow with a value equal to 1

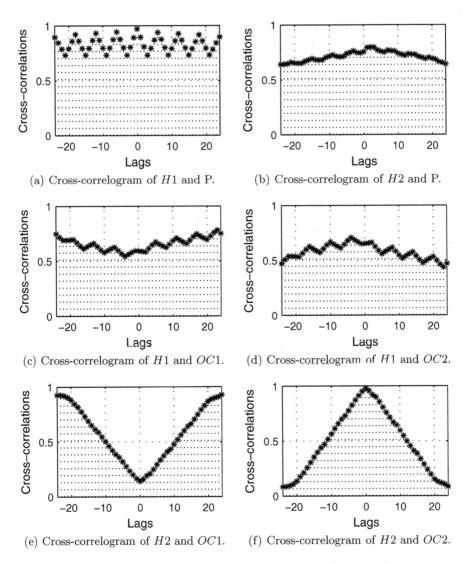

(a) Cross-correlogram of $H1$ and P.

(b) Cross-correlogram of $H2$ and P.

(c) Cross-correlogram of $H1$ and $OC1$.

(d) Cross-correlogram of $H1$ and $OC2$.

(e) Cross-correlogram of $H2$ and $OC1$.

(f) Cross-correlogram of $H2$ and $OC2$.

Fig. 11. Cross-correlograms between different variables of the models (see text) when the model perceives a repeated sequence of 1000 $P1/P2$ wall profiles, in an alternate fashion, with $\alpha(0) = 0$

(as shown by the thin black curves in Figure 10). With respect to the prediction capability, however, the network simply predicts that the signal at time $t + 1$ will be identical to the signal at time t. Although this simple strategy allows the network to produce the right answer in most of the cases, it fails to predict correctly in the cases in which the value of the signal suddenly varies from time t to $t + 1$ (Figure 10). Indeed, exactly predicting this sudden change is not possible due to the noise affecting the automaton's step-size.

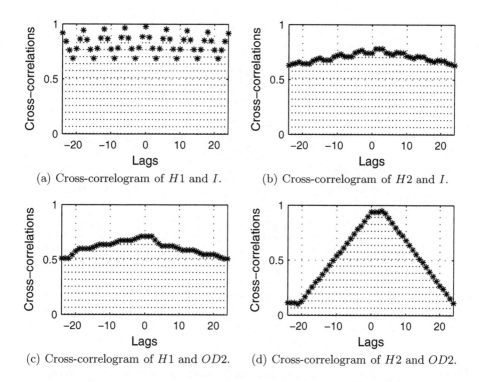

(a) Cross-correlogram of $H1$ and I.   (b) Cross-correlogram of $H2$ and I.

(c) Cross-correlogram of $H1$ and $OD2$.   (d) Cross-correlogram of $H2$ and $OD2$.

Fig. 12. Cross-correlograms between hidden units' activation and the input signal or desired output of the second categorization output unit, when the model perceives a repeated sequence of 1000 $P1/P2$ wall profiles, in an alternate fashion, with $\alpha(0) = 0$

The capabilities of the model are robust with respect to signal's random systematic transformations of the first type indicated in (2) (recall that these transformations are related to variations of the initial position of the automaton with respect to the object and to the step size): none of them prevents the system's capabilities to emerge.

In order to understand in detail how the system performs prediction and categorization, a test was run where the system was presented for 1000 times, in an alternate way, the two wall profiles, each time with $\alpha(0) = 0$ (that is, the automaton is placed at the beginning of the input time series). The data collected in this test were used to build cross-correlograms capturing correlations within couples of time series related to various variables of the network, namely, the input value, the hidden units' activations, and the output units' activations. Let us denote with I the input unit's activation (i.e. the perceived signal), with $H1$ and $H2$ the two hidden units' activation, with OP the activation of the output unit devoted to prediction, with $OC1$ and $OC2$ the activation of the two output units devoted to categorization, with $OD1$ and $OD2$ the desired output for the two categorization units, and with $P1$ and $P2$ the two wall profiles.

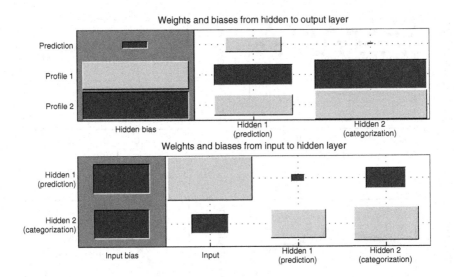

Fig. 13. Hinton plot of model's connection weights after training. The black and gray squares respectively correspond to negative and positive values of the weights, whereas their size is proportional to the weights' absolute value.

The cross-correlograms between the hidden and the output units' activation, reported in Figure 11, give important indications on the role played by the hidden units in the model's responses:

1. The comparison of cross-correlograms of Figure 11(a) and 11(b), related to the correlation between $H1/H2$ and OP, indicate that $H1$ has a strong correlation with OP whereas $H2$ has an almost null correlation with it.
2. The cross-correlograms of Figure 11(c) and 11(d) indicate that $H1$ has a very low anti-correlation with $OC1$ and a very low correlation with $OC2$.
3. The cross-correlograms of Figure 11(e) and 11(f) indicate that $H2$ has a strong anti-correlation with $OC1$ and a strong correlation with $OC2$.

Altogether, these data corroborate the suggestion that $H1$ mainly underlies the model's prediction capability, whereas $H2$ mainly underlies its categorization capability, and that a high and low activation of the latter tends to cause the model to categorize the input pattern respectively as $P2$ and $P1$.

These interpretations are further corroborated by the cross-correlograms related to the hidden units, the input unit, and the second wall category ($OD2$; the cross-correlograms with $OD1$ give similar information), reported in Figure 12, which show that:

1. The comparison of cross-correlograms of Figure 12(a) and 12(b), related to the correlation between $H1/H2$ and I, indicates that $H1$ has a strong correlation with I whereas $H2$ has an almost null correlation with it.

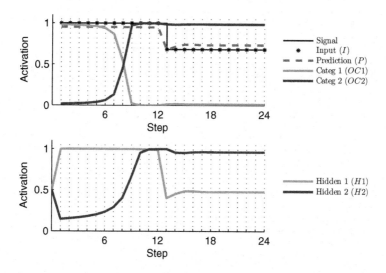

Fig. 14. Activations of the hidden and output units when the automaton first perceives a signal of 1 and then of 0.68

2. The comparison of cross-correlograms of Figure 12(c) and 12(d), related to the correlation between $H1/H2$ and $OD2$, indicate that $H1$ has no correlation with $OD2$ whereas $H2$ has a strong correlation with it.

Figure 13 reports the values of the model's weights that emerged with training. The analysis of the weights between the hidden and output units confirms the indications given by the cross-correlograms, and also allows formulating a more detailed explanation of the functioning of the system:

1. The weights from $H1$ and $H2$ to OP show that P depends only on $H1$, as the weight of the connection from $H1$ is positive (positive correlation) whereas the weight from $H2$ is close to zero (no correlation). Indeed, lesioning this weight, that is setting it to zero, has no effect on prediction performance (data not reported).
2. The high weights from $H2$ to $OC1$ and $OC2$ confirm that this hidden unit greatly contributes to determine the category of the wall profile, namely $P1$ when it is low and $P2$ when it is high. $H1$ also partially contributes to the categorization as its weights to $OC1$ and $OC2$ are different from zero (its high activation tends to cause a categorization of the input signal as $P2$).
3. The analysis of the weights between the input and the memory units on one side, and the output units on the other side, give other important indications on how the system solves the task.
4. Considering the connections to $H1$, the positive connection weight between I and $H1$ implies that $H1$ implements the prediction capabilities by "relaying" the input signal: the model tends to return a high or low prediction value

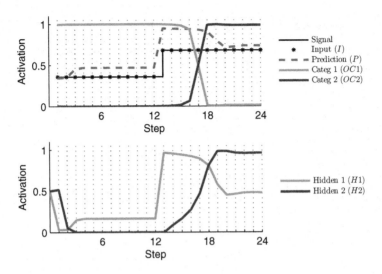

Fig. 15. Activations of the hidden and output units when the automaton first perceives a signal of 0.36 and then of 0.68

respectively for high or low input signal levels. Note that this implies that the model is not capable of returning an accurate prediction when the signal suddenly changes level one step in advance, as shown in Figure 10. The weights from $H1$ to itself do not play an important role. Indeed, lesioning them does not impair performance (data not shown).

5. Considering the connections to $H2$, the positive self-connection of $H2$ indicates that it has a strong inertia. The positive connection from $H1$ (that, as we have seen, positively correlates with the input pattern and generates P) indicates that $H2$'s categorization capacity strongly depends on $H1$'s activation. Lesioning the connection between the input and $H2$ indicates that it is also important for categorization (data not reported).

A further refinement of these interpretations is furnished by two other experiments where the input signal to the system is handcrafted in order to highlight particular aspects of its internal dynamics. In particular, Figure 14 shows the dynamics of the model's hidden and output units' activation when it first perceives a signal of 1 and then of 0.68 (recall that the level of the signal has been normalized in the range $[0, 1]$), whereas Figure 15 shows the dynamics of the same variables when the system first perceives a signal of 0.36 and then of 0.68.

With respect to the prediction capability, these figures confirm that prediction capability ($H1$) relies in part on categorization ($H2$). In fact, if $H2$ gives the category $P1$, then $H1$ gives $P = 1$ with both $I = 1$ (Figure 14) and $I = 0.68$ (Figure 15: note how after the signal abruptly changes, the prediction makes a mistake for about six cycles because the activation of $H2$ is incorrectly categorizing the input as $P1$). On the other hand, if $H2$ gives the category $P2$, $H1$ gives

$P = 1$ with $I = 1$ (Figure 14), but it gives $P = 0.68$ with $I = 0.68$ (Figure 14 and 15).

With respect to the categorization capability, it is interesting to see how the system can solve the $I = 1$ ambiguity. $H2$ moves slowly toward 1 (that implies $P2$) both when $I = 1$ (Figure 14) and when $I = 0.68$ (Figure 15), whereas it stays at 0 when $I = 0.36$ (Figure 15). This implies that $H2$ has the value of 1, (corresponding to $P2$) as a fixed-point attractor value when $I > 0.36$ or so, and 0 (corresponding to $P1$) when $I = 0.36$. For this reason, when the signal level $I = 1$ has been preceded by a signal $I = 0.36$ (corresponding to $P1$), $H2$ approaches 1 ($P2$) only slowly and so continues to give $P1$ for some time until the system perceives $I = 0.36$ again.

4.3   Three Objects Task: Experimental Setup

In this task, the automaton moves along a circular trajectory around three different "objects" and at each step detects the distance from them (see Fig. 16 and compare with [2]).

The various settings of the experimental setup can be summarized as follows:

1. *Objects.* The three objects are illustrated in Figure 17: a "square", a "thick cross" and a "thin cross". The number of objects is denoted with N_P. Notice that, given the shape of the objects, during one lap around an object the automaton experienced a signal formed by four succeeding equal waves, similarly to the wall task.

2. *Model.* The tested model was an Elman neural network with $N_I = 1$ input unit, $N_H = 3$ hidden units, and $N_O = 4$ output units. The input unit was activated by the sensor reading normalized in $[0, 1]$. The first output unit was devoted to predict the next input pattern while the remaining three output units were devoted to encode the categories of the three signal patterns. Such categories were locally encoded as $\{1, 0, 0\}$, $\{0, 1, 0\}$, and $\{0, 0, 1\}$) respectively for the three objects.

3. *Training.* Training was performed as in the wall task.

4. *Steps (Speed).* The number of steps #*Steps* the automaton took to scan an object's profile during a presentation was randomly varied from presentation to presentation as in the wall task.

5. *Starting Point.* The starting point $\alpha(0)$ was randomly varied from presentation to presentation as in the wall task.

6. *Compression/Expansion of the Signal Level (Object Size).* The size of the object is denoted by S (this represents half of the longest arm of the cross objects and half the size of the square's side) and half of the length of the shortest arm of the crosses is denoted with T. In the majority of experiments reported below, the size of the objects was set at fixed values, whereas in few other experiments it was randomly varied at each presentation (but kept fixed within it) within the range $[0.5, 1.8]$ (recall that S is related with the parameter a_y of equation 3).

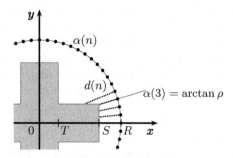

Fig. 16. Important parameters of the three object recognition task: the dots represent different positions in space occupied by the automaton from which it detects a certain distance d from the object. R is the automaton's distance from the object's center, S is half of the size of the object's longest axis, and T is half of the object's shortest axis.

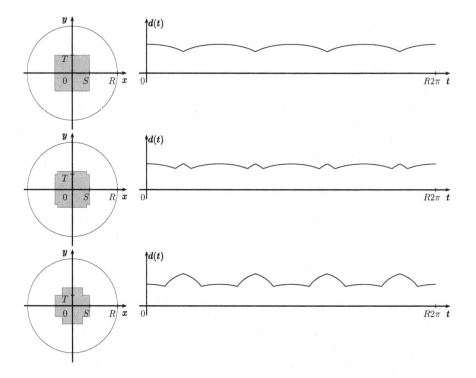

Fig. 17. Left graphs: the three crosses used in the object task, characterized by $(\rho_1, S_1) = (1, 0.95)$, $(\rho_2, S_2) = (0.70, 1.13)$, and $(\rho_3, S_3) = (0.41, 1.35)$: these values were set to similar maximum and minimum values of the sensor's reading (in few experiments was randomly varied). The radius R was set to 2.5. Right graphs: the sensor readings caused by the three objects.

7. *Absolute Signal Level (Radius).* In the majority of experiments reported below the radius R of the circular trajectory followed by the automaton around the objects was set at fixed values, whereas in some variants of the experiment it was randomly varied at each presentation (but kept fixed within it) in the range $[2,3]$ (recall that R is linearly related to the parameter b_y of equation 3).

8. *Noise.* The two sources of noise illustrated in Section 2.3 were both set to 5%.

Note that, since the setting implies that $0 \leq T \leq S \leq R$, the ratio $\rho = \frac{T}{S}$ has the following restriction: $\rho \in [0,1]$. Also note that parameter ρ uniquely identifies the three objects (see Figure 16).

4.4   Three Objects Task: Results

With the two sources of noise on and the four random systematic transformations off, the Elman network achieves a satisfying performance both for the prediction and for the categorization tasks. As far as the prediction capability is concerned, as in the wall task, the system adopts a strategy of input repetition. Nevertheless, the network had low quadratic errors when tested with various step sizes (see Figure 18): the graph shows that the error decreases as the steps grow.

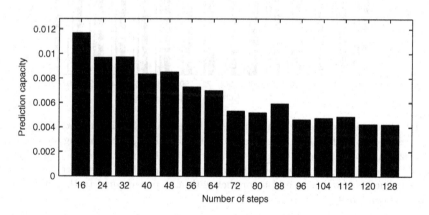

Fig. 18. Mean square error of the prediction unit with respect to #*Steps*

A further analysis of the system's prediction capabilities was obtained by presenting the patterns to the model for some time, and then by forcing the network to use its prediction as the next self-generated input. In this experiment, the quality of the prediction signal rapidly deteriorates (constant output), hence confirming that the system directly repeats the input as prediction.

The model also shows to be robust with respect to signal's random systematic transformations of the first type indicated in (2). In particular, the model is robust with respect to variations of the initial position of the automaton with respect to the object (corresponding to $\alpha(0)$), which does not deteriorate performance (data not shown). Moreover, and surprisingly, the model has an even higher performance when trained with step sizes that vary randomly between the objects' presentations. This can be seen by comparing the model trained in two different conditions:

1. In each presentation #*Steps* is randomly set within the values indicated in (9), and the object approach angle $\alpha(0)$ is randomly set in the range indicated in (10).
2. The #*Steps* is set at the same value used in the test of performance, whereas the object approach angle $\alpha(0)$ is randomly set as in the previous condition.

The white and black bars of the histogram reported in Figure 19, which refer to the two training conditions, respectively, indicate that the performance of the model is generally higher when it is trained with varying step sizes. Further experiments with one square and three different crosses corroborate this result (data not reported).

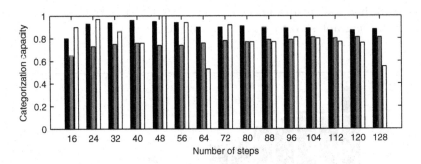

Fig. 19. Categorization performance of the model (y-axis) when it is tested with various step sizes (x-axis). The black bars refer to a model trained, with varying step size, to categorize the objects and to predict the next input. The gray bars refer to a model trained, with varying step size, only to categorize. The white bars refer to different models trained to categorize and to predict with a step size equal to the one used in the performance tests, reported on the x-axis.

Other tests showed that the signal's random systematic transformations of the second type, indicated in (3), completely disrupt the performance of the algorithm (data not reported). This result suggests that the capacity of the model both to categorize the object and to predict the next input heavily relies on the absolute and relative levels of the signals in time.

Another interesting result is that training the prediction capability of the model improves the model's capacity to categorize the objects. With this respect the gray histogram bars reported in Figure 19 show that the performance of the model deteriorates if the system is not trained to predict the next input. This

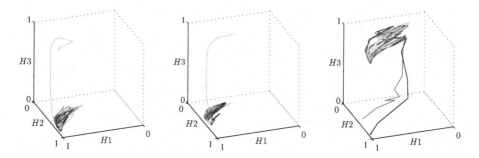

Fig. 20. Dynamics and attractors of the model's hidden units activations. Each of the three graphs represents the "history" of the three hidden units' activation when the model categorizes and predicts one of the three objects (respectively the square, the thick cross and the thin cross). Each graph reports the outcome of the experiment run in three different conditions: (a) the model categorizes the object after it has perceived the square (black line); (b) the model categorizes the object after it has perceived the thick cross (gray line); (c) the model categorizes the object after it has perceived the thin cross (light gray line).

result indicates that training the model prediction capability likely leads the system to develop internal representations that aid the categorization capability. Further analyses of the model's internal representations should be carried out to understand this outcome in further depth. These analyzes might be aided by some of the investigation methods suggested in Section 2.4. For example, Figure 20 shows the dynamics of the model's three hidden units when the system perceives the three objects. The graphs of the figure show that the model's object categorization and prediction capability is based on three different limit-cycle attractors: the model's internal state converges to a different limit-cycle attractor in order to categorize the different objects. Notice that when the system starts to perceive an object after it has perceived a different object, its internal state takes some time to settle to the limit-cycle attractor of the current object as its internal memory needs to synchronize with the dynamics of the new input time series. After the state has settled to the attractor corresponding to the object, then it follows a cyclic trajectory within it in order to predict the next input pattern.

5   Conclusions and Future Work

This paper presented a testbed that can be used to evaluate the capabilities of recurrent neural networks (and similar models) of integrating information in time, in particular the capabilities of categorizing different signals, of predicting future signals on the basis of past ones, and of doing so in the face of noise and systematic variations of the input signal. The paper also illustrated the potentialities of the testbed by exemplifying its functioning with two tests involving simple recurrent Elman networks engaged in solving two different prediction and categorization tasks.

The added value of the paper is manyfold. First, it highlights the need of building standard testbeds, metrics, and analysis tools to compare, and build taxonomies of, the increasing number of models proposed within the literature of the ABiALS community. Second, it presents a specific testbed that allows testing models' capabilities of categorization and anticipation. Third, it shows the potential utility of developing and using testbeds by showing some results obtained by applying the testbed proposed here to the study of the functioning of Elman neural networks. These applications showed that, even if a detailed understanding of the functioning of recurrent neural networks is very difficult, the dynamical principles that might underlie their capacity of integrating information in time are particularly interesting and make it worth designing and implementing testbeds and analysis tools, as those proposed here.

Future developments of this research will follow two main directions. On the one hand, it will continue to carry out systematic studies, in line with the preliminary experiments presented in Section 4, to understand the exact mechanisms that are developed by recurrent networks to integrate information in time, such as the formation of cyclic or fixed point attractors, units with progressive increases or decreases of activation, hierarchical abstract representations, etc. On the other hand, it will use the testbed to compare the capacities of the four neural networks described in Section 3 to capture different time regularities. This comparison could be important to highlight which particular features of temporal signals can be best integrated in time by the different models, and hence which types of tasks are more suitable for them.

References

1. Amari, S.I.: Dynamics of pattern formation in lateral-inhibition type neural fields. Biological Cybernetics 27, 77–87 (1977)
2. Cecconi, F., Campenní, M.: Recurrent and concurrent neural networks for objects recognition. In: Deved, V. (ed.): Proceedings of the International Conference on Artificial Intelligence and Applications (IASTED 2006), Innsbruck, Austria, pp. 216–221 IASTED/ACTA Press (2006)
3. Chakraborty, K., Mehrotra, K., Mohan, C.K., Ranka, S.: Forecasting the behavior of multivariate time series using neural networks. Neural Networks 5, 961–970 (1992)
4. Chappelier, J.C., Grumbach, A.: Time in neural networks. ACM SIGART Bulletin 5, 3–11 (1994)
5. Dorffner, G.: Neural networks for time series processing. Neural Network World 6, 447–468 (1996)
6. Doya, K.: Recurrent networks: learning algorithms. In: Arbib, M.A. (ed.) The Handbook of Brain Theory and Neural Networks, Second edn. pp. 955–960. The MIT Press, Cambridge, MA, USA (2003)
7. Elman, J.L.: Finding structure in time. Cognitive Science 14, 179–211 (1990)
8. Hellström, T., Holmström, K.: Predicting the stock market. Research and Reports Opuscula ISRN HEV-BIB-OP–26-SE, Department of Mathematics and Physics, Mälardalen University, Västerås, Sweden (1998)

9. Hochreiter, S., Schmidhuber, J.: Bridging long time lags by weight guessing and "Long Short-Term Memory". In: Silva, F.L., Principe, J.C., Almeida, L.B. (eds.) Spatiotemporal models in biological and artificial systems. Frontiers in Artificial Intelligence and Applications, vol. 37, pp. 65–72. IOS Press, Amsterdam (1996)

10. Hochreiter, S., Schmidhuber, J.: Long short-term memory. Neural Computation 9, 1735–1780 (1997)

11. Jaeger, H.: Tutorial on training recurrent neural networks, covering bptt, rtrl, ekf and the "echo state network". Gesellschaft für Mathematik und Datenverarbeitung Report 159, German National Research Center for Information Technology (2002)

12. Klapper-Rybicka, M., Schraudolph, N.N., Schmidhuber, J.: Unsupervised learning in LSTM recurrent neural networks. In: Dorffner, G., Bischof, H., Hornik, K. (eds.) ICANN 2001. LNCS, vol. 2130, pp. 684–691. Springer Verlag, Heidelberg (2001)

13. Maass, W., Natschläger, T., Markram, H.: Real-time computing without stable states: A new framework for neural computation based on perturbations. Neural Computation 14, 2531–2560 (2002)

14. Mitchinson, B., Pearson, M., Melhuish, C., Prescott, T.J.: A model of sensorimotor coordination in the rat whisker system. In: Nolfi, S., Baldassarre, G., Calabretta, R., Hallam, J.C.T., Marocco, D., Meyer, J.-A., Miglino, O., Parisi, D. (eds.) SAB 2006. LNCS (LNAI), vol. 4095, pp. 77–88. Springer, Heidelberg (2006)

15. Nolfi, S., Marocco, D.: Evolving robots able to integrate sensory-motor information over time. Theory in Biosciences 120, 287–310 (2001)

16. Nolfi, S., Tani, J.: Extracting regularities in space and time through a cascade of prediction networks: The case of a mobile robot navigating in a structured environment. Connection Science 11, 129–152 (1999)

17. Rumelhart, D.E., Hinton, G.E., Williams, R.J.: Learning representations by back-propagating errors. Nature 323, 533–536 (1986)

18. Schöner, G., Kelso, J.A.S.: Dynamic pattern generation in behavioral and neural systems. Science 239, 1513–1520 (1988)

19. Ulbricht, C., Dorffner, G., Canu, S., Guillemyn, D., Marijuán, G., Olarte, J., Rodríguez, C., Martín, I.: Mechanisms for handling sequences with neural networks. In: Dagli, C.H. (ed.): Intelligent Engineering Systems through Artificial Neural Networks (ANNIE 1992) New York, NY, USA, vol. 2, pp. 273–278 ASME Press (1992)

20. Williams, R.J., Zipser, D.: A learning algorithm for continually running fully recurrent neural networks. Neural Computation 1, 270–280 (1989)

21. Wilson, H.R., Cowan, J.D.: Excitatory and inhibitory interactions in localized populations of model neurons. Biophysical Journal 12, 1–24 (1972)

22. Ziemke, T., Jirenhedb, D.A., Hesslow, G.: Internal simulation of perception: a minimal neuro-robotic model. Neurocomputing 68, 85–104 (2005)

Construction of an Internal Predictive Model by Event Anticipation

Philippe Capdepuy[1], Daniel Polani[1,2], and Chrystopher L. Nehaniv[1,2]

[1]Adaptive Systems and [2]Algorithms Research Groups
School of Computer Science, University of Hertfordshire
College Lane, Hatfield, Herts, AL10 9AB, UK
{P.Capdepuy,D.Polani,C.L.Nehaniv}@herts.ac.uk

Abstract. We introduce information-theoretic tools that can be used in an autonomous agent for constructing an internal predictive model based on event anticipation. This model relies on two different kinds of predictive relationships: time-delay relationships, where two events are related by a nearly constant time-delay between their occurrences; and contingency relationships, where proximity in time is the main property. We propose an anticipation architecture based on these tools that allows the construction of a relevant internal model of the environment through experience. Its design takes into account the problem of handling different time scales. We illustrate the effectiveness of the tools proposed with preliminary results about their ability to identify relevant relationships in different conditions. We describe how these principles can be embedded in a more complex architecture that allows action-decision making according to reward expectation, and handling of more complex relationships. We conclude by discussing issues that were not addressed yet and some axis for future investigations.

1 Introduction

Designing agents that can act intelligently in a previously unknown environment is one of the most challenging issues in behavioral robotics. Such an agent must have the ability to construct an internal model describing the dynamics of the environment and the effect of its own actions on this environment. This can be mainly understood as extracting predictive relationships between events occurring in the perceptive field of the agent, whether these events are under its control (its actions) or if they are externally generated. This internal model allows the agent to predict forthcoming events, as well as the effect of its own actions on the environment. Such a predictive ability paves the way to anticipation and smart decision making by allowing the agent to decide which action to perform to obtain or avoid a given outcome. According to the classification of [1], these agents are said to perform *state anticipation*.

Our main focus in this paper is to define and evaluate tools that allow the construction of such an internal model regardless of any reinforcement. In this sense we are very close to latent learning and the concept of *expectancies* proposed

M.V. Butz et al. (Eds.): ABiALS 2006, LNAI 4520, pp. 218–232, 2007.

by Tolman [7]. We describe an architecture that uses these tools to effectively construct the internal model and we explain how this model can be used to anticipate events. The robustness of this architecture to length and variability of time-delays between relevant associations is evaluated in two experiments. A last experiment shows the temporal dynamic of the model and more especially the forgetting mechanism.

The paper is structured as follows: in Sec. 2 we formulate the problem of event anticipation along with some examples and what we would expect from our anticipation system. Section 3 introduces the main information-theoretic concepts used in our model and the two kinds of relationships they allow us to identify. Section 4 describes the anticipation architecture embedding these concepts. In Sec. 5 we describe preliminary results concerning the predictive efficiency of the proposed tools in a simple simulation experiment. Section 6 describe some possible extensions to the actual model, mainly considering the problem of action-selection and how our predictive model can be used in a reward-based behavior. We also introduce a possible mechanism for handling more complex situations involving sequences of events and non-occurrence. Section 7 summarizes the issues our model addresses and we discuss some of those that will be investigated in future work.

2   Event-Based Anticipation

2.1   Stating the Problem

Here we refer to anticipation in a very general way as the ability to predict, more or less accurately, the future occurrence of *perceptive events*. These events can be seen as different stimuli that the agent can encounter in its environment. We consider as a preliminary simplification that the agent is not allowed to act onto its environment, he can only observe it (handling of actions will be described in Sec. 6). Different from other approaches, the agent is not provided here with a continuous flow of sensoric values for different modalities. Instead we consider that the agent perceives discrete events in discrete time (0 to n different events can be observed at a given time-step). We will denote the set of possible events by $\mathcal{E}$. The agent is then observing a stream of events such as the one represented in Figure 1. The only relevant information that can be extracted from this stream are the relationships in time between similar or different events. The purpose of our work is to find an efficient way to identify these relationships in an anticipatory perspective.

2.2   Expected Properties

We want to infer a predictive model from observing the stream of events. According to a given recent past, the predictive model could then be used to anticipate what the next events should be, and when they will occur. One of the constraints we put on our model is that it should be robust to noise and variations in the

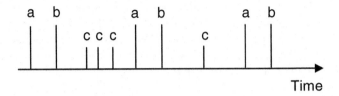

Fig. 1. Example of a stream of events $\mathcal{E} = \{a, b, c\}$ over time. The height is not relevant but just for clarity. In this particular example, we can observe that a is always directly followed by b with a fixed time delay. The event c seems to have a more complex pattern.

relationships. Also time-scale variations should have no effect on the efficiency of the predictive model construction (if for example all events have their delay multiplied by 2). Figure 2 shows three different cases where there exists a predictive relationship (a predicts b). One is rather obvious but different configurations of the time-delay between a and b and other noise events can lead to more difficult situations. To allow the extraction of these relationships, we will split our analysis in two different components. The first one is the relation from one event to all the others; the idea is to identify the most probable event that will occur shortly after another one (or shortly before if we look toward the past). The second component considers only pairs of events and its role is to measure the precision of the time-delay between these events.

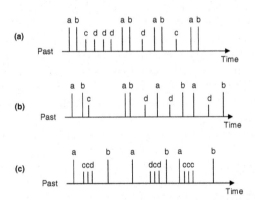

Fig. 2. Three different streams of events with $\mathcal{E} = \{a, b, c, d\}$. For each of them our aim is to identify the predictive relationship from a to b. Example **(a)** is quite obvious, events a and b follow each other very closely in time, and with a constant delay. Example **(b)** is more tricky as the delay between a and b varies, anyway it seems that b always follows a. In example **(c)** the delay between a and b is very large, providing room for many events to occur in between, nevertheless as this delay is constant we would like to identify such a relationship.

3   Information Theory and Anticipation

The goal of constructing an internal predictive model is to minimize the uncertainty of the predictions that the model will make. This construction can only be based on information acquired through experience, and therefore on a partial view of the environment, leading to probabilistic representations. Tools for dealing with such representations have been increasingly used in the context of sensorimotor coordination (for example Bayesian modeling in [6]), to analyze properties of the coupling between an agent and its environment (information-theoretic approach in [5]) and also to describe conditioning processes with information theory (see [3] and [4]). In our particular context, information theory is a very valuable tool because it is a natural framework to deal quantitatively with uncertainty.

3.1   Basis of Information Theory

Shannon's information theory is a mathematical framework that provides quantitative characterizations of probability distributions of events. We refer the reader to [2] for a complete introduction to the field. One of the main quantities we will be using is the entropy of a probability distribution. Consider a random variable X for which each event x can take a value in the set $\mathcal{X}$. The *entropy* of this random variable is defined as

$$H(X) = -\sum_{x \in \mathcal{X}} p(x) \log_2 p(x), \tag{1}$$

where $p(x)$ is the probability that event x occurs ($\sum_{x \in \mathcal{X}} p(x) = 1$ and $0 \leq p(x) \leq 1, \forall x \in \mathcal{X}$). This value reflects the uncertainty about the outcome of this random variable. The minimum is 0 for an absolutely predictable outcome (for example one outcome has a probability of 1) and the maximum is $\log_2(|\mathcal{X}|)$ if all outcomes are equiprobable.

The *information content* or self-information of one particular event x according to the given probability distribution is defined as

$$I(x) = -\log_2 p(x). \tag{2}$$

The minimum information content is 0 if this outcome has a probability of 1 and goes toward infinity as the probability approaches 0.

Our use of information theory in this model concerns the extraction of relationships between time-located events such as perceptions or actions. For understanding the tools described below, it is only necessary to keep in mind that high entropy H means high uncertainty, and high information content I means a low probability event (or surprising event).

3.2   Time-Delay Relationships

We will first focus on time-delay relationships between two events. For example, if an event b always occurs 50 timesteps after another event a, then we would like to identify this relationship. Also we would like the method to have some tolerance for variability, i.e. if b sometimes occurs 49 or 51 timesteps after a, we still consider that there exists a time-delay relationship between them.

For identifying these relationships, we will use information quantities. The principle used is based on the concept of *causal entropy* (see [8]) which in our case should be referred to as *predictive entropy*. The idea of predictive entropy is the following: let us consider that we want to identify a time-delay relationship between an event a and an event b always occurring after a. We will then use a random variable $D_{a,b}$ that represents the probability distribution of the observed time-delay for the next occurrence of b after a (i.e. the observed delay between an observation of a and the next subsequent observation of b). The entropy of this random variable $H(D_{a,b})$ reflects the strength of the relationship. The lower the entropy, the stronger the relationship. For example if b always occurs 50 timesteps after a, the entropy of $D_{a,b}$ will be 0 (only one event with a probability of 1, see Figure 3).

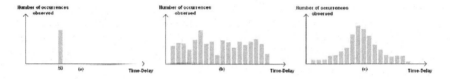

Fig. 3. Histograms of time-delay probability distribution, number of occurrences observed (vertical axis) for each possible time-delay (horizontal axis). **(a)** Histogram of an event b always occurring 50 timesteps after a, $H(D_{a,b}) = -\log_2(1) = 0$. **(b)** Example of a high entropy histogram. **(c)** Example of a low entropy histogram.

The original purpose of causal entropy is to determine whether there may be a relationship between two events from a to b or from b to a. This can be determined by comparing the entropies of $D_{a,b}$ and $D_{b,a}$. In our context, the goal is to identify relationships between many events. Therefore, we need a criterion for saying that there exists a time-delay relationship. In [3], the author states that the baseline from which the information provided by a conditional stimulus can be estimated is the prior estimate of the unconditional stimulus frequency. In our framework this can be translated as saying that the criterion for identifying a relationship from a to b is based on the self-relationship $D_{b,b}$, i.e. the distribution of observed time delays between two successive b events. We will therefore consider that there exists a relationship from a to b if a is a less uncertain predictor for b than b itself, i.e. if

$$H(D_{a,b}) < H(D_{b,b}). \tag{3}$$

Using causal entropy in our context leads to some problems that we need to solve. The first problem is that it is not robust at all to variability in time. If we consider for example two different conditions, in the first one, b occurred

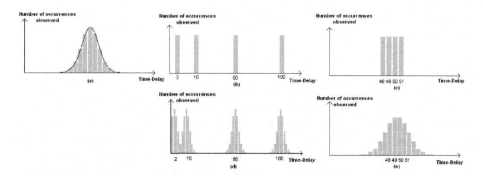

Fig. 4. Usefulness of adding Gaussian noise to time-delay events. **(a)** For one occurrence of a time-delay event, we add a discretized Gaussian distribution of realizations centered around the occurring event. **(b)** Example: histogram of the first condition without Gaussian noise, $H(D_{a,b}) = 2$. **(c)** Example: histogram of the second condition without Gaussian noise, $H(D_{a,b}) = 2$. **(d)** Example: histogram of the first condition with Gaussian noise, $H(D_{a,b})$ is high. **(e)** Example: histogram of the second condition with Gaussian noise, $H(D_{a,b})$ is low.

2,10,50 and 100 timesteps after a. In the second case, b occurred 48, 49 ,50 and 51 timesteps after a. For both conditions, $H(D_{a,b}) = 2$ (4 equiprobable outcomes, so $H(D_{a,b}) = \log_2(4) = 2$), therefore, we cannot identify which condition reflects a relationship. Obviously the second one seems to be a relationship where b occurs approximatively 50 timesteps after a, whereas the first condition doesn't seem to be a time-delay relationship.

To solve this problem, the idea is to introduce some variability in the probability distribution. Therefore rather than updating the statistics of $D_{a,b}$ by adding one realization of a given time delay t, we add a Gaussian distribution of time-delays centered around t, i.e. we add many realizations of t, then a bit less realizations of $t-1$ and $t+1$, even less for $t-2$ and $t+2$, and so on... Now if we get back to our example, adding Gaussian noise around the actual observed values of 48, 49, 50 and 51 will lead to overlapping Gaussians, and therefore to less variability than in the first condition, and consequently to a lower entropy (see Figure 4). For a given time-delay t, the number of realizations to add is computed for growing distances Δ_t as

$$\left\lfloor \left(\frac{\beta}{\sigma\sqrt{2\pi}} exp\left(- \frac{(\Delta_t - t)^2}{2\sigma^2} \right) \right) \right\rfloor \tag{4}$$

until this number reaches 0. The parameters β and σ of this function will be detailed in the Architecture section.

According to the quantity of information gained from using $D_{a,b}$ rather than $D_{b,b}$, we can compute a confidence value of the time-delay expectation as

$$\tau_{a,b} = \frac{H(D_{b,b}) - H(D_{a,b})}{H(D_{b,b})} \tag{5}$$

We can also compute the average expected time-delay between a and b as

$$\delta_{a,b} = \sum_{t \in D_{a,b}} p(t)t, \qquad (6)$$

where $p(t)$ is the observed probability of the time-delay t.

Another problem that has to be solved is the following. Let us suppose that after some time we have identified the time-delay relationship between a and b that has been described in the example above (Figure 4.e). Now if we consider that a new event c happened 10 timesteps before b, then the histogram of the random variable $D_{c,b}$ would be a perfect Gaussian centered on 10. The entropy of this random variable will be lower than the entropy of $D_{a,b}$ because of the small time variation between a and b. But obviously, if we had 4 realizations of b after a (48, 49, 50 and 51 timesteps), then we should be more confident in this relationship than for b after c, which had only 1 realization. Put another way, we should be more confident in a relationship that has occurred several times, even with some variation, than into a relationship that occurred only a few times, even with a perfectly constant time-delay. A way to solve this problem is to initialize any random variable $D_{a,b}$ with a uniform probability distribution of time-delays, e.g. an initial white noise. Then multiple realizations of a time-delay, even with some variability, will increase the probability of this time-delay and its neighborhood, and decrease the probability of the noise values, therefore the entropy of such a random variable will be lower than the entropy of a noisy random variable with only one realization of a time-delay.

3.3   Predictive Relationships Extracted from Contingency

Now we will focus on another type of relationship for which there is no precise delay between events a and b. We consider here relationships of the type "when a occurs, b is likely to occur soon". These relationships can be extracted from the contingency of events in the stream of perceptions. We will speak about them as *contingency relationships*, and we will consider that the closer b occurs after a, the stronger the relationship. Also we will consider that a predicts b if a *mainly* predicts b (relatively to predicting other events) and if b is *mainly* predicted by a (relatively to other events it is predicted by). The purpose of this criterion is the following: let consider an event a that happens all the time, and sometimes an event b, c or d happens. On one hand we can say that b, c and d are well predicted by a, because among all the possible predicting events, a is the most frequent. But on the other hand we cannot say that a usefully predicts b, c or d, because it predicts nearly everything (even itself), and therefore it is a useless predictor. That is why for establishing a predictive relationship from a to b, our criterion takes into account the future of a *and* the past of b.

We can translate these by the following principle: for each event e, we have two random variables, one is related to its past, i.e. it reflects the probability distribution of events that happened before e, we will refer to it as CP_e; and one is related to its future, i.e. the probability distribution of events that happened after e, we will refer to it as CF_e. In this context we will say that there is a relationship between a and b, such that b is a consequence of a if

$$I_{CF_a}(b) < H(CF_a) \tag{7}$$

and

$$I_{CP_b}(a) < H(CP_b). \tag{8}$$

This means that the information carried by b when occurring after a is less than the average information carried by an event that has occurred after a, thus b is more likely to occur after a than other events; and that a when occurring prior to b carries less information than the average information carried by an event in the past of b, i.e. a is more likely to have occurred before b than other events.

For each of these variables, event realizations are added according to their distance in time, i.e. when close in time, many realizations of the same event are added (for one actual occurrence), the number of realizations added decreasing with the distance. The exact number of realizations follows the same Gaussian equation 4, in which we replace t by 0, and Δ_t by the actual distance between the two events (negative values are discarded). Again we can define a confidence value of the contingency expectation, based on the loss of uncertainty, as

$$\kappa_{a,b} = \frac{1}{2}\left(\frac{H(CF_a) - I_{CF_a}(b)}{H(CF_a)} + \frac{H(CP_b) - I_{CP_b}(a)}{H(CP_b)}\right). \tag{9}$$

4 Architecture

The two information-theoretic tools described above are put together in an anticipation architecture. The main components of the architecture are shown in Figure 5. First saliency evaluation filters perceptive events, forwarding only the unusual events (those that carry most of the information). These perceptive events are used to update the internal model and their last observed occurrence is updated. The internal model and the last event occurrences are then used together to build expectations about forthcoming events.

4.1 Salience Filtering

We introduce a first mechanism that filters out some of the perceptions to avoid overloading the system with useless information. The precise criterion we use is that according to a distribution probability of perceptions E, which is constantly updated with new perceptions, we consider salient perceptions those that carry more information than the average information carried. Therefore the saliency criterion can be expressed as

$$I(e) > H(E) \tag{10}$$

where $e \in \mathcal{E}$.

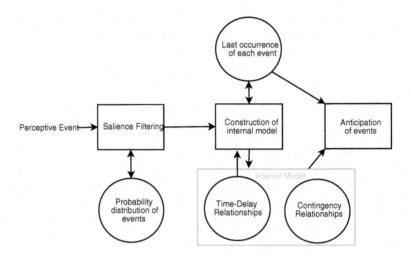

Fig. 5. Main architecture. Circles represent stored information, boxes are processes that generate information. See text for details.

4.2   Construction of the Internal Model

When an event is perceived, it is first stored into memory and replaces any previously stored occurrence of this event. The construction of the internal model is based on the two processes of finding time-delay and contingency relationships. When an event b is processed, for all events a that are in short-term memory, if it is the first occurrence of b since a occurred, we update the statistics of the random variables $D_{a,b}$, CF_a and CP_b. The parameters of the Gaussian used for updating the statistics are fixed for $D_{a,b}$ to β_0 and σ_0. For the two other random variables, these are adapted according to the event they concern, i.e. the longer the expected self time-delay between the concerned events, the more the Gaussian is flattened (hence the arrow from time-delay relationships to the construction of the internal model). The idea is to adapt to events that occur at very different timescales. Also the β parameter (the height of the Gaussian) is adapted according to the frequency of the added event, here the idea is to strengthen the association with rare events and to weaken associations with very common events. Therefore when adding an event b to the statistics of a, the parameters used are:

$$\sigma = \sigma_0(1 + \alpha\delta_{a,a}) \tag{11}$$

and

$$\beta = \beta_0(1 + \alpha\delta_{a,a} + \lambda\delta_{b,b}), \tag{12}$$

where α is the range adaptation coefficient and λ is the intensity adaptation coefficient (both low positive values). The higher α, the more the Gaussian is flattened for a given self time-delay. The higher λ, the more the added event is important for a given self time-delay.

4.3   Anticipation of Events

The constructed internal model, along with the memory of the last occurrences of events, can easily be used to determine the expected events using the following principles. For each past event a in memory, all the $D_{a,b}$ random variables are evaluated, and for each of them which validate the condition 3, the event b is added into the expectation list, along with its average time-delay $\delta_{a,b}$ and its confidence value $\tau_{a,b}$. Then for each possible event b, if we can find any event a in memory that is valid according to contingency conditions 7 and 8, then b is added to the expectations list, again with its average time-delay $\delta_{a,b}$ and its confidence value $\kappa_{a,b}$.

4.4   Forgetting Mechanism

We introduce a forgetting mechanism to allow for a quick replacement of relationships that are not relevant anymore. The principle of the forgetting mechanism is to define an upper bound to the total number of realizations of the random variables describing the internal model. When a new realization is added and increases the total number above the defined bound, one other realization is removed, by randomly choosing one of the events stored and removing one realization of this event.

5   Experiments

In this section we will evaluate the ability of the architecture described above to extract relevant predictive relationships from the stream of perceptions. The agent is not allowed to act, it can only passively perceive events coming from its environment. We first detail the experimental setup then we analyze the confidence value of relationships of interest.

5.1   Experimental Setup

Here we simulate some kind of Skinner box where the agent is situated. The perceptions of the agent are taken from the set $N1, N2, N3, N4, N5, N6, L1, Food$. The events from $N1$ to $N6$ are noise events that have no predictive value, whereas events $L1$ and $Food$ are causally associated, $L1$ predicting the $Food$ event ($L1$ stands for $Light$ 1, we consider than when the light is flashed, food will be given to the agent in a given delay). $L1 - Food$ sequence has a probability of 0.02 of being initiated at each timestep. The noise events are generated at each timestep with the respective probabilities ($N1 : 0.2, N2 : 0.1, N3 : 0.05, N4 : 0.025, N5 : 0.0125, N6 : 0.00625$). Other parameters of the simulation are the following. Gaussian parameters $\sigma_0 = 3$ and $\beta_0 = 100$. Range adaptation coefficient $\alpha = 0.25$. Intensity adaptation coefficient $\lambda = 0.1$. Random variables have an upper bound of 1000 realizations.

The first experiment measures the confidence values of the contingency and time-delay relationships after 10000 steps of simulation for different time-delay of the $L1 - Food$ association. The time-delays evaluated range from 1 to 80 timesteps with a variability of $+/-3$ timesteps.

For the second experiment we use the same procedure but the parameter investigated is the variability of the time-delay of the $L1 - Food$ association. The base time-delay used is 14 timesteps and with a variability ranging from $+/-0$ to 20 timesteps.

The third experiment aims at evaluating the dynamics of the internal predictive model over time. The $L1 - Food$ association has a time-delay of 14 timesteps and a variability of $+/-3$ timesteps. The experiment is running over 100000 timesteps, and during the range 40000 to 60000 $L1$ and $Food$ are not associated anymore, they are both presented at each timestep with the same probability of 0.01.

5.2   Results

Results of the first and second experiment are shown in Figure 6. We can see from these results that contingency relationships are successfully extracted for short time delays, less efficiently when the time delay increases, but they are robust to variability of this time delay. On the other hand, time-delay relationships have the opposite behavior, i.e. they are robust for long time delays, but they loose efficiency as the variability increases. These results confirm the expected behavior of these two anticipation mechanisms, which used together should allow the extraction of most relevant relationships.

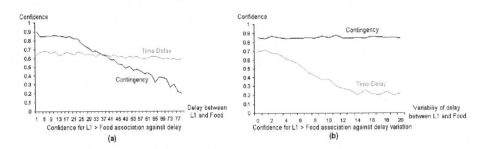

Fig. 6. Plotting of $\kappa_{L1,Food}$ (black) and $\tau_{L1,Food}$ (gray) after 10000 steps simulations. **(a)** Plotting against time delay between $L1$ and $Food$. Time-delay relationship is robust whereas contingency is not. **(b)** Plotting against variability of the time delay between $L1$ and $Food$. Contingency relationship is robust whereas time-delay relationship is not.

Results of the third experiment are shown in Figure 7. We can see that both relationships are quickly learned, correctly forgotten when the two stimuli are not associated anymore, and then their confidence value increases as soon as the events are paired again. These results show that the architecture correctly accounts for the forgetting mechanism. We can see that for a long enough time

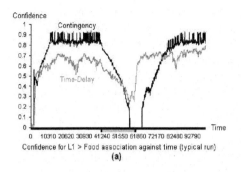
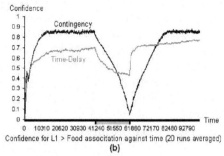

Fig. 7. Plotting of $\kappa_{L1,Food}$ (black) and $\tau_{L1,Food}$ (gray) against time during 100000 steps of simulation. In the range 40000 to 60000 $L1$ and $Food$ are not causally associated (shown in gray on the horizontal axis). **(a)** A typical run. **(b)** Average of 20 experiments.

of exposure to the unpaired events during a typical run, the agent can completely forget the contingency relationship. On the other hand, the time-delay relationship is maintained for a longer time and its original confidence value is recovered very quickly when the events are paired again, whereas the contingency relationship shows a slower recovery rate.

6   Possible Extensions

6.1   Actions and Rewards

Until that point we have only used the anticipation architecture in the context of an agent that can only passively observe the stream of perceptive events. But the point of such an architecture is to be used for action-selection. This aspect can be considered from two different perspectives: goal-oriented behavior and reinforcement learning. For both cases we will consider that actions are special perceptive events (e.g. proprioceptive events) that are generated when the agent performs the action. The particularity of these events is that they cannot be predicted by anything (from the agent's point of view) as they are dependent upon the will of the agent. By propagating these proprioceptive events into the architecture it becomes possible for the agent to extract predictive relationships between his actions and their effect in the environment.

In the case of goal-oriented behavior, we consider that the agent wants to reach a given goal whose definition is outside the scope of the architecture. For simplicity reasons, we can consider that the goal is a particular event. By chaining backward into the predictive model from this event toward possible actions, it is possible to identify which actions should lead to the occurrence of the goal event. It should also be in principle possible to plan more complex sequences of actions to reach intermediate events that will ultimately lead to the goal. If we now consider the case of action-selection based on reward expectation (as in reinforcement learning), the predictive model can be used in the opposite way. The idea would be to attach reinforcement values to particular events (such as the acquisition of food or an electric shock). When the agent must decide what action to perform, it is possible for him to estimate the effect of each possible action and moreover to compute an expected reinforcement value by chaining forward until reaching events with reinforcement. During the chaining, confidence values of the relationships can be used to estimate the probability of obtaining the reinforcement. The computed values can be used to select the action that will most probably lead to reinforcement. An advantage of this system over classical reinforcement learning is that it is possible to introduce a complex online modulation of reinforcement values (for example food events are rewarding only if the agent is hungry).

6.2   Handling Complex Predictive Relationships

One of the most difficult issues of anticipatory systems is to be able to identify complex phenomena involving many different events. An example of such a phenomena is that when an event a occurs, doing the action b will result in the event c occurring. One possible way to tackle this issue is to introduce sequence of events. The idea is to construct sequences of events that will be processed as normal events and that can therefore be used as predictors for other events. The problem here is to take care of the combinatorial explosion when grouping events. Therefore we need a criterion for creating new sequences, and also another one for discarding them when they have proved unsuccessful. The idea is to introduce a sequence generation probability p_{sg} that will be used each time an event b is processed to decide if a new sequence has to be created, another event a is then chosen randomly in the recent history and a new sequence a, b is registered. Subsequent occurrences of this sequence would then be recognized and the corresponding event generated and processed by the anticipation system. Using a sequence destruction probability p_{sd} evaluated at each time-step, a randomly chosen sequence may be destroyed if it has no predicting power, with a probability growing with the "age" of this sequence. Forwarding sequence events in the normal events' pathway allows for the construction of longer sequences by associating already existing sequences with other events.

Another case of complex relationship is when an event c predicted by a can be avoided if the action b is performed before c occurs. In this case we have to take into account the NON-occurrence of an expected event. The idea is that when an expected event did not happen after a sufficiently long time, a special event, opposite of the expected one, is generated and forwarded into the normal pathway. For example if an event a predicts an event c, and if after some time this event c still has not occurred, then we will generate an event $\bar{c}$ and forward it into the event processing pathway. This event can then be associated with another event b that caused this non-occurrence, or to the sequence of events a, b.

7   Conclusion

We have introduced two information theory based tools for extracting time-delay and contingency relationships in the stream of perceptions. These tools have been put together into an architecture that uses them for constructing an internal model of the environment. We have shown two distinct properties of contingency and time-delay relationships, the former is robust to variations of the delay between two stimuli, and the latter keeps its efficiency when the time-delay gets larger. An obvious advantage of these tools is that they allow simultaneous identification of relationships with completely different time scales without suffering from complexity increase. We have also shown the efficiency of this architecture for constructing a relevant internal model that is able to quickly adapt to a changing environment. Nevertheless some more extensive tests have to be carried out to evaluate the architecture in different conditions.

One of the advantages of this internal predictive model is that it can be used in two different ways. On the one hand it can be used to predict which events will occur and then perform appropriate actions to take advantage of this knowledge, such as avoiding a negative reinforcement. On the other hand it can also be used for goal-oriented behavior. In this case the goal would be a particular event (usually a positive reinforcement) the agent wants to obtain. Using the predictive model it can identify which events predict the goal and then chain back until it can find which actions can initiate the sequence of events leading to the goal. However this last part is a bit more complex as it involves not only predictive relationships between events, but also true causal relationships which are more difficult to identify. For example if we consider that an agent has learned that the sound of a bell predicts food delivery (by the experimenter), ringing the bell will not bring the food because the source of causality is upstream to both events and not from one to the other. Identification of causality requires the agent to actively inject information into the environment by acting upon it. In the example of the bell described above, if the agent can ring the bell by itself, then it would quickly realize that the bell and the food are not causally associated. Such a principle could also be used as a drive toward exploratory behavior. The idea would be that when a given predictive relationship has been identified, the agent could then try to more precisely evaluate this relationship by provoking the first event and then identify if the relation is causal or not.

One drawback of the architecture is that we use purely symbolic events, so no relation between them can be found apart from the predictive ones; it is impossible to define a notion of similarity between events and therefore impossible to generalize the predictive relationships. For this to be possible, events should not be only symbolic but they should possess a set of properties from which a notion of distance and subsets could be used.

Another problem is that the computational complexity of the model grows quickly with the number of different events that the agent can perceive. This was the reason for us to introduce a saliency filter so as to get rid of irrelevant events. Another possible way to avoid this problem and the previous one would be to map real events defined in a space of properties to a symbolic space by using categorization, i.e. grouping similar perceptions into one symbolic event, hence allowing for some generalization of relationships and also limiting the number of different events.

References

1. Butz, M.V., Sigaud, O., Gerard, P.: Internal models and anticipations in adaptive learning systems. In: Butz, M.V., Sigaud, O., Gérard, P. (eds.) Anticipatory Behavior in Adaptive Learning Systems. LNCS, vol. 2684, pp. 86–109. Springer, Heidelberg (2003)
2. Cover, T., Thomas, J.: Elements of Information Theory. John Wiley and Sons, New York (1991)
3. Gallistel, C.R.: Frequency, Contingency and the Information Processing Theory of Conditioning. In: Sedlmeier, P., Betsch, T. (eds.) Frequency Processing and Cognition, pp. 153–171. Oxford University Press, Oxford, UK (2002)
4. Gallistel, C.R.: Conditioning from an Information Processing Perspective. Behavioural Processes 61(3), 1234 1–13 (2003)
5. Klyubin, A.S., Polani, D., Nehaniv, C.L.: Tracking Information Flow through the Environment: Simple Cases of Stigmergy. In: Artificial Life IX. Proceedings of the 9th International Conference on the Simulation and Synthesis of Living Systems, pp. 563–568. The MIT Press, Cambridge (2004)
6. Kording, K.P., Wolpert, D.M.: Bayesian integration in sensorimotor learning. Nature 427, 244–247 (2004)
7. Tolman, E.C.: Principles of purposive behavior. In: Koch, S. (ed.) Psychology: A Study of Science, pp. 92–157. McGraw-Hill, New York (1959)
8. Waddel, J., Dzakpasu, R., Booth, V., Riley, B.T., Reasor, J.D., Poe, G.R., Zochowski, M.: Causal Entropies- A measure for determining changes in the temporal organization of neural systems. Journal of Neuroscience Methods (In Press) (2007)

The Interplay of Analogy-Making with Active Vision and Motor Control in Anticipatory Robots

Kiril Kiryazov[1], Georgi Petkov[1], Maurice Grinberg[1],
Boicho Kokinov[1], and Christian Balkenius[2]

[1] Central and East European Center for Cognitive Science, New Bulgarian University, 21 Montevideo Street, Sofia 1618, Bulgaria
[2] Cognitive Science, Lund University, Kungshuset, Lundagård SE-222 22 LUND, Sweden
kiryazov@cogs.nbu.bg, gpetkov@mail.nbu.bg, mgrinberg@nbu.bg,
bkokinov@nbu.bg, Christian.Balkenius@lucs.lu.se

Abstract. This chapter outlines an approach to building robots with anticipatory behavior based on analogies with past episodes. Anticipatory mechanisms are used to make predictions about the environment and to control selective attention and top-down perception. An integrated architecture is presented that perceives the environment, reasons about it, makes predictions and acts physically in this environment. The architecture is implemented in an AIBO robot. It successfully finds an object in a house-like environment. The AMBR model of analogy-making is used as a basis, but it is extended with new mechanisms for anticipation related to analogical transfer, for top down perception and selective attention. The bottom up visual processing is performed by the IKAROS system for brain modeling. The chapter describes the first experiments performed with the AIBO robot and demonstrates the usefulness of the analogy-based anticipation approach.

Keywords: Cognitive modeling, Anticipation, Analogy-making, Top-down Perception, Robots.

1 Introduction

Anticipation is an important function of human cognition – it makes human behavior more flexible than the behavior of other animals, it allows us to act before the incident has happened and thus to survive and actively change our environment knowing what the consequences will be. Obviously, anticipation is generated at various levels by various mechanisms. This chapter describes an attempt to demonstrate that analogy-making can be a useful mechanism for anticipation.

Analogy-making is thought to be a central mechanism in human cognition that underlies various other cognitive processes [1,2]. Thus it is natural to expect that anticipation could also be based on analogy. We believe, however, that the process is bidirectional: not only analogy plays an important role in anticipation, but also analogy-making benefits from human abilities to anticipate. That is why we explore the possibility to use the same basic mechanisms for both processes.

M.V. Butz et al. (Eds.): ABiALS 2006, LNAI 4520, pp. 233–253, 2007.
© Springer-Verlag Berlin Heidelberg 2007

A number of models of analogy-making have been proposed (see a review in [3]). One of the best known models is the Structure-Mapping Theory [4] and its implementation in SME [5] and MAC/FAC [6]. This model has introduced the structural focus of analogy-making, namely that analogy is about mapping of systems of relations. It is a kind of pipe-line model – it insists on the linearity of the processes and separates perception, mapping and retrieval in sequential steps. Although ACME, ARCS [7] and LISA [8] are connectionist types of models and thus are inherently parallel, they still rely on sequential processing and separate mapping from retrieval. None of these models is interested in how the representations are build – they fully ignore the perceptual processes and work on manually coded situations. On the other hand, COPY-CAT and TABLETOP [9, 10, 11] have focused on the process of perception and how it integrates with mapping. Moreover, the authors insist that perception cannot be separated from mapping [12]. These models are highly interactive and the processes of perception and mapping run in parallel and influence each other. Unfortunately, these models have no long-term memory and thus they do not explain how memory retrieval is performed and how it interacts with the rest of the processes. COPYCAT and TABLETOP have never been applied in a real world domain and their perceptual abilities are limited to "high-level perception". Thus, they cannot be directly applied in a real robot. However, the ideas behind COPYCAT and TABLETOP have been very instrumental in our research and in extending AMBR with perceptual abilities.

AMBR[1] [13, 14, 15, 16] is a model for analogy-making, based on the cognitive architecture DUAL [17]. AMBR models analogy as emergent phenomena – it emerges from the local interactions of a huge number of micro-agents. This model fully integrates retrieval and mapping processes and demonstrates how they interact and influence each other. The distributed representation of the episodes and the dynamic parallel nature of AMBR's mechanisms are a good starting point for modeling integration between perceptions and high-level cognitive processes. There are several main assumptions of AMBR. Firstly, context is not just a source of noise but is crucial for cognition. The context determines the relevant pieces of knowledge in long term memory (LTM) that can be used for solving the current problem. This context-sensitivity allows the system to be very flexible and at the same time very effective. Secondly, analogy, and particularly the ability for mapping is a central property for cognition. The cognitive system continuously maps the new information (coming from perceptions) with the old one (coming from memory) and adjusts both of them until they fit each other consistently. Thirdly, cognition is dynamic and all mechanisms run in parallel and interact with each other.

This chapter describes an extension of the AMBR model, building an integrated robot architecture that will try to cover all processes needed for real robot anticipation behavior: from visual perception through selective attention through high-level reasoning to actual physical movements in the environment. This is achieved by combining sparse bottom up visual perception performed by the IKAROS system with mapping this information to elements of LTM – concepts and old episodes – the retrieved portions of the episodes are immediately mapped onto the sparse target description and anticipatory relations are build that will be analogous to the ones in the old episode. These anticipations, about possible properties and relations, guide in a top-down

---

[1] AMBR is an acronym for Associative Memory-Based Reasoning.

fashion the further visual processing performed by the IKAROS system. Thus, the robot has an active vision system and builds only representations that are mapped to the known information (it does NOT build an extensive representation of ALL aspects of the environment). The extended representation further guides the retrieval and mapping process and finally all three together arrive at an analogy that projects back predictions which cannot be tested by the visual system but would require physical movement of the robot through the environment for actively exploring it.

A simple scenario where the robot anticipation abilities could be tested in reality involves an object hidden somewhere in the house and the task of the robot would be to find it. A robot without anticipation abilities would have to search exhaustively all possible places in all rooms. This will require probably several days or even weeks of search. Clearly, though, when people are confronted with a similar problem, they do not search blindly. They anticipate where the hidden object might be. In most cases we rely on our previous experience – we are spontaneously reminded about previous situations when we have searched for an object and try to transfer this experience to the new target situation. This is what our small dog-like AIBO will do – it will search for its bone hidden somewhere in the room. It is important to mention that this search is not based on a general regularity, such as that the object is always in the drawer, but actually the object can be at different places and still be analogous to one previous episode.

In this first step we have simplified the search by locating everything in one room, but this could be extended in the future. Unlike our previous attempts to build anticipatory mechanisms, this attempt is based on real robots acting in a real environment. This requires implementing all parts of the integrated architecture including visual perception, attention, reasoning, memory, and motor action. Thus even if it is a very simplified version and environment, it has all required capacities put together and working together and influencing each other.

2 Environment and Scenarios

We define our tasks and test the model in a house-like environment and in a "find-an-object" scenario.

The house-like environment consists of several rooms with doors between some of them. There are various objects like cubes, balls, and cylinders in the rooms. We used Sony AIBO robots (ERS-7). The goal of the robot is to find a bone (or bones) hidden behind an object. In a more complicated task there could be many robots: some of the robots should find and collect some 'treasures', whereas other robots play the role of 'guards' that try to keep the treasures and hide them dynamically or block the way of the treasure-hunters. Here, we start with the simplest case. We have one room, in which the robot itself and simple objects like cylinders and cubes with different colors (Figure 1) are located. The bone is hidden behind one of the objects. All objects (besides the hidden bone) are visible for the robot.

The AIBO robot has to predict where the bone is hidden based on analogies with past episodes and go for it. The episodes are manually built for the moment, but we plan to work on the learning process by which the newly perceived situations will remain in LTM.

Fig. 1. Simple scenario – "Where is my bone?"

In order to simplify the problems related to 3D vision we decided to have one camera attached on the ceiling having a global 2D view of the scene. There is a color marker on the top of the AIBO to facilitate its recognition. A web camera server sent the image data via TCP/IP to the network camera module of IKAROS. All software is installed on remote computers which communicate with the robot through wireless network.

3 Integrated Architecture

The integrated architecture of the system consists of several main modules (see Figure 2):

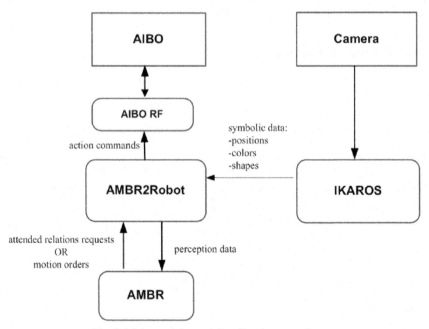

Fig. 2. Main modules and data flow between them

- AMBR – the core of the system – is responsible for attention and top-down perceptual processing, for reasoning by analogy, for decision making, and for sending a motor command to the robot controller.
- IKAROS module – a low-level perception module performing bottom up information processing.
- AMBR2Robot – a mediation module, the link between AMBR and IKAROS and the robot controller.
- AIBO robot.
- Camera attached to the ceiling.

A global camera takes visual information of the environment. It is received by the IKAROS module. The visual information is processed and symbolic information about objects in the environment is produced. This symbolic information is used from AMBR2Robot to provide AMBR with bottom-up perception information and also to handle the top-down requests which are described below. ABMR2Robot also waits for a "do-action" message from AMBR, which when received makes the module to control the robot and guide it to the target position using AIBO Remote framework. AMBR does the substantial job of making predictions about where the bone is hidden based on the representation of the current situation and making analogy with past situations. AIBO Remote Framework is a Windows PC application development environment which enables the communication with and control of AIBO robots via wireless LAN.

3.1 AMBR

The AMBR model for analogy-making is a multi-agent system, which combines symbolic and connectionist mechanisms [13, 14, 15]. Knowledge is represented by a large number of interconnected DUAL-agents. Each agent stands for an object, a relation, a property, a concept, a simple proposition, or a procedural piece of knowledge. The connectionist's activation of the agents represents their relevance to the current context. There are two special nodes that are the sources of activation – the INPUT and GOAL nodes – which are representations of the environment and the goals, respectively. Activation spreads from these two nodes to other nodes (typically instances of objects and relations from the target scene) then to their respective concepts and further up the concept hierarchy, then back to some of the concept instances and prototypes. There is no separation between the semantic and episodic memories – they are strongly interconnected.

The active part of the long-term memory forms the Working Memory of the model. Only active agents can perform symbolic operations like for example sending short messages to their neighbors, adjusting their weights, or creating new agents. This is a very important interaction between the symbolic and the connectionist parts of the model. The *speed* of the symbolic operations depends on the activation level of the respective agent. Thus, the most relevant (active) agents work faster, the less relevant – more slowly, and the irrelevant ones do not work at all.

Table 1. AMBR basic mechanisms

Spreading activation	The activation of the agents represents their relevance to the current context. It spreads just like in a neural network. The sources of the activation are two special nodes – INPUT and GOAL. The AMBR agents that represent the environment are attached to the INPUT, whereas the representation of the target is attached to the GOAL.
Marker emission and passing	Each *instance-agent* (representing a concrete token) emits a marker that spreads to the respective *concept-agent* (representing type) and then upward to the class hierarchy. When a marker from the target situation comes across a marker from a memorized situation, a *hypothesis-agent* between the two marker-origins is created. The hypothesis-agents always connect two agents and represent the inference that these two agents are analogical.
Structural correspondences	There are various mechanisms for structural correspondence that create new hypotheses on the basis of old ones. For example, if two relations are analogical, their respective arguments should also be analogical; if two instance-agents are analogical, their respective concepts should also be analogical, etc.
Constraint satisfaction network	The consistent hypotheses support each other, whereas the inconsistent ones compete with each other. Thus, dynamically, a constraint satisfaction network of interconnected hypotheses emerges. After its relaxation, a set of *winner-hypotheses*, which represent the performed analogy, is formed.

Table 2. Main types of AMBR agents

Instance-agent	Represents tokens, i.e., particular exemplars. The instance-agents can represent objects, as well as aspects and relations.	*Examples:* bone-1, red-21, behind-3…
Concept-agent	Represents types, i.e., classes of similar exemplars. Again, can represent objects or relations.	*Examples:* bone, color, behind…
Hypothesis-agent	Always connects two elements – one from the target situation and one from a memorized one. Represents an inference that there is something in common between the two elements – they have common super-class or they are the respective arguments of corresponding relations.	*Examples:* bone1<-->bone-3, left-of<-->right-of, red-12<->green-8…
Winner-hypothesis	Represents an already established analogical correspondence between two elements. The hypothesis-agents become winners or fizzle out.	The same form as the hypothesis-agents

Tables 1 and 2 summarize the main mechanisms and agent-types used in AMBR and describe the role and the routine of each of them. It is important to note, however, that all these mechanisms run in parallel and influence each other.

3.2 Extensions of AMBR for the AIBO Robot Application

In order to make AMBR a model of the mind of a real robot several new mechanisms were developed and implemented. Most importantly, several analogical transfer mechanisms have been developed which will allow robust predictions based on analogy. The present development is related to the extension of the model with a dynamic top-down perceptual mechanism, a mechanism for control of attention, mechanisms

Table 3. New ABMR mechanisms

Perceptual Anticipation (top-down influence on perception)	By a series of messages, the instance-agents from memorized situations inform the relevant relations in which they participate for all their hypotheses. If a certain relation collects the hypotheses for all its arguments, it creates an *anticipation-agent*. The anticipation-agents are copies of their mentor-relations but all their arguments are replaced with the respective analogical elements from the target situation.
Attention	The attention mechanism monitors all anticipation-agents, sorts them by their activation (i.e., relevance), and at fixed time intervals asks the perceptual system to check the relation represented by the most active one.
Goal-related Anticipation (Transfer of the solution)	When a certain hypothesis transforms itself into a winner-hypothesis, it informs its base element. The latter, in turn, informs the relations, in which it participates. The respective relations erase all anticipations and hypotheses that are inconsistent with the new winner. Thus, in reality, the anticipation mechanism creates many different possible solutions of the problem that compete with each other, whereas the transfer mechanism works by deleting most of them on the basis of the best analogy. As a final result of the transfer mechanism only the solution that is most consistent with the performed analogy remains.
Action	The *cause-agents* (representing causal relations) are equipped with a special routine. Via special messages, the agents, attached to the GOAL node inform the cause-relations, in which they participate, that the latter are close to the goal. After a period of time, if such 'close-to-goal' cause-agent receives information that it participates in a winner-hypothesis, it checks its antecedents for *action-agents* (representing description of a certain action or movement). If all these conditions are met, the action mechanism sends an order for executing the respective action.

for transferring parts from a past episode in memory towards the now perceived episode (analogical transfer), and mechanisms for planning and ordering actions based on that transfer. All new mechanisms are summarized in Table 3 while the new AMBR agent's types are given in Table 4. Note that all these mechanisms overlap in time and influence each other. It should be stressed that there is no central executive in AMBR. Instead, the AMBR agents interact only with their neighbors and perform all operations locally, with a speed, proportional to their relevance to the current context.

Table 4. Specialized AMBR agents

Anticipation-agent	Represents expectation that a certain relation is present in the environment.	*Examples:* ?red-cube-12?, ?behind-bone-cylinder-12?…
Cause-agent	Represents a certain casual relation. It always has antecedents and consequences. One cause-agent can be instance-agent or anticipation-agent.	*Example:* Cause1 -antecedents: move-12, behind-2 -consequences: find-8
Action-agent	Represents the description of a certain action or movement. The presence of an action-agent in the target situation does not mean that it will be executed. In order for AIBO to execute the respective action, a special procedure for this should be triggered.	*Examples:* Move (AIBO, cylinder-12),

More detailed description of the new AMBR mechanisms follows below.

Top-Down Perception
It is known that when executing a task in order to achieve a specific goal, top-down mechanisms are predominant [21, 22]. This finding is implemented by making AMBR the initiator of information acquisition actions.

At first, the robot looks at a scene. In order for the model to 'perceive' the scene or parts of it, the scene must be represented as an episode, composed out of several agents standing for objects or relations, attached to the input or goal nodes of the architecture. It is assumed that the construction of such a representation starts by an initial very poor representation (Figure 3) built by the bottom up processes. This includes, usually, only symbolic representations of the objects from the scene without any description of their properties and relations. These are attached to the input of the model (in the example, object-1, object-2, and object-3).

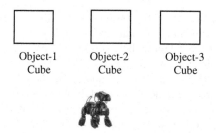

Object-1   Object-2   Object-3
Cube   Cube   Cube

Fig. 3. Initial representation of the scene (bottom-up perception)

The representation of the goal is attached to the goal node (usually `find-t`, `AIBO-t`, and `bone-t`). During the run of the system some initial correspondence hypotheses between the input (target) elements and some elements of the memory episodes (bases) emerge via the mechanisms of analogical mapping (Figure 4).

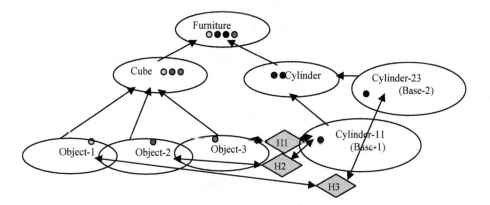

Fig. 4. Creation of hypotheses (H1, H2, H3) on the basis of marker intersections

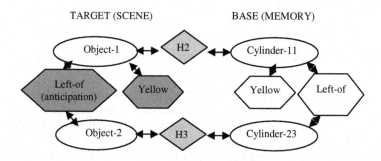

Fig. 5. Formation of anticipation agents in AMBR on the basis of missing in the scene arguments of already mapped relations. (Note that 'left-of' relation is asymmetric and the order of arguments is coded in AMBR although it is not shown in the picture).

The connected elements from the bases activate the relations in which they are participating. The implemented *dynamic perceptual mechanism* creates predictions about the existence of such relations between the corresponding objects in the scene. As shown in the example of Figure 5, Object-1 from the scene representation has been mapped onto Cylinder-11 in a certain old and remembered situation. The activation mechanism adds to working memory some additional knowledge about Cylinder-11 – e.g. that it is yellow and is positioned to the left of Cylinder-23, etc. (Figure 5) The same relations become anticipated in the scene situation, i.e. the system anticipates that Object -1 is possibly also yellow and could be on the left of the element, which corresponds to Cylinder-23 (if any), etc. Thus, various anticipation-agents emerge during the run of the system.

Attention

The attention mechanism deals with the anticipations generated by the dynamic perceptual mechanism, described above. With a pre-specified frequency, the attention mechanism chooses the most active anticipation-agents and asks the perceptual system to check whether the anticipation is correct (e.g. corresponds to an actual relation between the objects in the real scene). AMBR2Robot, as described earlier, simulates the perceptions of AMBR based on input from a real environment (using IKAROS). It receives requests from AMBR and simply returns an answer based on the available symbolic information from the scene.

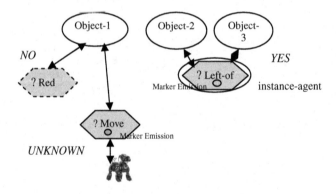

Fig. 6. Processing of the different types of answers of relation questions

The possible answers are three: 'Yes', 'No', or 'Unknown'. In addition to colors ('color-of' relations), spatial relations, positions, etc., it also generates anticipations like "the bone is behind 'object-1' ", or "if I move to 'object-3', I will find the bone". Those relations play a very important role for the next mechanism – the *transfer of the solution* (i.e. making a firm prediction on which an action will be based) – as explained below.

After receiving the answers, AMBR manipulates the respective agent (see Figure 6). If the answer is 'Yes' it transforms the anticipation-agent into an instance-agent. Thus the representation of the scene is successfully augmented with a new element, for which the system tries to establish correspondences with elements from old episodes in

memory. If the answer is 'No', AMBR removes the respective anticipation-agent together with some additional anticipation-agents connected to it. Finally, if the answer
is 'Unknown', the respective agent remains an anticipation-agent but emits a marker
and behaves just like a real instance, waiting to be rejected or accepted in the future. In
other words, the system behaves in the same way as if the respective prediction is correct. However, the perceptual system or the transfer mechanism (see below) can dis-
card this prediction.

Transfer of the Solution
Thus, the representation of the scene emerges dynamically, based on top-down processes of analogical mapping and associative retrieval and of the representation in
AMBR2Robot and its functioning. The system creates many hypotheses for correspondence that self-organize in a constraint-satisfaction network (see Figure 7)

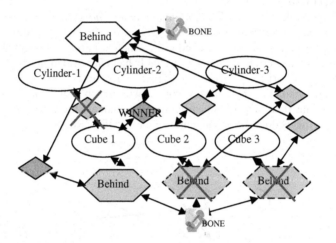

Fig. 7. Constraint satisfaction network between hypotheses. Winner hypotheses remove many
of the inconsistent anticipations until only few anticipation-agents remain.

Some hypotheses become winners as a result of the relaxation of that network and
at that moment the next mechanism, the *transfer of the solution*, does its job. In fact,
the transfer mechanism does not create the agents, which represent the solution. The
perceptual mechanism has already transferred many possible relations but now the
task is to remove most of them and to choose the best solution. As in the previous
examples, let's take a target situation consisting of three cubes and let the task of
AIBO be to find the bone. Because of various mappings with different past situations
the anticipation mechanism would create many anticipation-agents with a form similar to: "The bone is behind the left cube". This is because in a past
situation (sit-1 for example) the bone was behind the left cylinder and now the left
cylinder and the left cube are analogical. Because of the analogy with another situation, for example, the anticipation that "the bone is behind the middle
cube" could be independently created. Another reason might be generated due to
which the right cube will be considered as the potential location of the bone. Thus

many concurrent possible anticipation-agents co-exist. When some hypotheses win, it is time to disentangle the situation.

The winner-hypotheses care to propagate their winning status to the consistent hypotheses. In addition, the inconsistent ones are removed. In the example above, suppose that sit-1 happens to be the best candidate for analogy. Thus, the hypothesis 'left-cylinder<-->left-cube' would become a winner. The relation 'behind' from the sit-1 would receive this information and would care to remove the anticipations that the bone can be behind the middle or behind the right cylinder.

As a final result of the transfer mechanism, some very complex causal anticipation-relations like "if I move to the cube-1 this will cause finding the bone" become connected with the respective cause-relations in the episodes (bases) from memory via winner-hypotheses.

Action Execution

The final mechanism is *sending an action command* (see Figure 8). The cause-relations that are close to the GOAL node trigger it. The GOAL node sends a special message to the agents that are attached to it, which is in turn propagated to all cause-relations. Thus, at a certain moment, the established cause-relation "if I move to cube-1, this will cause finding the bone" will receive such a message and when one of its hypotheses wins, it will search in its antecedents for an action-agents. The final step of the program is to request the respective action to be executed and this is done again via a message to AMBR2Robot.

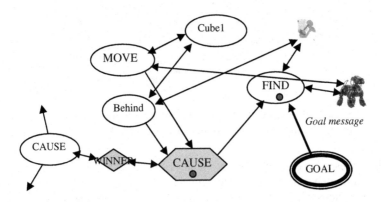

Fig. 8. Hypotheses of the cause-relations receive markers from the GOAL node. If the consequents satisfy the goal, then the actions from the conditions are executed.

3.3 Connecting ABMR with the Robot

In order to connect AMBR with the real world several new modules were developed. A major step was to build a perceptive mechanism with active vision elements based on the platform IKAROS [18, 19]. Several modules of IKAROS related to perception were successfully integrated in order to carry out the difficult task of bottom-up visual perception and object recognition. This results in a hybrid system where IKAROS performs the non-symbolic processes best suited for perceptual processing, while

AMBR performs the high level symbolic and connectionist operations best suited for analogy making Another module – AMBR2Robot – was developed as a general mediating layer between the perception modules of IKAROS and ABMR. AMBR2Robot supports the selective attention mechanisms, which were described above.

3.4 IKAROS

IKAROS is a platform-independent framework for building system-level cognitive and neural models [18, 19] (see also www.ikaros-project.org). The system allows systems of cognitive modules to be built. The individual modules may correspond to various cognitive processes including visual and auditory attention and perception, learning and memory, or motor control. The system also contains modules that support different types of hardware such as robots and video cameras. The modules to be used and their connectivity are specified in XML files that allow complex cognitive systems to be built by the individual modules in IKAROS. Currently, there are more than 100 different modules in IKAROS that can be used as building blocks for different models.

In the present work, IKAROS was used for visual perception and object recognition. An IKAROS module receives images from a camera while another module segments the image into different objects based on color. The result is sent to AMBR2Robot for further processing.

The object recognition proceeds in several stages. In the first stage, the color image is mapped onto the RG-chromaticity plane to remove effects of illumination and shadows. In parallel, the edges are extracted in the image. These edges are used as preliminary contours of the objects. In the second processing stage, the colors are normalized in-between the edges in x and y-direction to produce homogenous color regions (Figure 9). In the third stage, the individual color pixels are clustered into regions with similar color. At this stage, a template for each color is used to form only clusters for colors that are known to belong to the target objects. Each color is defined as a circle sector around the white-point in the RG-chromaticity plane. In the fourth stage, a rectangular region is formed around each cluster, which is used to delimit each object (Figure 10). Finally, a histogram of the edge orientations within each object regions is calculated which is then used to categorize the shape of the object.

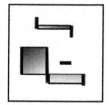 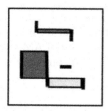 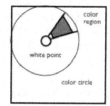

Fig. 9. Left. An initial image with found vertical edges. **Middle.** Colors after normalization between edges. **Right.** A color region in the RG-chromaticity plane.

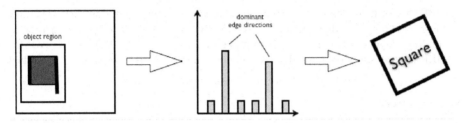

Fig. 10. The recognition of shapes. **Left**. A region is formed around the color cluster. **Middle**. A histogram of edge orientations is calculated. **Right**. The distribution of edge orientations is used to determine the shape.

Note that we are not trying to find the complete contours of the objects. Instead, the method is based on the distribution of different edge orientations which is a much more robust measure. The different processing stages were inspired by early visual processing in the brain but adapted to efficient algorithms. The mapping to the RG-chromaticity plane discards the illuminant and serves the same role as the interaction between the cones in the retina [23]. The detection of edges is a well known function of visual area V1 [24]. The color normalization within edge elements was inspired by theories about brightness perception [25] and filling-in [26].

3.5 AMBR2Robot

AMBR2Robot mediates between AMBR, IKAROS and the robot. It provides AMBR with perceptual information from IKAROS and also serves for implementing the selective attention mechanism in the model. The other main purpose of this module is receiving the action tasks from AMBR and executing them using AIBO-RF. We could say that it simulates the link between the mind (AMBR) and the body (perception system, action system).

The work of the module AMBR2Robot formally can be divided into three sub-processes:

1. Bottom-up perception
2. Top-down perception
3. Performing actions

Bottom-Up Perception
At this stage just a small part of the scene-representation is sent to AMBR. As described above information is further transformed into the form used for knowledge representation in AMBR by creating a set of AMBR agents with appropriate slots and links and connecting them to the so-called input of the architecture.

Top-Down Perception
As mentioned above AMBR sends top-down requests in the form of questions about the presence of properties and relations about the identified objects. These requests are received by AMBR2Robot and are answered based on visual symbolic information

provided by IKAROS. Relations represent the most important information for analogy-making and are extracted by AMBR2Robot from the scene description which does not contain them explicitly but only implicitly (e.g. in the form of coordinates and not spatial relations).

The main types of top-down perception requests are for:

- spatial relations: right-of, in-front-of, in-front-right-of, etc…
- sameness relations: same-color, unique-shape, etc…
- color properties: orange, blue, red, etc…

The spatial relations are checked based on the objects' positions as described by their coordinates and size and with respect to the gaze direction of the robot. Figure 11 shows how the above example relation request (left-of object-2 object-3) is processed. Positions of all objects are transformed in polar coordinates respective to a robot-centric coordinate system. Then some comparison rules are applied to the coordinates.

For processing the sameness relation the relevant properties (shape or color) of all the visible objects are compared.

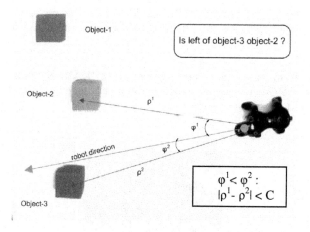

Fig. 11. Processing the spatial relation requests

Action

AMBR2Robot receives action commands from AMBR and, knowing the positions of the target object and the robot, navigates AIBO by sending movement commands via the AIBO Remote Framework (see Figure 2). During the executed motion, IKAROS is used to update the robot's position in the scene (the other objects in the scene are assumed to have fixed positions) and only the robot is actually being tracked.

The robot is guided directly to the target object without any object avoidance (to be implemented in the future in more sophisticated examples). After the robot has taken the requested position, it is turned in the appropriate direction to push and uncover the target object. At the end it takes the bone if it is there, otherwise it stops.

4 Results

In this chapter the results from a single run of the system are described.

There are two past situations in the robot's memory (Figure 12.a) The robot faces the situation showed in Fig 12.b

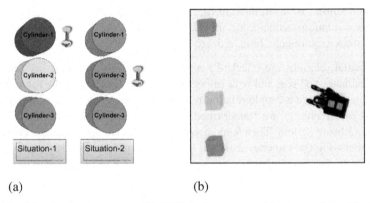

(a)   (b)

Fig. 12. (a) Old episodes in memory **(b)**AIBO is in a room with three cubes with different colors

The image (from the global camera) is sent to the IKAROS module. In Fig 14 the visual field after RG-chromaticity transformation and the edge histogram for the one of the recognized objects are shown.

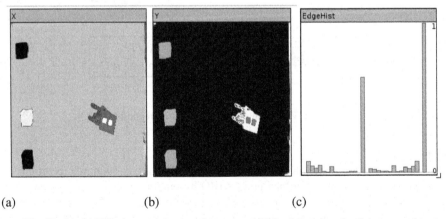

(a)   (b)   (c)

Fig. 13. (a),(b) RG-chromaticity transformation **(c)**Edge histogram for the upper cube

The IKAROS module recognizes the objects and produces the input:

```
object-1, shape=cube, position= (50,85)
object-2, shape=cube, position=(71,303)
object-3, shape=cube, position=(75,429)
aibo-I, position=(438,301), direction=3.387
```

Processing that information AMBR2Robot sends part of it as bottom-up perceptual information to AMBR:

```
object-1: cube, object-2: cube, object-3: cube
```

In the top-down perception part, lots of relation requests are generated from AMBR. Here we show some of them, including the answers from AMBR2Robot:

```
? behind: bone-t -> UNKNOWN
? left-of: object-1 object-2 -> YES
? blue: object-3 -> YES
? green: object-2 -> NO
? in-front-right-of: object-3 aibo-I -> NO
? move: aibo-I object-2 -> UNKNOWN
```

Some of the already created anticipation agents are turned into instance agents according to the answer:

```
anticip-blue-situation-1
anticip-left-of-1-situation-2
```

The name of the agents is formed as follows: 'anticip' stands for anticipatory. After that follows the name of the relation which it "anticipates". This relation can belong to one of the situations in robots memory (situation-1 or situation-2 in this case) Note that after transforming an anticipation agent into instance one its name remains the same.

After some time AMBR make an analogy with situation-1. Some of the winner-hypotheses (in the order they emerge) are:

```
object-3 ←→ right-cylinder-situation1
anticip-blue-situation-1 ←→blue-situation-1
aibo-I ←→ aibo-I-situation-1
anticip-move-situation-2 ←→move-situation-1
object-2 ←→ midle-cylinder-situation1
object-1 ←→ left-cylinder-situation1
```

Many other agents are mapped. After the mapping of the cause agent, the action mechanism is triggered, which sends a motion command to AMBR2Robot.

```
move-to object-3
```

AMBR2Robot guides the robot to the target. Once arrived, the robot uncovers the object and tries to get the bone. Figure 15 shows some images of the robot moving.

The model is tested in some other situations where other analogies are made. So far, we did not conduct any comparative experiments with other models because it is difficult to determine when an analogy "is right" in this scenario. But it is obvious that if the analogy is right, the performance will be much higher than a full-search method - checking all objects one by one.

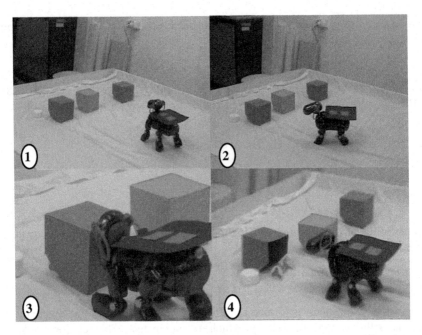

Fig. 14. 1. AIBO is starting to move. 2. approaching the target. 3. uncover object. 4. get the bone [2].

5 Summary

In this paper, a new approach of robot anticipatory behavior is presented which is based on predictions generated from an analogy with past experiences stored as episodes in memory. Anticipation is also used in a top-down perception mechanism, which is using predictions about relations between perceived objects based on hypotheses of correspondence with relevant objects from the past experience of the robot.

The AMBR model of analogy-making has been used as a system reasoning core and has been further extended. The visual information is handled by modules from the IKAROS system and transformed into a symbolic one. AMBR2Robots mediate between the high-level reasoning part of the system and the modules for perception and action. In this paper, we presented the experience of implementing a combination of low-level perceptual information and higher-level reasoning in a real robot. The results seem very promising and would allow us to take advantage of the strong side of both approaches, building a genuinely hybrid robot architecture able to deal with complex real life tasks

The architecture was successfully tested in a house-like environment with the tasks of finding a hidden object.

---

[2] This pictures are taken from demonstration at the second year review meeting of the MindRACES project.

6 Future Work

Our plans for the near future are to further develop the bottom-up perceptual mechanisms, improve the attention mechanism, and extend the model with various types of different actions.

We plan future integration with IKAROS system at the level of bottom-up perception system. IKAROS is able to create various salience maps on the basis of the perceptual input and the goals of the system. AMBR can use these salience maps in order to activate various concepts to a different degree. This contextual information will influence the retrieval processes to a higher degree.

The attention mechanism should be connected with the robot camera and particularly, with its gaze. Thus, both the salience maps and the top-down reasoning will influence the head-movement of the robot, and in turn, the order of checking the various predictions.

Finally, the repertoire of possible actions that AIBO can do should be expanded. In the current version of the model, AIBO is able only to move from its initial position to one predefined object. However, it is not yet implemented how it will do this if there are obstacles. We plan to implement A-star algorithms in the robot planning system to avoid obstacles and to move between rooms.

Finally, our long-term plans are to develop also learning and emotional mechanisms, as well as testing AIBO's behavior in more complex situations that include social interaction. For instance emotions can control some global parameters of the model like the capacity of working memory. The latter could lead to more superficial and less consistent analogies but which are done much faster.

We also plan to develop various mechanisms for 'task understanding', e.g. breaking up an abstract goal into several smaller sub-goals on the basis of analogy with past situations.

In the current implementation all memorized situations are manually predefined and stay static. Learning at different levels must be implemented to account for the experiences of the robot and allow the change of old episodes and the addition of new episodes in LTM.

Modeling social interactions including at least two robots is also very important to us. For instance one of them will hide the bone, whereas the other one will seek it. Both robots will have various anticipations (and possibly meta-anticipations) about the behavior of the other one which will give rise to interesting collective behavior.

Acknowledgments. This work is supported by the Project Mind RACES: from Reactive to Anticipatory Cognitive Embodied Systems (Contract No 511931), financed by the FP6. We would like to thank Marina Hristova for proofreading the text.

References

1. Hofstadter, D.: Analogy as the Core of Cognition. In: Gentner, D., Holyoak, K., Kokinov, B. (eds.) The Analogical Mind: Perspectives from Cognitive Science, MIT Press, Cambridge, MA (2001)
2. Holyoak, K., Gentner, D., Kokinov, B.: The Place of Analogy in Cognition. In: Gentner, D., Holyoak, K., Kokinov, B. (eds.) The Analogical Mind: Perspectives from Cognitive Science, MIT Press, Cambridge, MA (2001)

3. Kokinov, B., French, R.: Computational Models of Analogy-Making. In: Lynn Nadel (ed.) Encyclopedia of Cognitive Science. London: Macmillan, Nature Publishing Group, pp. 113–118 (2002)

4. Gentner, D.: Structure-mapping: A theoretical framework for analogy. Cognitive Science 7, 155–170 (1983)

5. Falkenhainer, B., Forbus, K., Gentner, D.: The structure mapping engine: Algorithm and examples. Artificial Intelligence 41(1), 1–63 (1989)

6. Forbus, K., Gentner, D., Law, K.: MAC/FAC: A model of similarity-based retrieval. Cognitive Science 19(2), 141–205 (1995)

7. Holyoak, K., Thagard, P.: Analogical mapping by constraint satisfaction. Cognitive Science 13, 295–355 (1989)

8. Hummel, J., Holyoak, K.: Distributed representation of structure: A theory of analogical access and mapping. Psychological Review 104, 427–466 (1997)

9. Hofstadter, D. and The Fluid Analogies Research Group: Fluid concepts and creative analogies. Basic Books, New York (1995)

10. Mitchell, M.: Analogy-making as Perception: A computer model. MIT Press, Cambridge, MA (1993)

11. French, R.: The Subtlety of Sameness: A theory and computer model of analogy making. MIT Press, Cambridge, MA (1995)

12. Charlmers, D., French, R., Hofstadter, D.: High- Level Perception, Representation, and Analogy: A Critique of Artificial Intelligence Methodology. Journal of Experimental and Theoretical AI 4(3), 185–211 (1992)

13. Kokinov, B.: A hybrid model of reasoning by analogy. In: Holyoak, K., Barnden, J. (eds.) Advances in connectionist and neural computation theory. Analogical connections, vol. 2, pp. 247–318. Ablex, Norwood, NJ (1994a)

14. Kokinov, B., Petrov, A.: Dynamic Extension of Episode Representation in Analogy-Making in AMBR. In: Proceedings of the 22nd Annual Conference of the Cognitive Science Society. Erlbaum, Hillsdale, NJ (2000)

15. Kokinov, B., Petrov, A.: Integration of Memory and Reasoning in Analogy-Making: The AMBR Model. In: Gentner, D., Holyoak, K., Kokinov, B. (eds.) The Analogical Mind: Perspectives from Cognitive Science, MIT Press, Cambridge, MA (2001)

16. Petrov, A., Kokinov, B.: Processing symbols at variable speed in DUAL: Connectionist activation as power supply. In: Proceedings of the Sixteenth International Joint Conference on Artificial Intelligence, (IJCAI-99), pp. 846–851. Morgan Kaufman, San Francisco, CA (1999)

17. Kokinov, B.: The DUAL cognitive architecture: A hybrid multi-agent approach. In: Proceedings of the Eleventh European Conference of Artificial Intelligence, (ECAI-94), John Wiley & Sons, London (1994b)

18. Balkenius, C., Moren, J.: From Isolated Components to Cognitive Systems. ERCIM News, No. 53 (2003)

19. Balkenius, C., Moren, J., Johansson, B.: System-Level Cognitive Modeling with Ikaros. Lund University Cognitive Studies, 133 (2007)

20. Petkov, G., Naydenov, C., Grinberg, M., Kokinov, B.: Building Robots with Analogy-Based Anticipation. In: Proceedings of the KI 2006, 29th German Conference on Artificial Intelligence, Bremen, (in press) (2006)

21. Duncan, J.: Selective attention and the organization of visual information. Journal of Experimental Psychology: General 113, 501–517 (1984)

22. Chalmers, D., French, R., Hofstadter, D.: High-level perception, representation, and analogy: A critique of artificial intelligence methodology. Journal for Experimental and Theoretical Artificial Intelligence 4, 185–211 (1992)
23. Dacey, D.M.: Circuitry for Color Coding in the Primate Retina. In: Proceedings of the National Academy of Sciences, 93, 582–588 (1996)
24. Hubel, D.H., Wiesel, T.N.: Receptive fields and functional architecture of monkey striate cortex. J Physiol 195(1), 215–243 (1968)
25. Paradiso, M.A., Nakayama, K.: Brightness perception and filling-in. Vision Research 31, 1221–1236 (1991)
26. Grossberg, S.: The complementary brain: unifying brain dynamics and modularity. Trends in Cognitive Sciences 4, 233–246 (2000)

An Intrinsic Neuromodulation Model for Realizing Anticipatory Behavior in Reaching Movement Under Unexperienced Force Fields

Toshiyuki Kondo[1] and Koji Ito[2]

[1] Department of Computer, Information and Communication Sciences,
Tokyo University of Agriculture and Technology,
2-24-16 Naka-cho, Koganei, Tokyo 184-8588, Japan
tkondo@ieee.org
http://www.livingsys.lab.tuat.ac.jp/
[2] Department of Computational Intelligence and Systems Science,
Tokyo Institute of Technology,
4259 Nagatsuta, Midori, Yokohama, 226-8502, Japan
ito@dis.titech.ac.jp
http://www.ito.dis.titech.ac.jp/

Abstract. Regardless of complex, unknown, and dynamically-changing environments, living creatures can recognize situated environments and behave adaptively in real-time. However, it is impossible to prepare optimal motion trajectories with respect to every possible situations in advance. The key concept for realizing the environment cognition and motor adaptation is a context-based elicitation of constraints which are canalizing well-suited sensorimotor coordination. For this aim, in this study, we propose a polymorphic neural networks model called CTRNN+NM (CTRNN with neuromodulatory bias). The proposed model is applied to two dimensional arm-reaching movement control under various viscous force fields. The parameters of the networks are optimized using genetic algorithms. Simulation results indicate that the proposed model inherits high robustness even though it is situated in unexperienced environments.

1 Introduction

Living creatures are information structuring systems which have enormous sensorimotor degrees of freedom. External environments can be recognized based on spatiotemporal integration of their sensorimotor information (e.g. vision, tactile, somatosensory stimulus). Given a task goal in addition to the environment cognition, smooth limb movements are immediately planned and executed in spite of huge DOF of our musculoskeletal systems [2]. However the detailed mechanisms of cognition and motor adaptation are still open questions [6,16].

Thus, several computational models for cognition and motor adaptation have been proposed. Most of them are based on *internal model* theory in which an adequate inverse model (i.e. sensorimotor mapping or controller) would be selected according to a prediction derived from forward models (e.g. [22]). In these

M.V. Butz et al. (Eds.): ABiALS 2006, LNAI 4520, pp. 254–266, 2007.

localist models, a novel pattern can be incrementally learned by allocating an additional module. But owing to this, they have less ability for dealing with unknown environments without the additional learning process.

In contrast, recently much attention has been focused on *dynamical systems* (or *distributed*) approach to the cognition and motor adaptation problem [17,21,1,20,15]. In this approach, the state prediction and motor generation are represented in accordance with the concept of attractors in dynamical systems theory.

On the other hand, recent neurophysiology has revealed that those environment cognitions and real-time adaptations can be observed in real nervous systems of insects and crustaceans. It is known that a variety of chemical substances called neuromodulators (NMs) play crucial roles to regulate the dynamic characteristics of the neural networks (e.g. activating/blocking/changing of synaptic connections) [5,14,7,13].

Even in the case of higher level animals, the ability of environment cognition and motor adaptation should be influenced by internal/external *hormones*. It is widely known that these intrinsic chemical conditions highly affect the anticipatory behaviors of animals. For example, if we have sufficient information about the situated environment (i.e. forward model of the external world), we can behave optimally. Otherwise, we would constrain our redundant musculoskeletal systems based on the anticipation given by the intrinsic chemical conditions.

Based on the physiological findings, various confectionist models of neuromodulation have been proposed [8,19,3,12,11,4,9]. In [11], a polymorphic neural networks with self reconfigurable ability was proposed, and it was applied to real robot control. In their work, it was argued that the neuromodulatory neural networks evolved in a computer simulation can be seamlessly applicable to real robot control. In [3], the homeostatic networks in which the ability of neural plasticity is the target to be evolved was proposed. In [19], the *GasNet* model was proposed, which was the first model that considered the spatiotemporal distribution of neuromodulators, and reported that it could show highly evolutionary performance compared with NoGas model.

In this paper, we explore a possible neuronal mechanism of environment cognition and motor adaptation in unknown (i.e. unexperienced) environments. For this aim, we developed a polymorphic neural networks model called CTRNN+NM (CTRNN with neuromodulatory bias). The proposed model was applied to two dimensional arm-reaching movement control in various viscous curled force fields. The parameters of the proposed model were optimized using genetic algorithms. The robustness of the optimized neural controller has been evaluated.

2 Proposed Method

2.1 CTRNN with NM Bias

In this study, a polymorphic neural networks model has been proposed. As the proposed model is based on a continuous time recurrent neural network (CTRNN)(e.g. [1]), it is named CTRNN+NM (CTRNN with NM bias).

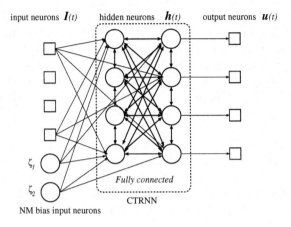

Fig. 1. CTRNN with NM bias

Figure 1 schematically shows the proposed model. The network basically consists of fully-connected hidden neurons, each of which has the following dynamics (leaky-integrator dynamics):

$$T_i\frac{ds_i(t)}{dt} = -s_i(t) + \sum_{j=1}^{N_h} w_{ij}h_j(t) + \sum_{k=1}^{N_s} w_{ik}I_k(t) \tag{1}$$

$$h_j(t) = \frac{1}{1 + \exp\left[-(s_j(t) - \theta_j)\right]} \tag{2}$$

where $s_i(t)$ and $h_i(t)$ are the internal state and output of the neuron i, respectively. N_s and N_h are the number of sensors and hidden units. T_i is the time constant of the neuron, w_{ij} and w_{ik} are the synaptic weights, and θ_j is the threshold of the neuron. These parameters ($[\boldsymbol{T}, \boldsymbol{w}, \boldsymbol{\theta}]$) are the target to be optimized.

As can be seen in the figure, the proposed model has additional bias inputs named NM bias, ζ_i. The characteristics of the CTRNN can be altered by modulating the NM bias ζ_i just like RNNPB (recurrent neural networks with parametric bias) proposed by Tani [20]. Generally, in those models, the crucial point to be noted is how the bias inputs can be regulated.

2.2   Diffusion of NM

In neurophysiology, self-recursive network modulation is known as "intrinsic neuromodulation" [13]. In the proposed model, we adopted the self-regulation method proposed in [11], in which the NM bias is controlled by the network itself.

As schematically shown in Figure 2, each hidden neuron has the capability to diffuse its specific (i.e. genetically-determined) type of NM ($\lambda_j = \{1, 2, \cdots, M\}$) in accordance with its activity and the diffusing function given by Equation 3, which can also be genetically modified.

$$\zeta_i(t) = \max_{j \in \{\forall \lambda_j = i\}} \left[\exp \left(-\frac{(h_j(t) - \mu_j)^2}{2\sigma_j^2} \right) \right] \tag{3}$$

In this example, the instance hidden neuron (shaded in the figure) can diffuse NM2 with the concentration ζ_2 depending on its activity.

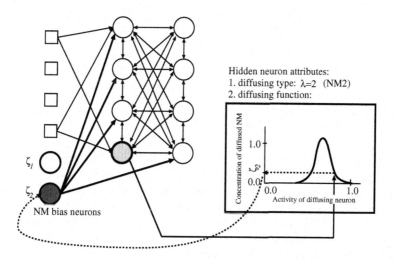

Fig. 2. Diffusion of NMs

The parameters for NM diffusion ($[\boldsymbol{\lambda}, \boldsymbol{\mu}, \boldsymbol{\sigma}]$) are also the target to be optimized.

2.3   Evolution of CTRNN with NM Bias

In this study, the parameters to be optimized (i.e. $[\boldsymbol{T}, \boldsymbol{w}, \boldsymbol{\theta}, \boldsymbol{\lambda}, \boldsymbol{\mu}, \boldsymbol{\sigma}]$) are evolutionarily determined by using genetic algorithms.

In the evolutionary process, each parameter is encoded as a real number within the range $[0, 1]$. On the contrary, in the decoding process each parameter is linearly transformed into the correspondingly defined range (Table 1) except for the diffusing type of NM, which is allocated discrete values with uniform probability according to the real value of the corresponding gene, e.g. $\lambda = 1$ is allocated if the gene is larger than 0.5, otherwise $\lambda = 2$.

In this study, we used tournament selection (# of candidates are two) with an elitist strategy as the selection mechanism in order to create the candidates in the next generation. After the selection process, genetic operators, i.e. crossover

Table 1. Parameter range

Time constant:	$T_j \in [0.01, 2.0]$
Synaptic weights (intra-neuron):	$w_{ij} \in [-5.0, 5.0]$
Synaptic weights (sensor neuron):	$w_{ik} \in [-5.0, 5.0]$
Threshold of neuron:	$\theta_j \in [-1.0, 1.0]$
NM type:	$\lambda \in \{1, 2\}$
NM diffusing function (center):	$\mu \in [0.0, 1.0]$
NM diffusing function (width):	$\sigma \in [0.01, 0.1]$

Table 2. Simulation conditions

Generations:	100000
Population (elitist strategy):	50
Crossover rate (uniform crossover):	0.5
Mutation rate:	0.04

and mutation are applied to the selected candidates. The crossover operator selects a pair of parents from the candidates, and uniform crossover is executed with a certain probability (crossover rate). On the other hand, in the mutation operation each gene locus of the candidates are simply replaced by a random real value with a certain probability (mutation rate). The parameters of the genetic algorithms are listed in Table 2. The individuals are evaluated based on the optimization criterion given in Section 3.3.

3   Experiments

3.1   Arm-Reaching Movement in Various Force Fields

To investigate the validity of the proposed model, it is applied to two dimensional arm-reaching movement control in various viscous curled force fields. Figure 3 schematically illustrates the task. As shown, a human arm can be modeled as a planar two-link manipulator.

The equations of motion of the two link arm are described by:

$$M(q)\ddot{q} + h(q, \dot{q}) = \tau + J(q)^T f_{Env} \tag{4}$$

where q, $M(q)$, $h(q, \dot{q})$, $J(q)^T$ are joint angle vector, inertia matrix, Coriolis' force, Jacobian matrix in joint coordinate, and f_{Env} is external force in Cartesian coordinate, respectively.

In this model, the joint torque τ can be derived from activities of antagonist muscles, and each muscle is contracted based on motor command u which corresponds to the output of the CTRNN (cf. Equation 12):

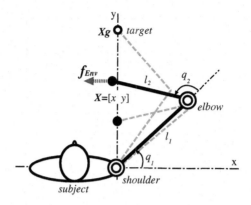

Fig. 3. Reaching task

$$\boldsymbol{\tau} = \boldsymbol{G}^T \cdot diag\left[F_1, F_1, F_2, F_2\right] \cdot \boldsymbol{u}, \tag{5}$$

$$G = \begin{bmatrix} 0.04 & -0.04 & 0.0 & 0.0 \\ 0.0 & 0.0 & 0.025 & -0.025 \end{bmatrix}^T, \tag{6}$$

$$F_j = 1 - k(q_j - q_j^{(0)}) - b\dot{q}_j, \tag{7}$$

where G and F are matrix of the moment arm and the maximum force of each muscle, respectively. This equation implies that shoulder muscles have higher gains than elbow.

In this experiments, we can simulate arbitrary external force in the hand coordinate as changes of environments. For instance, $\boldsymbol{f}_{Env}$ described by Equation 8 is a viscous curled force field (hereafter VF), in which the hand suffers an orthogonal force in proportion to the hand velocity $\dot{\boldsymbol{X}}$ (see Figure 4). In the study, the two-link arm dynamics are numerically calculated using open dynamics engine (ODE) provided by R.Smith [18].

In order to develop robust neural controller which has "how to adapt" instead of "how to move", we assumed two different force fields (i.e. $\boldsymbol{f}_{Env_1}$ and $\boldsymbol{f}_{Env_2}$) as the training environments in the evolutionary optimization experiments [10]. Here, $\boldsymbol{f}_{Env_1}$ is a null field (hereafter NF), in other words n=0.0 in Equation 10). On the other hand, $\boldsymbol{f}_{Env_2}$ corresponds to VF, and also n=5.0.

$$\boldsymbol{f}_{Env} = B\dot{\boldsymbol{X}} \tag{8}$$

$$\dot{\boldsymbol{X}} = \begin{bmatrix} \dot{x} & \dot{y} \end{bmatrix}^T \tag{9}$$

$$B = \begin{pmatrix} b_{11} & b_{12} \\ b_{21} & b_{22} \end{pmatrix} = n \begin{pmatrix} 0.0 & -1.0 \\ 1.0 & 0.0 \end{pmatrix} \tag{10}$$

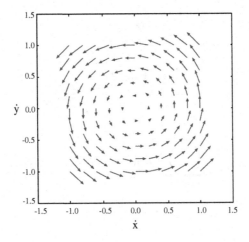

Fig. 4. Viscous curled force field

3.2   Neural Controller

The sensory inputs to the neural controller is I, and the outputs of the controller is motor command u. In equations 11 and 12, SH, EL, f, and e represent *Shoulder*, *Elbow*, *flexor*, and *extensor*, respectively. The other parameters for the CTRNN+NM model used in the following experiments are listed in Table 3.

$$I = [q^{SH} \; q^{EL} \; \tau^{SH} \; \tau^{EL} \; \zeta_1 \; \zeta_2]^T \tag{11}$$

$$u = [u_f^{SH} \; u_e^{SH} \; u_f^{EL} \; u_e^{EL}]^T \tag{12}$$

Table 3. Parameters for CTRNN+NM

# of sensor neurons N_s:	4
# of hidden neurons N_h:	10
# of NM type:	2

3.3   Evaluation Criteria

The evaluation criteria for the arm-reaching control task are given by the following equations:

$$E = \alpha E_1 + E_2, \tag{13}$$

$$E_1 = \frac{1}{T} \int_T \exp\left[-\frac{(\boldsymbol{X}_g - \boldsymbol{X})^T (\boldsymbol{X}_g - \boldsymbol{X})}{2\sigma_g^2} \right] dt \tag{14}$$

$$E_2 = \frac{1}{T} \int_T \frac{1}{1 + \boldsymbol{u}^T \boldsymbol{u}} dt \tag{15}$$

where $\boldsymbol{X}$ and $\boldsymbol{X}_g$ are hand and target positions, respectively. In the following experiments, the start position and target position are fixed ($\boldsymbol{X}_g = [0.0 \quad 0.5]^T$). Therefore, the criterion E_1 represents averaged position errors, and E_2 indicates the averaged energy consumptions of muscles. To determine the priority between the criteria E_1 and E_2, a scaling coefficient $\alpha=10$ is used.

4   Results

Because we are interested in the robustness of the proposed CTRNN+NM-based controller, it will be compared with a normal CTRNN-based controller. As has been noted, the neural controllers (i.e. CTRNN and CTRNN+NM) were optimized under two environments (i.e. NF($n=0.0$) and VF($n=5.0$)) using genetic algorithms with the above mentioned evaluation criteria (cf. Equation 13). To verify the evolvability of both networks statistically, we executed five runs with different random seed for each network structure. Figure 5 shows the average and best performance (i.e. evaluation criterion E) in the final generation. As can be seen, both neural controllers were optimized at the same level under the two environments.

After the optimization, the evolved neural controllers were evaluated in the following four environments: $n=0.0$, 2.5, 5.0, and 7.5 (cf. Equation 10). Here, $n=0.0$ and $n=5.0$ are the training environments whereas $n=2.5$ and $n=7.5$ are unexperienced (i.e. unknown) environments. Figure 6 (a) and (b) demonstrate the resultant hand trajectories in the four kinds of viscous curled force fields. Also Figure 6 (c) and (d) illustrate the resultant hand velocity curves in the four environments.

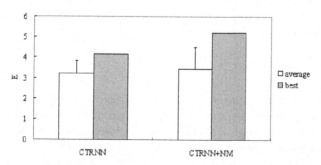

Fig. 5. Average and best performance (E) in the final generation

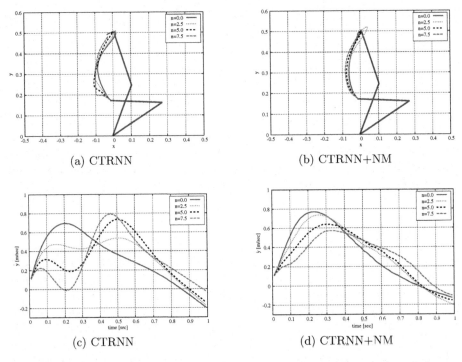

Fig. 6. Resultant hand trajectories and velocity curves in different size of viscous curled force fields. Here n=0.0 and n=5.0 are the training environments, in contrast, n=2.5 and n=7.5 are unexperienced (i.e. unknown) environments. (a,c) In the case of CTRNN, the neural controller learned different trajectories and velocity curves with respect to environments, since it has to store different sensorimotor mappings in the monolithic neural networks. (b,d): On the other hand, CTRNN+NM can recognize the environmental change via its sensorimotor feedback, and it can appropriately modulate the sensorimotor mapping so as to keep the optimal trajectory in spite of the changes.

For further investigation of the robustness of the optimized neural controllers, we measured the performance (i.e. E) of them (i.e. CTRNN and CTRNN+NM) while the viscous parameters (b_{12} and b_{21} in Equation 10) are exhaustively changed in a range ($b_{12} \in [-10.0, 0.0]$, $b_{21} \in [0.0, 10.0]$). Figure 7 shows the results of the exhaustive evaluation experiments.

5   Discussions

In this paper, a possible neuronal mechanism of environment cognition and motor adaptation in unknown (i.e. unexperienced) environments has been investigated.

Based on the physiological findings, we proposed a polymorphic neural networks model with self-reconfigurable feature, called "CTRNN with NM bias." The proposed neural networks model was applied to a planar two-link arm-reaching movement control in various (e.g. partly unexperienced) viscous curled force fields.

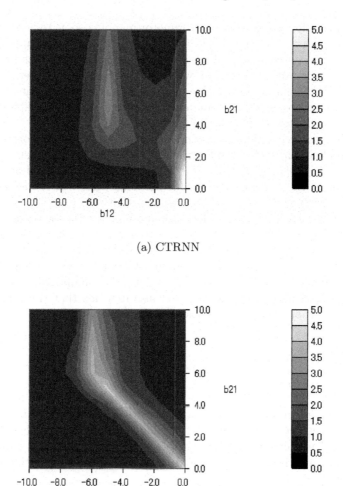

(a) CTRNN

(b) CTRNN+NM

Fig. 7. Robustness of the evolved controllers ((a) CTRNN and (b) CTRNN+NM) while viscous parameters (b_{12} and b_{21}) are exhaustively changed. (a) In the case of CTRNN, the controller shows high robustness against b_{21}, but it is brittle under b_{12} changes. (b) In contrast, CTRNN+NM demonstrates high robustness against not only b_{12}, but also b_{21} (Note that the gradation specifies different range). Especially, the diagonal line keeps high performance.

As can be seen in Figure 6 (a) and (c), the neural controller based on a normal CTRNN learned different trajectories and hand velocity curves with respect to environmental changes. This is because the normal CTRNN has to store them as different dynamics in a monolithic neural networks.

On the contrary, Figure 6 (b) and (d) indicate that the proposed CTRNN+NM model can recognize environmental change via its sensorimotor feedback, and it can appropriately modulate the sensorimotor mapping so as to keep an optimal hand velocity curve (i.e. bell-shape) in spite of environmental changes. The crucial point to be noted here is that the diffusing conditions of the NM bias are dependent on neuron state vectors within the identical neural networks. Thus, there are self-referential loops with a wide variety of time constants. These multiple feedback loops would contribute to a real-time adaptation ability through continuing interactions with infinite environments.

According to verification experiments (Figure 7), the neural controller based on the CTRNN is specialized to the training environments and it showed brittleness against b_{12} changes. Because the y-directional component of hand velocity is dominant in the reaching movements assumed here, less robustness against b_{12} is considered fatal compared with b_{21} sensitivity.

In contrast, the CTRNN+NM demonstrates higher robustness against changes not only in b_{12} but also in b_{21}. Especially, the diagonal line keeps high performance. This implies that the proposed CTRNN with NM bias model evolved "adaptation strategy" (i.e. sensorimotor constraints and their elicitation procedure) instead of "sensorimotor mapping" as it is (i.e. an optimal inverse dynamics of the experienced environment). Because it seems that the CTRNN+NM extracts the dynamic structure of the external environments ($\boldsymbol{f}_{Env}$), which has following form:

$$B = \begin{pmatrix} 0 & -\Gamma \\ \Gamma & 0 \end{pmatrix} \tag{16}$$

Due to these considerations, we confirmed that the proposed model has high robustness even though it is situated in unexperienced environments. Therefore the proposed model should be a simple solution to explain the self-referential adaptation, which is essential to work out the environment cognition and motor adaptation.

In addition, the condition of the NM bias neurons reflects the contextual interaction between the environments, because the activities are dependent on the dynamics of internal leaky-integrator neurons. Therefore, the behavior of the controlled system with unconstrained redundancy should be canalized by the bias. Even if the controlled system with the proposed network method is located in unexperienced environments, it would adapt to the situation by modulating the NM bias itself. Thus, the system is implicitly anticipatory due to the optimization of its internal network structure with evolutionary optimization techniques. And, it is explicitly anticipatory due to the recurrent modulatory neural biases that proved useful for the stabilization of arm reaching movements in dynamic environments.

Acknowledgments

This research has been partially supported by a Grant-in-Aid for Scientific Research on Priority Areas (No.454, 2005–2009) and Young Scientists (B)

(No.18700195, 2006–2007) from the Japanese Ministry of Education, Culture, Sports, Science, and Technology.

References

1. Beer, R.D.: Dynamical approaches to cognitive science. Trend in Cognitive Sciences 4-3, 91–99 (2000)
2. Bernstein, N.: The Coordination and Regulation of Movements. Pergamon, Oxford (1967)
3. Di Paolo, E.A.: Homeostatic adaptation to inversion of the visual field and other sensorimotor disruptions. In: From animals to animat 6, pp. 440–449. MIT Press, Cambridge (2000)
4. Eggenberger, P., Ishiguro, A., Tokura, S., Kondo, T., Uchikawa, Y.: Toward Seamless Transfer from Simulated to Real Worlds: A Dynamically-Rearranging Neural Network Approach. In: Demiris, J., Wyatt, J.C. (eds.) Advances in Robot Learning. LNCS (LNAI), vol. 1812, pp. 44–60. Springer, Heidelberg (2000)
5. Getting, P.A., Dekin, M.S.: Tritonia swimming: A model system for integration within rhythmic motor systems. In: Selverston, A.I. (ed.) Model neural networks and behavior, pp. 3–20. Plenum Press, New York (1985)
6. Haken, H.: Principles of Brain Functioning. Springer-Verlag, Berlin Heidelberg (1996)
7. Hasselmo, M.: Neuromodulation and cortical function: modeling the physiological basis of behavior, Behavioral Brain Research, vol. 67, Elsevier Science B.V, pp. 1–27 (1995)
8. Husbands, P., Smith, T.M.C., Jakobi, N., O'Shea, M.: Better Living Through Chemistry: Evolving GasNets for Robot Control, Connection Science, vol.10 (3-4) Carfax, pp. 185-210 (1998)
9. Ishiguro, A., Fujii, A., Eggenberger, P.: Neuromodulated Control of Bipedal Locomotion Using a Polymorphic CPG Circuit. Adaptive Behavior 11-1, 7–17 (2003)
10. Karniel, A., Mussa-Ivaldi, F.A.: Sequence, time, or state representation: how does the motor control system adapt to variable environments? Biological Cybernetics 89-1, 10–21 (2003)
11. Kondo, T.: Evolutionary design and behavior analysis of neuromodulatory neural networks for mobile robots control. In: Applied Soft Computing, vol. 7-1, pp. 189–202. Elsevier, North-Holland, Amsterdam (2007)
12. Kondo, T. Ishiguro, A. Tokura, S. Uchikawa, Y. and Eggenberger, P., Realization of Robust Controllers in Evolutionary Robotics: A Dynamically-rearranging Neural Network Approach, In: Proceedings of the 1999 Congress of Evolutionary Computation, vol.1, pp. 366–373 (1999)
13. Marder, E., Thirumalai, V.: Cellular, synaptic and network effects of neuromodulation. Neural Networks 15, 479–493 (2002)
14. Meyrand, P., Simmers, J., Moulins, M.: Construction of a pattern-generating circuit with neurons of different networks. NATURE 351, 60–63 (2 may 1991)
15. Paine, R.W., Tani, J.: Motor primitive and sequence self-organization in a hierarchical recurrent neural network. Neural Networks 17, 1291–1309 (2004)
16. Poggio, T., Bizzi, E.: Generalization in vision and motor control Insight Review Articles. Nature 431, 768–774 (2004)
17. Schöner, G., Kelso, J.A.S.: Dynamic pattern generation in behavioral and neural systems. Science 239, 1513–1520 (1988)

18. Smith, R.: Open dynamics engine v0.5 user guide, (2004) http://ode.org/
19. Smith, T.M.C., Husbands, P., Philippides, A., O'Shea, M.: Neuronal Plasticity and Temporal Adaptivity: GasNet Robot Control Networks. Adaptive Behavior 10 (3–4), 161–183 (2002)
20. Tani, J.: Learning to generate articulated behavior through the bottom-up and the top-down interaction processes. Neural Networks 16, 11–23 (2003)
21. Thelen, E., Smith, L.B.A: A Dynamic Systems Approach to the Development of Cognition and Action. MIT Press, Cambridge, MA (1994)
22. Wolpert, D.M., Kawato, M.: Multiple paired forward and inverse models for motor control. Neural Networks 11, 1317–1329 (1998)

Anticipating Rewards in Continuous Time and Space: A Case Study in Developmental Robotics

Arnaud J. Blanchard and Lola Cañamero

Adaptive System Research Group
School of Computer Science
University of Hertfordshire
College Lane, Hatfield, Herts AL10 9AB, UK
{A.J.Blanchard, L.Canamero}@herts.ac.uk

Abstract. This paper presents the first basic principles, implementation and experimental results of what could be regarded as a new approach to reinforcement learning, where agents—physical robots interacting with objects and other agents in the real world—can learn to anticipate rewards using their sensory inputs. Our approach does not need discretization, notion of events, or classification, and instead of learning rewards for the different possible actions of an agent in all the situations, we propose to make agents learn only the main situations worth avoiding and reaching. However, the main focus of our work is not reinforcement learning as such, but modeling cognitive development on a small autonomous robot interacting with an "adult" caretaker, typically a human, in the real world; the control architecture follows a Perception-Action approach incorporating a basic homeostatic principle. This interaction occurs in very close proximity, uses very coarse and limited sensory-motor capabilities, and affects the "well-being" and affective state of the robot. The type of anticipatory behavior we are concerned with in this context relates to both sensory and reward anticipation. We have applied and tested our model on a real robot.

1 Introduction

A very important problem for autonomous robots is to be able to explore and learn about their world on their own while interacting with objects and other agents in it. The use of "rewards" that satisfy survival-related needs (see e.g., [9,10]) is a natural approach to this problem.

Related to this view, Reinforcement Learning (RL) aims to make an agent learn which actions it should perform in order to maximize the acquisition of rewards. This learning paradigm is interesting as it permits to "program" agents easily to make them carry out particular tasks in an efficient way by emitting different signals (reinforcement) according to the relevance of their actions. Moreover, animals and humans can be very efficient in this type of learning, such as evidenced for example by the way dogs can be trained, or how rats can learn to move in a maze in order to find a source of food. Biological agents

M.V. Butz et al. (Eds.): ABiALS 2006, LNAI 4520, pp. 267–284, 2007.

can thus anticipate the reward associated with their actions, and biologically-inspired learning models aim to contribute to the understanding of how this is done. However, good models for this type of behavior are complex to design, and the main difficulty is to identify the cues predicting rewards.

Although the main emphasis of our work is not reinforcement learning, it presents some commonalities with this approach to learning and can be interpreted in this light. Our work presents a different cognitive system that is capable of setting its own goals (i.e. sensation it should try to reach) on the basis of reinforcement, probability, and recency. Our focus is on modeling cognitive development in a small autonomous robot interacting with an "adult" caretaker—typically a human—in the real world. This interaction occurs in very close proximity (we could say that, for the most part, it takes place within the robot's "personal space"), uses very coarse and limited sensory-motor capabilities, and affects (increases or decreases) the "well-being" of the robot, not only modifying its current "affective" state but also the memories and goals of the robot on different time scales.

The architecture we are using follows a Perception-Action approach [13], in which perception and action form a tightly coupled loop. The type of anticipatory behavior we are concerned with in this context relates to both sensory and state anticipation. It is linked to sensory anticipation, as predictions of the "reward" guide the robot's sensory processing and what it attends to. It also relates to state anticipation, since decision making, regarding the action to execute, is led by the goal of improving or maintaining the internal well-being. Anticipatory behavior is triggered by taking into account simple "expectations" that are grounded in the remembered affective effects that perception-action couplings had on the "well-being" of the model. The main interest of our work is the development of a system that is capable to autonomously form "desired sensations", that is, goals in continuous spaces (although not addressed herein, the mechanism may also be extended to discrete spaces.

The remainder of the paper is structured as follows. In Section 2 we briefly describe the scenario that sets the framework for our work—the type of problems we are addressing and the robotic architecture we are using. Section 3 presents the main ideas of classical approaches of reinforcement learning and discusses the problems raised by them, particularly those having to do with their arbitrary discretization of the environment and the use of computational resources. In Section 4 we propose a new architecture able to handle these problems and present the results of experiments we have carried out on a real robot. Finally, Section 5 draws some conclusions and perspectives to continue this work.

2   The Context of Our Work: Scenario and Architecture

The scenario in which we are working focuses on modeling the development of attachment bonds between a robot and an adult or expert agent—typically a human, although it could also be another, more "knowledgeable" robot—which acts as its caretaker and with the help of whom it can safely satisfy its needs,

Fig. 1. Robot used in the experiments

perform tasks, and learn how to interact with the environment and what the important things are.

Attachment bonds develop naturally in animals and humans, but how can they originate and develop in robots? Why would a robot *want* to interact with other robots and humans and what would make it *like* such interactions? To bootstrap attachment bonds and the affective interactions that stem from them, we previously took inspiration from a naturally occurring imprinting phenomenon (see [3], [4]). Imprinting is a phenomenon in which many animals (particularly birds and mammals) form special attachment bonds with objects to which they are exposed to very early in life. Imprinting is a very important mechanism favoring adaptation and learning within the developmental process, in particular filial imprinting, in which the imprinting object is treated as a parent, giving rise to affiliated behaviors such as approaching and following. Such behaviors are highly beneficial to very young animals and humans who still cannot act autonomously in the world, and provide a basis to obtain needed resources and security, for social facilitation and learning (e.g., young animals encounter new situations and learn new things to which they would not be exposed if they were not following their caretakers around), and for emotional development (e.g., by matching the emotional state of the caretaker one can learn appropriate emotional and be-havioral reactions to different types of situations). Imprinting could provide the same advantages to an autonomous robot having to inhabit a dynamic, unknown, and social environment.

To implement this scenario, we have used a Perception-Action (PerAc) archi-tecture implemented and tested in a Koala robot (k-team.com/robots/koala, see Figure 1). The PerAc approach is rooted in psychology [15] and in robotics [13], and postulates that perception and action are tightly coupled and encoded at the same level. Perception-action loops can be considered in terms of homeosta-tic control, which leads to the hypothesis that behavior is executed to correct perceptual errors. Actions that allow the correction of different perceptual errors are selected on the grounds of sensorimotor associations, which can be learned or hard-coded in the architecture of the robot. According to this view, action is thus executed as a "side-effect" of wanting to achieve, improve, or correct some perception and, in the case of imprinting, the robot "approaches", "follows" or "avoids" the imprinted object as it moves around. Figure 2 summarizes the PerAc architecture we have used to model imprinting—we refer the reader to [8] for details of the architecture and the experiments we carried out to test it. We used two types of imprinting objects that produced equally satisfactory results: a human and a cardboard box. To detect and learn about the imprinting object, the only feature that the robot used was distance, as measured by infrared sensors located at its front. With our architecture, the robot can learn, memorize, and retrieve at various time scales (see [8]) different "positive" and "negative" sensations produced during environmental interactions. The sensations are retrieved based on the effects they had on the internal well-being or comfort of the robot.

The problem addressed in this paper concerns solely the autonomous learning of goals (desired sensations in our case). However, from a developmental perspective imprinting is only adaptive at an early time in life, and young animals, humans, and robots, must sooner or later start trying other behaviors, such as exploring the world on their own and learning new things and their relevance, exploiting knowledge and skills already acquired, imitating what others do and, most importantly, making decisions to switch among all those behaviors in an adaptive way. We thus extended the previous PerAc architecture to give rise to and switch among all these behaviors depending on the affective state of the robot. To do so, we use the notion of comfort in terms of satisfaction of parameters to model the robot's internal needs as the key mechanism that modulates the underlying basic architecture—we refer the interested reader to [5] for details.

3   Relation to Reinforcement Learning

Although our model and work are not explicitly about Reinforcement Learning (RL) proper, it can be regarded in the light of this learning paradigm, as has been pointed out to us, notably through the relation that our notion of "comfort" bears to the use of "rewards" in RL. Our work presents a different cognitive system that is capable of setting its own goals (i.e. sensations it should try to reach) on the basis of reinforcement, probability, and recency.

3.1   Classical Reinforcement Learning

The temporal-difference model [18] is a very common and efficient reinforcement learning method. Its principle is to discretize the inputs (from the sensors and the internal states of an agent) in order to obtain a finite number of possible states (inputs).

The expected reinforcement for each state is evaluated using the actual reinforcement of the state in addition to the reinforcement expected in states

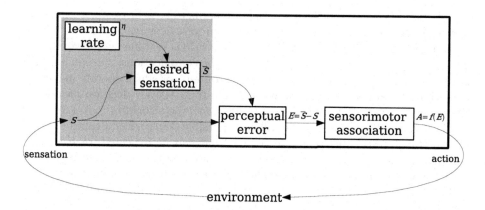

Fig. 2. Architecture used to model imprinting. The gray area highlights the problem addressed in this paper.

immediately accessible. The agent then acts in order to reach the states maximizing the expected reinforcement. Even if the convergence of the algorithm is proved, learning is very slow because the agent needs to try each state several times, and it strongly depends on the discretization used, which can lead to a huge number of different states. It is therefore also very demanding in terms of memory, since it needs to be able to store all the expected reinforcements for each possible state.

Q-learning [19] uses similar principles but it works even if the agent does not know which action to execute in order to reach a given state. The agent learns the expected reinforcement for each possible state-action couple. This increases again the time of learning because there are many more possibilities to explore, and the quantity of memory needed is multiplied by the number of different possible actions.

3.2 The Problem of Discretization

In artificial intelligence, numerous powerful algorithms have been designed to learn, anticipate and decide. However, they are often inappropriate when applied to robots in the real world, particularly if the robots are not pre-programmed to detect specific stimuli. For example, many models of classical or instrumental conditioning need to predefine the set of possible stimuli to consider. Information theory [11] provides powerful tools to statistically measure the temporal correlation between events and anticipate them. However, this approach also relies on discretization, and the problem is again to define the set of events by discretizing the world.

Discretization can be adaptive, for example by grouping together events that carry the same predictive information. To do this, we can use classification algorithms like the k-mean, Kohonen's maps, Estimation-Maximization algorithm, etc. (see [7]). Many of these algorithms need strong assumptions on the distribution of classes and the discretization needs to be arbitrarily or randomly initialized, so that the quality of the learning process depends on random initializations. When developing the Q-learning algorithm, Watkins was aware of the difficulty to cope with continuity: "To avoid the complications of systems which have continuous state-spaces, continuous action sets, or which operate in continuous time, I will consider only finite, discrete-time Markov decision processes" [19, page 38]. Even after the discretization is done, the algorithm converges quite slowly because it needs to try the different possible states several times in order to statistically estimate the reinforcement that can be expected for each one. Once the reinforcement can be reliably anticipated for each state, the agent can act in order to reach the state with the highest expected reinforcement.

These approaches are very powerful when they are used in simulation, since the environment is often discrete (e.g. a grid where the agent is moving) and it is easy to make an agent try different situations a large number of times. They can be well adapted to robotics when the elements of the environment are predefined, and there are obvious salient cues that the robot can consider as classes of events (e.g. salient color or pattern).

However, in the case of robots in real environments without specific features, the robots have to find by themselves the cues predicting rewards. These cues are not necessarily salient, and can for example be a specific light intensity, a range of sound frequencies or a specific position, rather than binary signals associated with the presence or absence of light, sound, shape, etc, as is usually the case in discretized environments. Humans and animals are very efficient at discriminating similar stimuli if they have distinctive predictive values. In this case, using the salience of sensations can be misleading, since for example a light being turned on or off might not have any predictive value, whereas a small change in the intensity of a light at a specific level can be significant.

Most algorithms involving discretization are not able to cope efficiently with this kind of situation because they waste vast amounts of memory storing the predictions of expected rewards for many different values of the sensory input, even though most of them are not relevant or are redundant. Moreover, there is usually no difference between the effect of a small reward obtained immediately and the promise of an important reward to be obtained later. However, in some cases it is very important to make such a distinction: for example, if a robot is about to "die" it should go where it is sure and quickly find at least a small reward (e.g. partially satisfy a need by consuming a small resource), whereas it should try to maximize the long-term reward when it has more time (e.g. go to a farther location with larger quantities of that resource and where it could satisfy that need until satiated).

4   Our Approach to Learning

As there is no free lunch [20], there is no general algorithm that, on average, performs better than any other one without further and more suitable assumptions about the structure of the environment. We make and use assumptions about

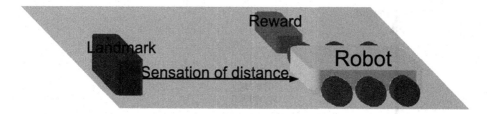

Fig. 3. Using the distance to a landmark detected by its distance sensor, the robot must be able to anticipate the presence of the reward on its side

the world in order to improve our architecture. In our work, we assume that the world is continuous since physical robots acting in the real world have to deal with a continuous, rather than discrete, environment: there are continuous variations of rewards with continuous variations of sensory inputs and the relations between rewards and sensory inputs are consistent. Consequently, if the agent—a robot in our case—receives a high reward for a specific sensory input (sensation), it can anticipate a good reward for other close sensations. Therefore, instead of estimating the expected reward for all the many possible states and trying to reach the state anticipating the maximum reward, we propose to make the robot memorize only the sensation associated with the best reward, and we will call it *desired sensation*. As previously mentioned, our robot can memorize and recall desired sensations at different time scales, depending on its affective state (see [8]). However, for the sake of clarity, in this paper we focus only on one time scale.

4.1   Desired Sensations

To illustrate the various possibilities, we consider a continuous environment—our usual environment in which a robot interacts with a human and objects in the real world—and we use sets of real variables: $S = \{s_1, s_2, \ldots\}$ for sensory input (light intensity, pressure, distance to obstacles, etc.), $A = \{a_1, a_2, \ldots\}$ for actions (velocity, rotation angle, etc.) and a real variable r to represent immediate reward. To simplify, we focus on one dimension of sensory input ($S = \{s\}$) and we consider the problem depicted in Figure 3, where a robot can move only forwards and backwards. It receives the distance to a landmark (e.g. a caretaker) as sensory input (S) and it receives and must be able to anticipate the presence of a reward (e.g. a source of energy) located at its side (r).

In order to make the robot learn the sensation associated with the highest reward, we could simply set the desired sensation ($\hat{S}$) to be equal to the current sensation (S), but only when the reward (r) is higher than the highest remembered reward ($\hat{r}$) (see also Figure 4).

$$\text{if } r > \hat{r} \text{ then } \begin{cases} \hat{r} = r \\ \hat{S} = S \end{cases} \qquad (1)$$

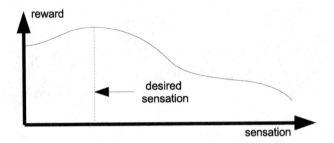

Fig. 4. Desired sensation depending on the reward associated with the sensation

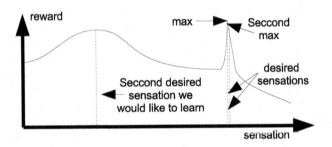

Fig. 5. Impossibility of learning a local maximum

The problem with this equation is that if the reward is very high once and is never high again, or if the sensation is very hard to obtain, the desired sensation learned would be useless. Moreover, the robot would not be able to learn more than one sensation associated with a reward. Actually, even if it memorizes another desired sensation associated with a slightly smaller reward, the principle of continuity makes this desired sensation infinitely close to the previous one learned as we can see in Figure 5. Therefore, to be reliable and robust the robot should not only memorize the sensations associated with the highest reward, but also the sensations associated with a positive reward at a high probability.

We have shown in Equation 1 how to memorize the sensation associated with the maximum reward. Equation 2 shows how the robot could compute the most probable sensation ($\overline{S}$) as the average of all sensations at each point in time (t).

$$\overline{S}_t = \frac{S_0 + \ldots + S_t}{t + 1} \tag{2}$$

However, to implement this the robot would need to store all the sensations at all the times, which is not only biologically implausible but also virtually impossible. A more plausible and realistic approach would involve memorizing the current average sensation, which summarizes the "history" of the robot's sensations over its lifetime, and also avoids the above problem. This amounts to the use of an incremental rule such as the one shown in Equation 3, which is similar to the learning rule of Rescorla and Wagner [16], used for conditioning.

$$\overline{S}_t = \frac{S_0 + \ldots + S_{t-1} + S_t}{t+1}$$
$$= \frac{\frac{S_0 + \ldots + S_{t-1}}{t} \times t + S_t}{t+1}$$
$$= \frac{\overline{S}_{t-1} \times t + S_t}{t+1}$$
$$= \frac{\overline{S}_{t-1} \times (t+1) - \overline{S}_{t-1} + S_t}{t+1}$$
$$= \overline{S}_{t-1} + \frac{1}{t+1}\left(S_t - \overline{S}_{t-1}\right)$$
$$= \overline{S}_{t-1} + \eta_t \cdot \left(S_t - \overline{S}_{t-1}\right) \tag{3}$$

The learning rate is $\eta_t = \frac{1}{\widetilde{T}_t}$, and in this case we only need a variable that increases with time ($\widetilde{T}_t = \widetilde{T}_{t-1} + 1; \widetilde{T}_0 = 1$) and another variable to memorize the current average sensation ($\overline{S}$). The complexity of the calculus is very low and biologically plausible.

Now the agent can learn two extreme cases: the sensation associated with the best reward ($\hat{S}$), and the average sensation ($\overline{S}$), regardless what the reward is.

It is nevertheless not very useful to learn only those extreme cases. The first one indicates the sensation associated with the best reward, but this memory might not be reliable as it may have happened only once. The second case indicates which are the sensations that happen more often, but this does not mean that they are good things for the robot, only that they appear often in its environment. However, all the intermediate cases are very important because in order to maximize the cumulative reward, the agent should balance the effect of the reward and the effect of the probability. If a robot urgently needs a reward (for example consuming a resource to avoid dying), it should focus on the sensations promising small rewards with high probabilities (easy to obtain), but if the situation is not urgent, it should focus on sensations promising higher rewards in order to maximize the cumulative reward and also to learn more about these high rewards. The robot must thus be able to memorize a range of desired sensations, from those obtained often but predicting small rewards, to those rarely obtained but predicting high reward.

In [4] we have shown how a robot can learn the average "best" sensation by weighting each sensation with the associated reward [1] simply by modifying the function of the learning rate η_t with $\eta_t = \frac{r_t}{\widetilde{r}_t}$ with $\widetilde{r}_t = \widetilde{r_{t-1}} + r_t; \widetilde{r}_0 = r_0$. However, the probability of the robot obtaining the reward did not reflect the importance of the reward. Moreover, past experiences with highly positive and

---

[1] In [4] the desired sensations were called "desired perceptions", and what we call here "reward", to follow the more usual terminology in machine learning, corresponds to the notion of comfort used there and in other related papers.

negative rewards would have the same consequences as past experiences with an average constant reward.

We propose in Equation 4 a solution for learning different desired sensations $(\overline{S^k})$ where the relation between the importance of the reward and its probability is controlled by the parameter k:

$$\overline{S_t^k} = \frac{e^{k.r_0}.S_0 + \ldots + e^{k.r_t}.S_t}{e^{k.r_0} + \ldots + e^{k.r_t}}$$

$$= \overline{S_{t-1}^k} + \frac{e^{k.r_t}}{e^{k.r_0} + \ldots + e^{k.r_t}} \cdot \left(S_t - \overline{S_{t-1}^k}\right) \qquad (4)$$

For extreme values of k (namely 0 and $+\infty$) we obtain, respectively, the same results as in Equation 2 because $e^0 = 1$, and Equation 1 because:

$$\lim_{k \to +\infty} \frac{e^{k.r_0}.S_0 + \ldots + e^{k.r_t}.S_t}{e^{k.r_0} + \ldots + e^{k.r_t}} = S_{argmax(r_0,\ldots,r_t)}$$

Another advantage of this formula is that only the *variation* of the reward— i.e. of the *comfort* in our model—and not its absolute value, influences learning; therefore, we do not need to define *a priori* which level of reward value has to be considered a good reward. We can actually add any constant (c) to the reward without changing the learning rate:

$$\eta_t^k = \frac{e^{k.(r_t+c)}}{e^{k.r_0+k.c} + \ldots + e^{k.r_t+k.c}}$$

$$= \frac{e^{k.r_t}.e^{k.c}}{e^{k.r_0}.e^{k.c} + \ldots + e^{k.r_t}.e^{k.c}}$$

$$= \frac{e^{k.r_t}}{e^{k.r_0} + \ldots + e^{k.r_t}}$$

$$= \frac{e^{k.r_t}}{\widetilde{r_t^k}} \qquad (5)$$

with $\widetilde{r_t^k} = \widetilde{r_{t-1}^k} + e^{k.r_t}$.

4.2   Avoided Sensations

Our previous work had only taken into account desired sensations. We have shown how a robot can learn sensations predicting rewards, but it can also be useful to learn sensations predicting danger or negative reward (punishment) in order to avoid them. With our model, such negative sensations are easy to compute as they are equal to the sensations S_t^k for negative values of the parameter k. We will call them *avoided sensations*.

The main problem when computing the desired sensations is that they can be between two local maxima and therefore predict a reward where there is no reward, i.e. a "false positive", see Figure 6.

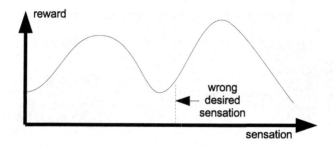

Fig. 6. Wrong desired sensation, resulting from the average of multiple local maxima

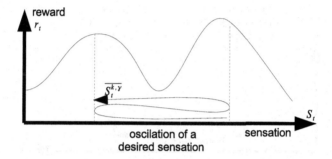

Fig. 7. Oscillation of a desired sensation between local maxima ($\gamma < 1$)

The solution is to make the robot partly forget the past and consequently have its desired sensations moving among local maxima but not staying in between.

In [4] we raised the learning rate η_{t^k} to the power of γ, with γ taking values between 0 and 1, in order to make the robot continuously learn and partly forget the effect of the old experiences:

$$\overline{S_t^{k,\gamma}} = \overline{S_{t-1}^{k,\gamma}} + \left(\frac{e^{k.r_t}}{\widetilde{r_t^k}} \right)^{\gamma} \left(S_t - \overline{S_{t-1}^{k,\gamma}} \right) \tag{6}$$

The smaller γ is, the higher the learning rate and the faster the desired sensation changes; therefore, the desired sensation oscillates between local maxima depending on the exploration of the robot, as depicted in Figure 7.

The main problem with partly forgetting the past is that the robot will not be able to remember a sensation associated with a good reward if it did not experience it recently. However, desired sensations oscillate between local maxima, and avoided sensations oscillate between local minima; therefore, if the robot memorizes the extreme values ($\widehat{S}$) of the successive desired and avoided sensations (see Figure 8 desired sensation), it can remember two (minimum and maximum valued) sensations anticipating a positive reward and two sensations anticipating a negative reward (punishment). In the conclusions we discuss how to deal with the memorization of an unlimited number of sensations.

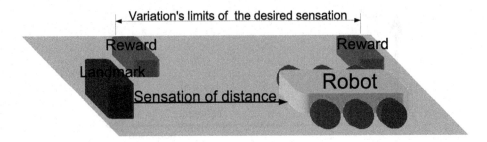

Fig. 8. The desired sensation strictly oscillates between two rewards

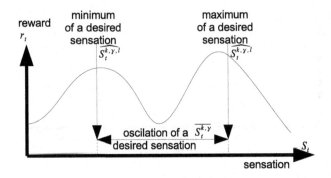

Fig. 9. Extreme values of a desired sensation $(k > 0)$. Values for l are $l < 0$ at the leftmost extreme, and $l > 0$ at the rightmost extreme.

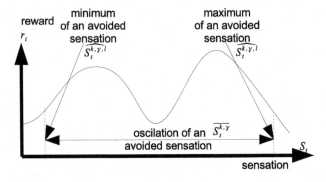

Fig. 10. Extreme values of an avoided sensation $(k < 0)$. Values for l are $l < 0$ at the leftmost extreme, and $l > 0$ at the rightmost extreme.

In order to remember these extreme values, we use an equation similar to (4) but this time the robot memorizes the desired sensations $(S^{k,\gamma})$ with extreme values of the sensations themselves, instead of memorizing the sensations associated with extreme rewards! The weight in the exponential function is therefore the desired sensation itself multiplied by the parameter l, which determines whether the agent memorizes the minimum ($l < 0$) or the maximum ($l > 0$) value of the desired sensation (see Figure 9 and Figure 10).

We can see in Equation 7 how these extreme values are defined with $\widehat{S_t^{k,\gamma,l}} = e^{l \cdot S_0^{k,\gamma}} + \ldots + e^{l \cdot S_t^{k,\gamma}}$:

$$\widehat{S_t^{k,\gamma,l}} = \widehat{S_{t-1}^{k,\gamma,l}} + \frac{e^{l \cdot \overline{S_t^{k,\gamma}}}}{\widetilde{S_t^{k,\gamma,l}}} \cdot \left(\overline{S_t^{k,\gamma}} - \widehat{S_{t-1}^{k,\gamma,l}} \right) \tag{7}$$

5   Experiments and Results

We have tested this algorithm on a Koala robot that had to memorize sensations associated with reward or punishment. The robot could move forward and backwards—towards or away from a "caretaker"[2] that was also used as a landmark. The experimental setup is depicted in Figure 8. The sensory input (S) used was the frontal distance sensor, which measures the distance of the caretaker in front. The right distance sensor was used to detect rewards (r) in the form of an object located in close proximity within the range of its right sensor that provides comfort to the robot, that is, positive reward r. Figure 11 shows the reward obtained by the robot as a function of the sensation of distance to the landmark (e.g. the caretaker). As we are only interested here in the learning system of the robot, we make (it is hard-coded) the robot alternatively move forward and backward, and we observe how it creates its desired sensations (i.e. its goals), shown in Figure 12.

In Figure 11, we can observe how the desired sensations of the robot evolve over time and the experiences of the robot. We compute the desired sensation as ($\overline{S^{k,\gamma}}$ with $k = +400; \gamma = 0.9$), and the avoided sensation as ($\overline{S^{k,\gamma}}$ with $k = -400; \gamma = 0.9$). If k or γ differ, the curves are more or less smooth and qualitatively similar.

We present the results in Figure 12. The desired sensation oscillates between sensations 75 and 425, which correspond to the presence of the reward (object to the right of the robot). The avoided sensation oscillates between the two rewards at the beginning and then around them, which indicates that the robot should avoid being between the objects providing reward or behind them.

Desired and avoided sensations are constantly changing; therefore, the robot cannot remember anything for a long time. However, the next step for the robot is to memorize the extremes of these desired and avoided sensations. We present in Figure 13 the evolution of these extremes ($\widehat{S^{k,\gamma,l}}$) for the same values of k and γ and -0.1 and 0.1 for l (l is small because the amplitude of the sensation is high); however, this does not have a strong effect on the qualitative result. The extremes of the avoided sensations quickly converge (almost at the first cycle) to the sensations corresponding to the rewards on the side (the reward 75 and

---

[2] We used different types of stimuli as caretaker, notably humans and cardboard boxes that could be static or moved by humans.

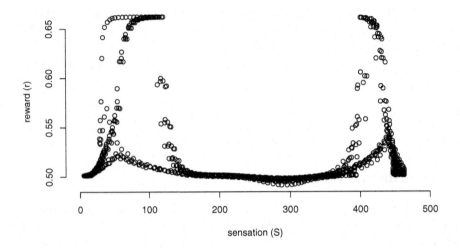

Fig. 11. Value of the reward (r) as a function of the sensation (S). We see that the maximum reward is for sensations of about 75 and 425 (the unit is not important), which correspond to the presence of an object on the right of the robot.

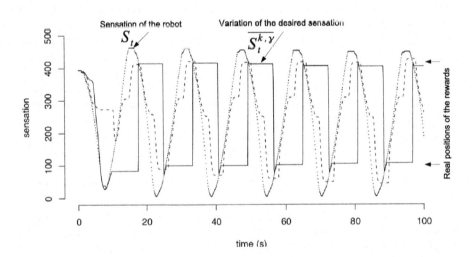

Fig. 12. Evolution of the sensation (S_t) of the robot in dotted line, the desired sensation $(\overline{S^{k,\gamma}}$ with $k = +400; \gamma = 0.9)$ in solid line and the avoided sensation in dashed line $(\overline{S^{k,\gamma}}$ with $k = -400; \gamma = 0.9)$. The desired sensation oscillates between sensations 75 and 425, which correspond to the presence of the reward. The avoided sensation oscillates between the two rewards at the beginning and then around them, indicating that the robot should avoid being between or around the objects providing the rewards.

425). The extremes of the avoided sensations correspond at the beginning to the sensation obtained when the robot is located between the rewards, and at the end to the sensation obtained when it is located behind the rewards. This means

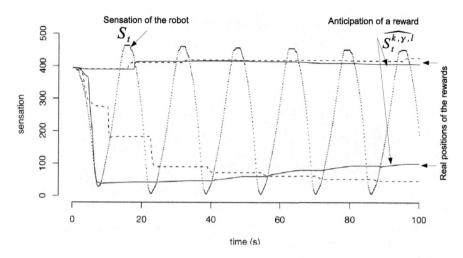

Fig. 13. Evolution of the extreme values ($S^{k,\gamma,l}$) of desired and avoided sensations using the same parameter values as previously for k and γ, but with a value of 0.1 for l in the curves at the top, and -0.1 in the curves at the bottom. The extremes of the desired sensations are in solid line and the extremes of the avoided sensations are in dashed line.

that the robot should avoid staying between or behind the objects providing rewards, since in those two cases no reward is obtained.

6 Conclusion and Perspectives

We have presented the first basic principles and implementation of what could be regarded as a new approach to reinforcement learning, where agents—robots interacting with objects and other agents in the (continuous) real world—can learn to anticipate rewards using their sensory inputs.

Doya, for example, proposed approximating the reward function in order to process reinforcement learning in continuous time and space [12], but we argue that it is enough to only memorize where the rewards are even if the robots can not know what these rewards are. The advantages of our approach are that the robot memorizes only the relevant information and does not require much memory or computation time. It does not use notions of events or discretization, and this strongly reduces the effects of a priori choices and decreases learning time. Our model also requires virtually no a priori knowledge about the world, consisting of a couple of parameters that need to be set: k to balance the relation between the importance of the reward's value and its probability, and γ and l to vary the average speed (adaptability) of learning. However, these parameters only have quantitative effects in our case, and we have already proposed in [4] ways to modulate such parameters as a function of the affective state (see also e.g. [1,2] for related work using affective modulation).

Moreover, our approach permits learning with only one presentation of the reward, which is very useful in robotics where exploration is expensive—not

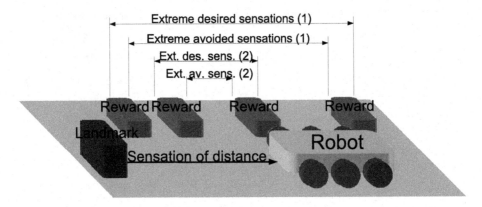

Fig. 14. Using successive detections of desired and avoided rewards, the robot could anticipate many rewards

only computationally but also in terms of the potential threats the robot can encounter while exploring, use of its batteries, and disadvantages that neglecting other activities the robot has to perform might bring to its performance and survival. Autonomous robots thus need to decide whether they should explore their environment or exploit their current knowledge about it in order to use what they learn efficiently. A number of strategies to achieve this have been proposed, e.g., [17,14,6].

In this paper we have shown how a robot can predict the presence of two rewards only; however, we can extend our approach to have the robot learn many more rewards looking for the two extreme desired sensations in between two extreme avoided sensations and so on (see Figure 14).

We are also currently working to expand this algorithm to more dimensions, so that the robot can learn about and recognize the landmark (caretaker) and the objects providing reward using different dimensions and sensory modalities. We will use these more elaborate learned desired and avoided sensations in a complete architecture making our robot act in order to respectively reach and avoid these sensations. For example, it will move forward when the closest desired sensation is to be closer to the landmark (or caretaker), and move backward when being closer to the landmark is the closest avoided sensation. External anticipatory mechanisms can be useful in helping our learning system to focus on unexpected or surprising situations, and therefore on what is likely to be the most relevant information. Moreover, it may be useful to cluster desired sensations with the context, or to treat them at a more abstract level in order to elaborate plans.

Acknowledgments

We would like to thank Carol Britton for feedback on an earlier draft of this paper. Arnaud Blanchard is funded by a research studentship of the University of Hertfordshire. This research is partly supported by the EU-funded Network of Excellence HUMAINE (FP6-IST-2002-507422).

References

1. Arkin, R.: Behavior-Based Robotics. The MIT Press, Cambridge (1998)
2. Avila-Garcia, O., Cañamero, L.: Using hormonal feedback to modulate action selection in a competitive scenario. In: Schaal, S., Ijspeert, J., Billard, A., Vijayakumar, S., Hallam, J., Meyer, J.A. (eds.) From Animals to Animats 8. Proceedings of the 8th International Conference on Simulation of Adaptive Behavior, pp. 243–252. The MIT Press, Cambridge, MA (2004)
3. Bateson, P.: What must be know in order to understand imprinting? In: Heyes, C., Huber, L. (eds.) The Evolution of Cognition, pp. 85–102. The MIT Press, Cambridge, MA (2000)
4. Blanchard, A., Cañamero, L.: From imprinting to adaptation: Building a history of affective interaction. In: Proc. of the 5th Intl. Wksp. on Epigenetic Robotics, pp. 23–30 (2005)
5. Blanchard, A.: and Cañamero, L.: Developing Affect-Modulated Behaviors: Stability, Exploration, Exploitation, or Imitation?. In: Kaplan, F., et al. (eds.) In: Proc. Sixth Intl. Conf. on Epigenetic Robotics: Modeling Cognitive Development in Robotic Systems, Lund University Cognitive Studies 128, 17–24 (2006)
6. Blanchard, A., Cañamero, L.: Modulation of exploratory behavior for adaptation to the context. In: Kovacs, T. J. M.(ed.) Biologically Inspired Robotics (Biro-net) in AISB'06: Adaptation in Artificial and Biological Systems. vol. II, pp. 131–139 (2006)
7. Butz, M.V., Sigaud, O., Gérard, P.: Internal models and anticipations in adaptive learning systems. In: Butz, M.V., Sigaud, O., Gérard, P. (eds.) Anticipatory Behavior in Adaptive Learning Systems. LNCS (LNAI), vol. 2684, Springer-Verlag, Heidelberg (2003)
8. Cañamero, L., Blanchard, A., Nadel, J.: Attachment Bonds for Human-Like Robots. International Journal of Humanoid Robotics 3(3), 301–320 (2006)
9. Cos-Aguilera, I.: Cañamero, L., and Hayes, G.: Learning Object Functionalisites in the Context of Action Selection. In: Nehmzow, U., Melhuish, C. (eds.) Towards Intelligent Mobile Robots, TIMR'03: 4th British Conference on Mobile Robotics. University of the West of England, Bristol, UK, August 28–29 (2003)
10. Cos-Aguilera, I.: Cañamero, L., and Hayes, G.: Motivation-Driven Learning of Action Affordances. In: Cañamero, L. (ed.) Agents that Want and Like: Motivational and Emotional Roots of Cognition and Action. Papers from the AISB'05 Symposium. University of Hertfordshire, UK, April 14–15, pp. 33–36 AISB Press, (2005)
11. Cover, T.M., Thomas, J.A.: Elements of Information Theory. Wiley-Interscience (1991)
12. Doya, K.: Reinforcement learning in continuous time and space. Neural Computation 12(1), 219–245 (2000)
13. Gaussier, P., Zrehen, S.: Perac: A neural architecture to control artificial animals. Robotics and Autonomous Systems 16, 291–320 (1995)
14. Oudeyer, P.Y., Kaplan, F.: Intelligent adaptive curiosity: a source of self-development. In: Berthouze, L., Kozima, H., Prince, C.G., Sandini, G., Stojanov, G., Metta, G., Balkenius, C. (eds.) Proc. of the 4th Intl. Wks. on Epigenetic Robotics. Lund University Cognitive Studies. vol. 117, pp. 127–130 (2004)

15. Prinz, W.: Perception and action planning. European journal of cognitive psychology 9(2), 129–154 (1997)
16. Rescorla, R., Wagner, A.: A.: A theory of pavlovian conditioning: Variations in effectiveness of reinforcement and nonreinforcement. In: Black, A., Prokasy, W. (eds.) Classical Conditioning II, pp. 64–99. Appleton-Century-Crofts, New York (1972)
17. Steels, L.: The autotelic principle. In: Iida, F., Pfeifer, R., Steels, L., Kuniyoshi, Y. (eds.) Embodied Artificial Intelligence. LNCS (LNAI), vol. 3139, pp. 231–242. Springer, Heidelberg (2004)
18. Sutton, R., Barto, A.: A temporal-difference model of classical conditioning. In: Proceedings of the Ninth Annual Conference of the Cognitive Science Society, 355–378 (1987)
19. Watkins, C.: Learning from Delayed Rewards. PhD thesis, King's College (1989)
20. Wolpert, D., Macready, W.: No free lunch theorems for optimisation. IEEE Trans. on Evolutionary Computation 1, 67–82 (1997)

Anticipatory Model of Musical Style Imitation Using Collaborative and Competitive Reinforcement Learning

Arshia Cont[1,2], Shlomo Dubnov[2], and Gérard Assayag[1]

[1] Ircam - Centre Pompidou - UMR CNRS 9912, Paris
{cont,assayag}@ircam.fr
[2] Center for Research in Computing and the Arts, UCSD, CA
sdubnov@ucsd.edu

Abstract. The role of *expectation* in listening and composing music has drawn much attention in music cognition since about half a century ago. In this paper, we provide a first attempt to model some aspects of musical expectation specifically pertained to short-time and working memories, in an anticipatory framework. In our proposition *anticipation* is the mental realization of possible predicted actions and their effect on the perception of the world at an instant in time. We demonstrate the model in applications to automatic improvisation and style imitation. The proposed model, based on cognitive foundations of musical expectation, is an active model using reinforcement learning techniques with multiple agents that learn competitively and in collaboration. We show that compared to similar models, this anticipatory framework needs little training data and demonstrates complex musical behavior such as long-term planning and formal shapes as a result of the anticipatory architecture. We provide sample results and discuss further research.

1 Introduction

About half a century ago, the musicologist Leonard Meyer drew attention to the importance of *expectation* in the listener's experience of music. He argued that the principal emotional content of music arises through the composer's choreographing of expectation [18]. Despite this significance, musical expectation has not enjoyed its cognitive importance in existing computational systems, which mostly favor prediction-driven architectures without enough cognitive constraints. In this paper, we will introduce a first attempt towards modeling musical systems with regards to the psychology of musical expectations. For modeling these constraints, we use anticipatory systems where several accounts of musical expectation are modeled explicitly. We claim that such cognitive modeling of music constitutes complex musical behavior such as long-term planning and generation of learned formal shapes. Moreover, we will show that the anticipatory approach greatly reduces the dimensions of learning and allows satisfactory performance when little data is available.

M.V. Butz et al. (Eds.): ABiALS 2006, LNAI 4520, pp. 285–306, 2007.
© Springer-Verlag Berlin Heidelberg 2007

We start the paper by studying the cognitive foundations of music as the core inspiration for the proposed architecture. In Section 2, we discuss important aspects of the psychology of music expectation such as auditory learning, mental representations of expectations and auditory memory. Our hope is that such studies create motivations for modeling complex musical behavior as proposed.

We demonstrate our system in applications to automatic music improvisation and style imitation as a direct way to showcase complex musical behavior. Musical style modeling consists of building a computational representation of the musical data that captures important stylistic features. Considering symbolic representations of musical data, such as musical notes, we are interested in constructing a mathematical model, such as a set of stochastic rules, that allows for the creation of new musical improvisations by means of intuitive interaction between human musicians or music scores and a machine. In Section 3, we study some of the important approaches in the literature for the given problem. We will be looking at these systems from two perspectives: that of representation and memory underlying the challenge of dimensionality of music information, and that of modeling addressing learning and grasping stylistic features.

Section 4 provides background on anticipatory modeling used in the proposed system. It also contains the main idea behind our anticipatory modeling of musical expectation. Our design explicitly addresses two types of anticipatory models introduced in [4]: state anticipation and payoff anticipation. In this work, we tend to model two aspects of musical expectations, namely dynamic adaptive and conscious expectations as discussed in Section 2.

The proposed architecture features reinforcement learning as an interactive module between the system and an outside environment and addresses adaptive behavior in auditory learning. In general, our model is a modular system that consists of three main modules: *memory, guides* and *learning*. The memory module serves compact representations and future access to music data. Guides are reinforcement signals from the environment to the system or from previous instances of the system onto itself that guide the learning module to relevant places in memory for updates and learning. The learning module captures stochastic behavior and planning through interactive learning. In Section 5, we overview the general architecture and each module will then be presented separately in sections 6 to 8. After the design, we demonstrate some generation results in Section 9. Results show evidence of long-term planning achieved through learning as an outcome of anticipatory modeling of working memory in music cognition. Furthermore, the system requires much less training data compared to similar systems, again due to the proposed anticipatory framework. We end this chapter by discussing the complexity of the proposed anticipatory architecture and future works.

2   Cognitive Foundations

The core foundations of the proposed model in this chapter are based on the *psychology of musical expectation*. In his recent book, David Huron vastly studies various aspects of music expectation [15]. Here, we highlight important aspects of his work along other cognitive facts pertinent to our proposal.

2.1 Auditory Learning

There is extensive evidence for the *learning* aspect of musical expectation through auditory learning and in opposition to innate aspects of these behaviors. One of the most important discoveries in auditory learning, has been that listeners are sensitive to the probabilities of different sound events and patterns, and these probabilities are used to form expectations about the future. An important research landmark in favor of this claim is the work in [24]. On the other hand, the brain does not store sounds. Instead, it interprets, distills and represent sounds. It is suggested that the brain uses a combination of several underlying presentations for musical attributes [15]. A good mental representation would be one that captures or approximates some useful organizational property of an animal's actual environment.

But how does the brain know which representation to use? Huron suggests that expectation plays a major role [15]. There is good evidence for a system of rewards and punishments, which evaluates the accuracy of our unconscious predictions about the world. Our mental representations are being perpetually tested by their ability to usefully predict ensuing events, suggesting that *competing and concurrent representations* may be the norm in mental functioning [15]. This view is strongly supported by the neural Darwinism theory of Edelman [11]. According to this theory, representations compete with each other according to Darwinian principles applied to neural selection. Such neural competition is possible only if more than one representation exists in the brain. In treating different representations and their expectation, each listener will have a distinctive listening history in which some representations have proved more successful than others.

2.2 Mental Representations of Expectation

According to Huron, memory does not serve for recall but for *preparation*. In chapter 12 of his book, Huron tries to address the structure rather than content of mental representations and introduces a taxonomy for auditory memory that constitutes at least four sources of musical expectations as follows:

Veridical Expectation: Episodic Memory is an explicit memory and a sort of autobiographical memory that holds specific historical events from our past. Episodic memory is easily distorted and in fact, the distortion occurs through repeated retelling or recollection. Most importantly, our memories for familiar musical works are episodic memories that have lost most of their autobiographical history while retaining their accuracy. This sense of familiarity or expectation of familiar works is refereed to, by Huron and Bharucha, as *Veridical expectation*.

Schematic Expectation: Schematic expectation is associated with *Semantic memory*; another type of explicit memory which holds only declarative knowledge and is distinguished from episodic by the fact that it does not associate the knowledge to any historical past but as stand-alone knowledge. This kind of memory is most active in first-exposure listening (when we do not know the piece) where our past observations and learned schemas are generalized. These sort of auditory generalizations are reminiscent of the learned categories characteristic of semantic memory.

Dynamic Adaptive Expectation: Expectation associated with Short-term memory is *Dynamic Adaptive Expectation*. It occurs when events do not conform with expectations that have been formed in the course of listening to the work itself. These expectations are updated in realtime especially during exposure to a novel auditory experience such as hearing a musical work for the first time.

Conscious Expectation: All the three types of expectations discussed above are unconscious in origin. Another important class of expectations arise from conscious reflection and prediction. Such explicit knowledge might come from external sources of information (such as program notes) or as part of a listener's musical expertise, or even arise dynamically while listening to a novel musical work. An argument for this last type of expectation, and most important for this work, is the perception of musical form during listening. This form of expectation comes from the mental desktop, which psychologists refer to as working memory.

All these expectation schemes operate concurrently and in parallel. Schematic expectations are omnipresent in all of our listening experiences. When listening to a familiar work, the dynamic-adaptive system remains at work – even though the veridical expectation anticipates exactly what to expect. Similarly, when listening for the first time to an unfamiliar work, the veridical system is constantly searching for a match with familiar works. The veridical system is essential for catching the rare moments of musical quotation or allusion. In short, an anticipatory effect such as *surprise* is a result of various types of interactions among these lower-level components of music expectation in cognition. For a thorough discussion see [15].

An ideal anticipatory model of music cognition should address all four types of expectations addressed above. However, for this work as a first attempt, we focus on *dynamic adaptive expectation* and *conscious expectation* and will address the rest in future works. With respect to conscious expectations, we are interested in expectations that arise dynamically while listening to a "new" musical work.

2.3   Memory and Reinforcement

The role of memory in the brain for music might hint to us how musical representations are stored and how they interact within themselves in the brain and with an outside environment. In the previous section we looked at the representational aspects of memory with regard to music expectation and here we briefly introduce the interactive level. This interactive level should guide us on how we can model memory access and learning for our purpose.

Snyder in [26] proposes an auditory model of memory that consists of several stages, from which we consider feature extraction, Long Term Memory (LTM) and Short Term Memory (STM). Feature extraction is a sort of perceptual categorization and grouping of data. Events processed at this stage can activate those parts of LTM evoked by similar events in the past. Activated LTM at this point forms a context for current awareness. This context takes the form of expectations that can influence the direction that current consciousness takes. Memory also acts like a filter determining which aspects of our environment we are aware of at a given time. LTM that reaches higher states of activation can then persist as current STM. Information in STM might be repeated or rehearsed. This rehearsal greatly *reinforces* the chances that the information being circulated in STM will cause modifications in permanent LTM. We consider both activation and reinforcement processes in our design of guide and learning modules.

Besides this unconscious level of reinforcement, like sensory representations, conscious thinking also requires some guidance and feedback to ensure that thinking remains biologically adaptive [15]. Useful thinking needs to be rewarded and encouraged, while useless thinking needs to be suppressed or discouraged.

3   Background on Stochastic Music Modeling

In this section, we look at several prior attempts at modeling music signals either for generation (automatic improvisation or style imitation) or modeling long-term dependencies observed in music time series. In this work, we are interested in *automatic* systems where there are no rules or a priori information abducted into the system by experts and everything is learned through the life-span of the system. Moreover, we are interested in systems which address directly the complexity of music signals as will be clear shortly.

Earlier works on style modeling employed information theoretical methods inspired by universal prediction. In many respects, these works build upon a long musical tradition of statistical modeling that began in the 50s with Hiller and Isaacson's "Illiac Suite" [14] and Xenakis using Markov chains and stochastic processes for musical composition [29]. In what follows, we will review some of the state-of-the-art systems proposed in the literature from two standpoints: musical representation and stochastic modeling.

3.1   Musical Representation

Music information has a natural componential and sequential structure. While sequential models have been extensively studied in the literature, componential or multiple-attribute models still remain a challenge due to complexity and explosion in the number of free parameters of the system. Therefore, a significant challenge faced by music signals arises from the need to simultaneously represent and process many attributes of music information. The ability (or inability) of a system to handle this level of musical complexity can be revealed by studying its ways of musical representations or memory models both for storage and learning. Here, we will compare different memory models used and proposed in the literature for systems considering this complex aspect of music signals. We will undertake this comparison by analytically looking at each model's complexity and its modality of interaction across attributes, which in term determine its power of (musical) expressivity. We will be looking at *cross-alphabets* [9,2,20], *multiple-viewpoints* [6] and mixed memory *Factorial Markov models* [25].

In order to better understand each model in this comparison, we use a toy example demonstrated in Figure 1 containing the first measure of J.S.Bach's *two-part invention No. 5* (Book II). The music score in figure 1 is parsed between note onsets to obtain distinct events through time as demonstrated. In this article, we consider discrete MIDI signals as is clear from the figure. For the sake of simplicity, we only represent three most important attributes, namely pitch, harmonic interval and beat duration of each parsed event as shown in Table 1. This table represents 15 time series vectors I_t corresponding to 15 parsed events in Figure 1, where each event has three components (i_t^ℓ). Let k_t denote the number of components for each vector I_t and n_t^ℓ denote the dictionary size for each attribute i_t^ℓ. Later in Section 6, we will use the same example to demonstrate our alternative representation scheme.

Table 1. Music attributes for distinct events of parsed score in Figure 1

Event Number I_t	I_1	I_2	I_3	I_4	I_5	I_6	I_7	I_8	I_9	I_{10}	I_{11}	I_{12}	I_{13}	I_{14}	I_{15}
MIDI Pitch (i_t^1)	0	51	63	62	63	0	65	0	67	67	63	68	68	58	60
Harmonic Interval (i_t^2)	0	0	0	0	24	0	0	0	4	5	0	8	6	0	0
Duration (i_t^3)	1	1	1	1	1	1	1	2	1	1	1	1	1	1	1

Cross Alphabets. The simplest model used so far in music applications is *cross-alphabet* where a symbol is a vector of multiple attributes. Therefore, cross-alphabet models are very cheap but they do not model interaction among components in any ways. To overcome this shortcoming, researchers have considered various membership functions to allow for these context dependencies through various heuristics [2,20]. Such heuristics might make the system dependent upon the style of music being considered or reduce generalization capabilities. Moreover, as the number of components (or dimensions) increase this representation becomes less informative of the underlying structure.

In our toy example each symbol of the alphabet is a unique 3-dimensional vector. Note that in this specific example, there are 15 alphabets since none of them is being reused despite considerable amount of modal interactions among components and high autocorrelations of each independent component.

Multiple Viewpoints. The *multiple viewpoints model* [6] is obtained by deriving individual expert models for each musical attribute and then combining the results obtained from each model. This means that in the multiple viewpoint model of the above example, three other rows for two-dimensional representations of <pitch, harmonic interval>, <pitch, duration>, etc. and one row

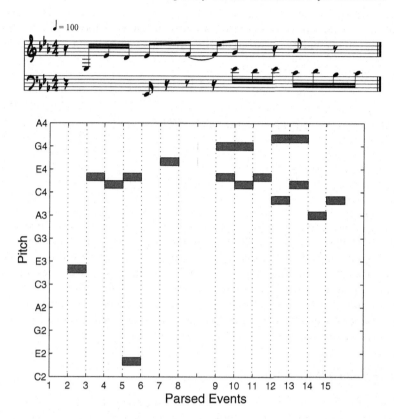

Fig. 1. Parsed PianoRoll presentation for the first measure of J.S.Bach's *two-part Invention No.5* (Book II) with quantization of $\frac{1}{16}$ beats

of three-dimensional vectors are added to the representation. At this stage, the model's *context* is constructed.

Multiple viewpoint models are more expressive than cross-alphabet models since by combining models we allow modal interactions among components. Moreover, the system can reach parts of the hypothesis space that the individual models would not be able to reach. However, the context space is obviously too large and hence, learning requires a huge repertoire of music for training data to generate few musical phrases [6]. In our toy example, with 9 distinct pitches, 6 distinct harmonic intervals and 2 durations, the state-space of this model amounts to $9 + 6 + 2 + 54 + 18 + 12 + 108 = 209$.

Factorial Markov Models. Mixed memory models are geared towards situations where the combinatorial structure of state space leads to an explosion of the number of free parameters. But unlike the above methods, the alphabets of the dictionary are assumed to be known instead of them being added online to the system. Factorial Markov models, model the coupling between components in a compact way.

To obtain a compact representation, we assume that components at each time t are conditionally independent given the previous vector event at $t-1$, or

$$P(I_t|I_{t-1}) = \prod_{\nu}^{k} P(i_t^{\nu}|I_{t-1}) \tag{1}$$

and that the conditional probabilities $P(i_t^{\nu}|I_{t-1})$ can be expressed as a weighted sum of "cross-transition" matrices:

$$P(i_t^{\nu}|I_{t-1}) = \sum_{\mu=1}^{k} \phi^{\nu}(\mu)a^{\nu\mu}(i_t^{\nu}|i_{t-1}^{\mu}) \tag{2}$$

where $\phi^{\nu}(\mu)$s are positive numbers that satisfy $\sum_{\mu} \phi^{\nu}(\mu) = 1$ and measure the amount of correlation between the different components of the time series. A non-zero $\phi^{\nu}(\mu)$ means that all the components at *one time step* influence the νth component at the next. The parameters $a^{\nu\mu}(i'|i)$ are $n \times n$ transition matrices, which provide a compact way to parameterize these influences [25].

The number of free parameters in Equation 2 is therefore upper-bounded by $O(k^2 n^2)$ (where n denote $\max n_i{}^1$) and the state-space size is $\prod_i n_i$. In our toy example the state-space size of the system would be $9 \times 6 \times 2 = 108$.

3.2   Stochastic Modeling

In this section, we review the systems mentioned above in terms of their ways of learning stochastic rules or dependencies from given musical sequences in order to generate new ones in the same style of music.

The most prevalent type of statistical model encountered for music are *predictive* models based on *context* implying general Markov models [5]. Universal prediction methods improved upon the limited memory capabilities of Markov models by creating context dictionaries from compression algorithms, specifically using the Lempel-Ziv incremental parsing [30], and employing probability assignment according to Feder et al. [12]. Music improvisation was accomplished by performing a random walk on the phrase dictionary with appropriate probabilistic drawing among possible continuations [10,9,20]. Later experiments explored Probabilistic Suffix Tree (PST) [22], and more recently in [2] using Factor Oracle (FO) [1]. Other methods include the use of Genetic Algorithms [3] and neural networks [13] just to name a few.

The inference and learning structures for Multiple Viewpoint Models (Section 3.1) can be categorized as *Ensemble Learning* algorithms and have had multiple manifestations [21,6]. One advantage of this type of modeling is the explicit consideration of long-term dependencies during learning where they combine the viewpoint predictions separately for long-term and short-term models

---

[1] In the original paper on factorial Markov models, the authors assume that the dictionary sizes are all the same and equal to n. For the sake of comparison we drop this assumption but keep n as defined above to obtain the coarse definition in Equation 2.

[21]. Due to the explosion of parameters, results of learning are hard to visualize and assess. Their generation samples are usually only a few monophonic bars out of learning on an entire database of music (e.g. all Bach chorals).

Despite the explicit componential representation of Factorial Markov Models, the correlation factors $\phi^\nu(\mu)$ model only *one step* dependencies and lack modeling long-term behavior, essential in computational models of music. Correspondingly, authors use this method to analyze correlations between different voices in componential music time series without considering generation [25].

Another main drawback of the above methods is lack of responsiveness to changes in musical situations that occur during performance, such as dependence of musical choices on *musical form* or changes in *interaction* between players during improvisation. Interaction has been addressed previously in [20] for PST based improvisation by means of a fitness function that influenced prediction probabilities according to an ongoing musical context, with no consideration of planning or adaptive behavior. Statistical approaches seem to capture only part of the information needed for computer improvisation, i.e. successfully modeling a relatively short term stylistics of the musical *surface*. Although variable Markov length and universal methods improve upon the finite length Markov approach, they are still insufficient for modeling the true complexity of music improvisation.

4   Background on Anticipatory Modeling

All of the systems reviewed in the previous section are based on predictions out of a learned context. In this work, we extend this view by considering *musical anticipation*, in accord with the psychology of musical expectation. Anticipation is different from both prediction and expectation. An anticipatory system, in short, is "a system containing a *predictive model* of itself and of its *environment*, which allows it to change state at an instant in accord with the model's predictions pertaining to a later instant" [23]. More concretely, *Anticipation* is the mental realization of possible predicted actions and their effect on the perception and learning of the world at an instant in time. Hence, anticipation can be regarded as a marriage of actions and expectations. In this framework, an anticipatory system is in constant interaction with an outside environment, for which it possesses an internal predictive model. In an anticipatory system, action decisions are based on future predictions as well as past inference. It simulates adaptive frameworks in the light of different behaviors occurring in interaction between the system with itself and/or its environment. In this view, the anticipatory effect can be described as a reinforcing feedback as a result of the interaction between the system and the environment onto the system.

Butz et al. [4] draw distinctions between four types of anticipation for modeling: *Implicit, Payoff, Sensory,* and *State* anticipations. We did not find a direct correspondence between those mentioned in Section 2. The proposed system in this chapter is both a *payoff anticipatory system* and *state anticipation system*. Figure 2 shows the diagrams for both models separately and how they use future predictions for decision making. The system proposed hereafter is state anticipatory because of the explicit use of prediction and anticipation during both learning and decision making. It is also a payoff anticipatory system because of the selective behavior caused by the collaborative and competitive learning and generation discussed in Section 8. From a musical standpoint following our introduction in Section 2, we attempt implicit modeling of short-term and working memories responsible for dynamic adaptive expectation and long-term planning.

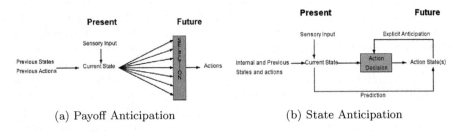

(a) Payoff Anticipation   (b) State Anticipation

Fig. 2. Anticipatory Modeling diagrams used in the proposed system

Davidsson in [7] proposes a framework for preventive anticipation where he incorporates collaborative and competitive multiple agents in the architecture. While this has common goals with our proposal, ours is different since Davidsson uses rule-based learning with ad-hoc schemes for collaboration and competition between agents. Recently, in the computer music literature, Dubnov has introduced an anticipatory information rate measure that, when run on non-stationary and time varying data such as audio, can capture anticipatory profile and emotional force data that has been collected using experiments with humans [8].

5   General Architecture

After the above introduction, it is natural to consider a *reinforcement learning (RL)* architecture for our anticipatory framework. The reinforcement learning problem is meant to be a straightforward framing of the problem of learning from interaction to achieve a goal. The learner and decision-maker is called the agent. The thing it interacts with, comprising everything outside the agent, is called the environment. These interact continuously, the agent selecting actions and the environment responding to those actions and presenting new situations to the agent. The environment also gives rise to rewards, special numerical values that the agent tries to maximize over time. This way, the model or agent is interactive in the sense that the model can change through time according to reinforcement signals sent by its environment. Any RL problem can be reduced to three signals passing back and forth between an agent and its environment: one signal to represent the choices made by the agent (the actions), one signal to represent the basis on which the choices are made (the states), and one signal to define the agent's goal (the rewards)[27]. In a regular RL system, rewards are defined for goal-oriented interaction. In musical applications, defining a *goal* would be either impossible or would limit the utility of the system to a certain style. In this sense, the rewards used in our interaction are rather *guides* to evoke or repress parts of the learned model in the memory, as discussed in Section 7.

In our system, agents are learners and improvisers based on the model-based RL *Dyna* architecture [27]. Here, the environment is anything that lies outside the agent, or in this case, a human performer or a music score fed sequentially into the system. Each agent has an internal model of the environment and adapts itself based on new musical phrases and rewards it receives at each interaction. For our purpose, we propose two execution modes as demonstrated in Figure 3. In the first, referred to as *Interaction mode*, the system is interacting either with a human performer (live) for machine improvisation or with music score(s) for style imitation and occurs when external information is being passed to the system from the environment. During the second mode, called *self listening mode*, the system is in the generation phase and is interacting with itself.

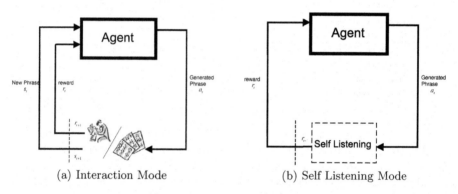

(a) Interaction Mode   (b) Self Listening Mode

Fig. 3. Machine Improvisation modes diagram

The internal models in agents play the role of *memory* and *mental representations* of input sequences from the environment and will be detailed in the following section. At each instance of interaction, the agents update their models and learn strategies as discussed in Section 7, using guides or rewards presented in Section 8.

6   Musical Representation

Representation of musical sequences in our system serves as musical memory, mental representation of music signals and internal models of the agents. A single music signal has multiple attributes and as stated earlier, each attribute is responsible for an individual mental representation which *collaborates* and *competes* with others during actions and decision making. This collaboration and competition is handled during learning and is discussed in Section 8. For now, it suffices to say that the agent in both modes of interaction in Figure 3 consists of multiple agents, each responsible for *one* musical attribute. This feature is of great importance since it reduces the dimensionality of the system during learning, allowing it to interact when small data is available and in a fast way. The number of attributes and nature of them are independent of the agent architecture. For this experiment, we hand-picked 3 different attributes (pitch, harmonic interval and quantized duration in beats) along with their first order difference, hence a total of 6. Upon the arrival of a MIDI sequence, it is quantized, cut into polyphonic "slices" at note onset/offset positions, and then different viewpoints are calculated for each slice. Slice durations are represented as multiples of the smallest significant time interval that a musical event would occupy during a piece (referred to as *Tatum*). For demonstration, Table 2 shows these features as time series data calculated over the score of figure 1.

Table 2. Time series data on the score of figure 1 showing features used in this experiment

Event Number	1	2	3	4	5	6	7	8	9	10	11	12	13	14	15
Pitch (MIDI)	0	51	63	62	63	0	65	0	67	67	63	68	68	58	60
Harmonic Int.	0	0	0	0	24	0	0	0	4	5	0	8	6	0	0
Duration	1	1	1	1	1	1	1	2	1	1	1	1	1	1	1
Pitch Diff..	0	0	12	-1	1	0	0	0	0	0	1	-3	0	-4	2
Harm. Diff.	0	0	0	0	0	0	0	0	0	1	0	0	-2	0	0
Dur. Ratio	1	1	1	1	1	1	1	2	0.5	1	1	1	1	1	1

After the data for each viewpoint is gathered it has to be represented and stored in the system in a way to reflect principles discussed in Section 2. Of most importance for us are the expressivity of the model, least computational complexity and easy access throughout the memory model. There are many possible solutions for this choice. In our multiple-agent framework, we have chosen to store each attribute as a *Factor Oracle (FO)* [1]. In this paper, we give a short description of the properties and construction of FO and leave out the formal definitions and musical interests [1,2]. Basically, a factor oracle is a finite state automaton learned incrementally in linear time and space. A learned sequence of symbols $A = a_1 a_2 \cdots a_n$ ends up in an automaton whose states are $s_0, s_1, s_2 \cdots s_n$. There is always a transition arrow labeled by symbol a_i going from state s_{i-1} to state s_i. Depending on the structure of A, other arrows may appear: *forward transitions* from a state s_i to a state s_j, $0 \leq i < j <= n$, labeled by symbol a_j; *suffix links*, directed backward and bearing no label. The forward transitions model a factor automaton, that is every factor $p = a_i a_{i+1} \cdots a_{j-1} a_j$, $1 \leq i \leq j \leq n$ in A corresponds to an unique transition path labeled by p, starting in s_0 and ending in state s_j. Suffix links connect repeated patterns of A, i.e. states sharing large common suffixes. In general, given a sequence, the constructed FO returns two *deterministic* functions: a transition function $F_{trn} : S \times \Sigma \to \{S \cup \emptyset\}$ and suffix links $F_{sfx} : S \to \{S \cup \emptyset\}$, where S is the set of states and Σ is the alphabet on which A is constructed. Figure 4 shows four instances of FO construction over data presented in Table 2.

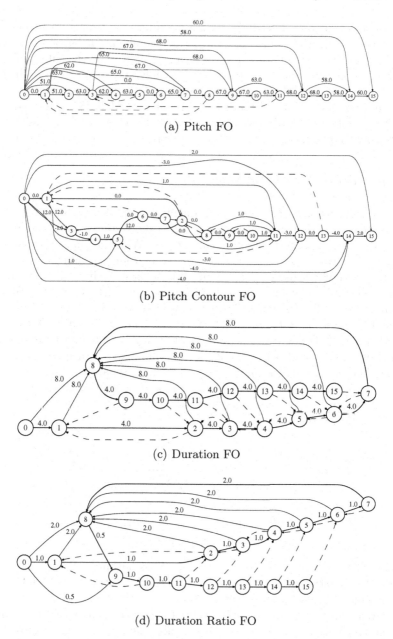

(a) Pitch FO

(b) Pitch Contour FO

(c) Duration FO

(d) Duration Ratio FO

Fig. 4. Learned Factor Oracles over pitch and duration sequences of Table 2. Each node represents a state, each solid line a *transition* and dashed line a *suffix link*.

An important property of FO for this work is their power of generation. Navigating the oracle and starting in any place, following forward transitions gener-ates a sequence of labeling symbols that are repetitions of portions of the learned sequence; following one suffix link followed by a forward transition generates an alternative path in the sequence, creating a recombination based on a shared suffix between the current state and the state pointed at by the suffix link. This shared suffix link is called context in context-inference models. In addition to completeness and incremental behavior of this model, the best suffix is known at the minimal cost of just following one pointer. By following more than one suffix link before a forward jump or by reducing the number of successive factor link steps, we make the generated variant less resemblant to the original.

7   Environmental Interactions

Guide signals received from the environment are an essential part of the proposed system since they define the sensitivity of the system to the outside world, directions it can take and its musical capabilities. When at time t the new music sequence $A^t = a_1 a_2 \cdots a_N$ is received from the environment, an ideal reward signal should reinforce those parts of memory that most likely evoke the sequence received to be able to generate recombinations or musically meaningful sequences thereafter. In the RL framework, this means that we want to assign numerical rewards to *transition states* and *suffix states* of an existing Factor Oracle with internal states s_i. Guide computation occurs using the previously learned FOs (defined by FO^{t-1}) and before incorporating the new sequence into the model.

After different attributes of A^t are extracted as separate sequences each in form $\{x_1 \ldots x_N\}$, we use a *probability assignment function* P from $S^* \to [0,1]$ (where S^* is the set of all n-tuples of states available to FO) to assign rewards to states in the model as follows:

$$P(s_{1*} s_{2*} \ldots s_{N*} | FO^{t-1}) = \left[\sum_{i=1}^{N} p(x_i | s_{i*}) \right] / N \qquad (3)$$

where

$$p(x_i | s_i^*) = \begin{cases} 1 & \text{if } F_{trn}(s_{i-1}^*, x_i) = s_i^* \\ 0 & \text{if } F_{trn}(s_{i-1}^*, x_i) = \emptyset \end{cases} \qquad (4)$$

Of course the exploration of the search space S^* is optimized by not considering all possible n-tuples. Instead a simple forward checking strategy is used to reduce S^* to significantly rewarded subsets.

Rewards out of Equation 3 reinforce the states in the memory that are factors of the new sequence A^t. In other words, it will *guide* the learning described later to the places in the memory that should be mostly regarded during learning and generation. For example, the reward for $\{s_{1*} s_{2*} \ldots s_{N*}\}$ would be equal to 1 if the state/transition path $s_{1*} \ldots s_{N*}$ regenerate literally the sequence A^t.

To assign rewards to suffix links, we recall that they refer to previous states with the largest common suffix. Using this knowledge, a natural reward for a suffix link would be proportional to the *length* of the common suffix that the link is referring to. Fortunately, using a factor oracle structure, this measure can be easily calculated online and has been introduced in [16]. Note that the process defined above assigns numerical values to states pertaining to associate paths (transitions or suffix links) in each FO. This value is an *immediate* reward, noted by $r(s_t, a_t)$ for emission of a symbol a_t while at state s_t.

Rewards or guides are calculated the same way for both modes of the system described before with an important difference. We argue that the rewards for the *interaction mode* (Figure 3(a)) correspond to a *psychological attention* towards appropriate parts of the memory and guides for the *self-listening mode* (Figure 3(b)) correspond to a *preventive* anticipation scheme. This means that while interacting with a live musician or sequential score, the system needs to be attentive to input sequences and during self-listening it needs to be preventive so that it would not generate the same path over and over. Moreover, these schemes provide the conscious reinforcement required to encourage or discourage useful and useless thinking as mentioned in Section 2.3. This is achieved by treating environmental rewards with positive and negative signs appropriately.

8 Interactive Learning

Reinforcement Learning techniques are mostly studied within Markov Decision Process (MDP) framework. An MDP in general is defined by a set of states-action pairs $S \times A$, a reward function $R : S \times A \to \mathbb{R}$ and a state transition function $T : S \times A \to S$. Given this MDP, RL techniques aim to find the policy as a mapping probability $Q(s, a)$. To conform the representational scheme presented before to this framework, we define MDP state-action pairs as FO states and emitted symbol from that state. The transition function would then be the deterministic FO transition functions as defined before. This way the policy can be represented as a matrix Q which stores values for each possible state-action pair in a given FO.

In a regular reinforcement learning session, the system simulates itself up to a fixed number of episodes with terminal states, in order to maximize the overall reward during each episode by learning a Q matrix. At each interaction cycle with their environments, depending on their mode of interaction (Figure 3), the agents receive guides, update their existing models, learn new ones (as FOs and only during the *interaction mode*), and learn policies through some *Q-learning* algorithm by simulating episodes of system run. In this view, one can say that during each learning episode the system is *practicing* or improvising fixed length pieces using what it has learned so far in order to adapt itself to new musical situations defined by the newly arrived sequence and to learn and update policies. The main cycle of the interactive learning is shown in algorithm 1. This architecture is based on Dyna [27] with multiple agents and FOs as models. This cycle happens at each interaction between the system and the environment. During the self-listening mode, the algorithm is the same except that FOs are not updated.

Hereafter, we focus on the policy learning algorithm. At this stage, the algorithm simulates episodes of improvisations using previously learned policies and

1 Receive the new sequence A^t from the environment;
2 Compute guides on FO^{t-1}s;
3 Update FO^{t-1}s to FO^ts using A^t;
4 Learn policies (Q matrices);

Algorithm 1. Interactive Learning

updates the Q matrices in order to maximize the environmental rewards. This RL module must conform to cognitive foundations presented in Section 2, i.e. agents should be collaborative, competitive, and memory-based.

8.1   Competitive and Collaborative Learning

As discussed in Section 2.1, different mental representations of music work in a collaborative and competitive manner based on their predictive power to make decisions. This can be seen as kind of a *model selection* where learning uses all the agents' policies available and chooses the best one for each episode. This winning policy would then become the *behavior policy* with its policy followed during that episode and other agents being influenced by the actions and environmental reactions from and to that agent.

At the beginning of each episode, the system selects one agent using the probability in Equation 5, with positive parameter β_{sel}, and M as the total number of agents or attributes. Low β_{sel} causes equiprobable selection of all modules and vice versa. This way, a behavior policy π^{beh} is selected *competitively* at the beginning of each episode based on the value of the initial state s_0 among all policies π^i as demonstrated in Equation 5.

$$Pr(i|s_0) = \frac{e^{\beta_{sel} \sum_k Q^i(s_0, a_k)}}{\sum_{j=1}^{M} e^{\beta_{sel} \sum_r Q^j(s_0, a_r)}} \quad , \quad \pi^{beh} = \underset{i}{\mathrm{argmax}}\, Pr(i|s_0) \quad (5)$$

During each learning episode, the agents would be following the behavior policy. For update of π^{beh} itself, we can use a simple Q-learning algorithm but in order to learn other policies π^i, we should find a way to compensate the mismatch between the target policy π^i and the behavior policy π^{beh}. Uchibe and Doya [28] use an *importance sampling* method for this compensation and demonstrate the implementation over several RL algorithms. Adopting their approach, during each update of π^i when following π^{Beh} we use a compensation factor $IS = \frac{Q^i(s_m, a_m)}{Q^{Beh}(s,a)}$ during Q-learning as depicted in Equation 6, where (s_m, a_m) are *mapped* state-action pairs of (s, a) in behavior policy to attribute i, and α is the learning rate.

$$Q^i(s_m, a_m) = Q^i(s_m, a_m) + \alpha \left[R(s_m) + \gamma \cdot IS \cdot \max_{a'}(Q^i(s_m, a')) - Q^i(s_m, a_m) \right] \quad (6)$$

$R(.)$ in the above equation is different from the immediate reward $r(.,.)$ introduced in Section 7. In an anticipatory system, we are interested in the impact of future predictions on the current state of the system. This means that the reward for a state-action pair would correspond to future predicted states. With this regard, Equation 7 calculates $R(s_t)$ with γ as a discount factor. Future predicted states and actions (s_{t_i}, a_{t_i}) are obtained by applying an ϵ-greedy algorithm on the current policy matrix and starting from s_t.

$$R(s_t) = \sum r(s_t, a_t) + \gamma r(s_{t+1}, a_{t+1}) + \cdots + \gamma^m r(s_{t+m}, a_{t+m}) + \cdots \quad (7)$$

This scheme defines the *collaborative* aspect of interactive learning. For example, during a learning episode, pitch attribute can become the behavior policy Q^{beh} and during that whole episode the system follows the pitch policy for simulations and other attributes' policies $Q^i(.,.)$ will be influenced by the behavior of the pitch policy as shown in Equation 6.

8.2   Memory-Based Learning

In the Q-learning algorithm above, state-action pairs are updated during each episode through an ϵ-greedy algorithm on previously learned policies and using updated rewards. This procedure updates one state-action pair at a time. In an ideal music learning system, each immediate change should evoke previous related states already stored in the memory. In general, we want to go back in the memory from any state whose value has changed. When performing updates, the value of some states may have changed a lot while others rest intact, suggesting that the predecessor pairs of those who have changed a lot are more likely to change. So it is natural to prioritize the backups according to measures of their urgency and perform them in order of priority. This is the idea behind *prioritized sweeping* [19] embedded in our learning with the priority measure in Equation 8 for a current state s and next state s', leading to a priority queue of state-action pairs (chosen by a threshold θ) to be traced backwards for more updates.

$$p \leftarrow |R(s) + \gamma \max_{a'}(Q^{Beh}(s', a')) - Q^{Beh}(s, a)| \quad (8)$$

9   Generation Results

There are many ways to generate or improvise once the policies for each attribute are available. We represent one simple solution using the proposed architecture. At this stage, the system would be in the *self listening mode* (Figure 3(b)). The agent would generate *phrases* of fixed length following a behavior policy (learned from the previous interaction). When following the behavior attribute, the system needs to *map* the behavior state-action pairs to other agents in order to produce a complete music event. For this, we first check and see whether there are any common transitions between original attributes and, if not, we would follow the policy for their derivative behavior. Once a phrase is generated, its (negative) reinforcement signal is calculated and policies are updated as in Section 8 but without updating the current models (FOs).

Improvisation Session after learning on Invention No.3 by J.S.Bach

Fig. 5. Style imitation sample result

Audio results of automatic improvisation sessions on different styles can be heard at the following URL:

http://www.crca.ucsd.edu/arshia/ABiALS06/

As a sample result for this paper, we include analysis of results for style imitation on a polyphonic piece, *two-part Invention* No.3 by J.S. Bach. For this example, the learning phase was run in *interaction mode* with a sliding window of 50 events with no overlaps over the original MIDI score. After the learning phase, the system entered *self listening* mode where it generates sequences of 20 events and reinforces itself until termination. Parameters used for this session were $\alpha = 0.1$ (in Equation 6), $\gamma = 0.8$ (in Equation 7), $\theta = 2$ for prioritized sweeping threshold, and $\epsilon = 0.1$ for the *epsilon*-greedy selection of state-action pairs. Number of episodes simulated during each RL phase was 100. The generated score is shown in Figure 5 for 240 sequential events where the original score has 348. For this generation, the *pitch* behavior has *won* all generation episodes and direct mappings of *duration* and *harmonic* agents have been achieved 76% and 83% in total respectively leaving the rest for their derivative agents.

While both voices follow a polyphonic structure, there are some formal musicological structures that can be observed in the generated score. Globally, there are *phrase* boundaries in measures 4 and 11 which clearly segment the score into three formal sections. Measures 1 to 4 demonstrate some sort of exposition of musical material, which are expanded in measures 7 to the end with a transition phase in measure 5 and 6 ending at a week cadence on G (a fourth in the given key). There are several thematic elements which are reused and expanded. For example, the repeated D notes appearing in measures 2 appear several times in

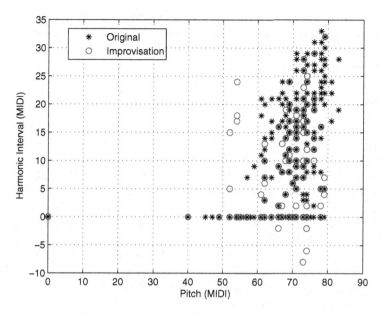

Fig. 6. Improvisation Space vs. Original Space

the score notably in measure 7 as low A with a shift in register and harmony and measure 9 and 15. More importantly, these elements or their variants can be found in the original score of Bach.

Figure 6 shows the pitch-harmony space of both the original MIDI and the generated score. As is seen, due to the collaborative and competitive multi-agent architecture of the system, there are new combinations of attributes which do not exist in the trained score.

10 Discussions

In this chapter we presented an anticipatory model of music cognition with application to automatic improvisation and style imitation. The proposed model covers short-term and working memory processes introduced in music cognition literature that result in dynamic adaptive expectations and long-term planning. The anticipatory model, in ABiALS terms, is a payoff and state anticipatory system which provides attentive and preventive frameworks during computation. We show that generation results demonstrate long-term and complex behavior thanks to this anticipatory and cognitive model.

Before any discussion, we would like to bring forth the difficulty of *evaluation* in case of automatic music generation. As should be clear to any musical reader, assessing a music generator in an objective manner, if not impossible, would set along disputable measures of goodness. On the other hand, in most music practices and styles, what is considered as *wrong* can be constituted as a *feature* depending on the context. Therefore we do not discuss the outcome of our design in aesthetic terms. Such considerations might become possible by careful design of perceptual experiments with human subjects, which we will address in our future work. Here we discuss further issues such as complexity of the proposed model and further research.

10.1 Model Complexity

In the architecture introduced above, because of the concurrent and competitive multiple agent structure, each component or attribute is modeled separately and the state-space size increases linearly with *time* as $k \times T$ coming down to 45 for the toy example. Modal interaction is not modeled by directed graphs between components but rather by influence of each attribute on others through the IS term in Equation 6 as a result of collaboration and competition between agents. Note that this choice comes from cognitive foundations of music and was not made for mere simplicity. The complexity of the system depends linearly on T, n_is and an adaptive environmental factor. This is because the arrows of the state-space model are inferred on-line and are dependent on the context being added to previous stored knowledge. We could say that it has an upper-bound of $O(nkT)$ but is usually much sparser than that.

The fact that T is a factor of both state-space size and complexity has advantages and shortcomings. The main advantage of this structure is that it easily enables us to access long-term memory and to calculate long-term dependencies, induce structures, and go back and forth in the memory at ease. However, one might say that as T grows, models such as Factorial Markov would win over the proposed model in terms of complexity since n_i would not change too much after some time T. This shortcoming is partially compensated by considering the phenomena of *finite memory process*. A finite memory process in our application is one that, given a factor oracle with N states and an external sequence A^t, can generate the sequence through a finite window of its history without using all its states [17]. More formally, this means that there exist a nonnegative number m such that the set $\{s_n \in FO : n \in \mathbb{N} \text{ and } n \in [N - m, N]\}$ would suffice for regeneration of a new sequence A^t. This notion is also supported by the fact that music data in general is highly repetitive [15] and not considering this would cause high reinforcement of earlier states in the memory through time. The parameter m is usually dependent on the style of music but for this presentation we keep it fixed.

Besides observing results, compared to similar systems, an anticipatory model reduces the complexity of the representation and learning. The proposed model and shown result need much less training data for learning (a single piece of music as training data to generate a rather long polyphonic sample) and is currently being developed for realtime improvisation.

10.2 Further Developments

As mentioned earlier, an ideal anticipatory model of music should consider all expectation processes in music perception mentioned in Section 2. In our first experiment, we attempted two and left the more difficult semantic and episodic processes for later works. To compliment the system, we would need more intelligent modules for music semantic learning and better representational schemes. Note that the sample result in figure 5 is a result of automatic interactive reinforcement learning without explicit consideration for semantic notions such as harmonic progressions or counterpoints. Adding these two notions to the system should further improve local consistencies in the results.

The interactive learning module can still be more efficient in each episode by considering directly relevant states for updates. This will bring us to the notion of Active Learning for future work. Also, note that the representation module using Factor Oracles does not in any way represent the complexity of feature extraction and perceptual bindings of the auditory system in the brain. It was rather chosen as a very efficient way to gather repetitive factors and structures in a sequence. Further alternatives should be studied for enhancement of this model.

References

1. Allauzen, C., Crochemore, M., Raffinot, M.: Factor oracle: A new structure for pattern matching. In: Bartosek, M., Tel, G., Pavelka, J. (eds.) SOFSEM 1999. LNCS, vol. 1725, pp. 295–310. Springer, Heidelberg (1999)
2. Assayag, G., Dubnov, S.: Using factor oracles for machine improvisation. Soft. Computing 8-9, 604–610 (2004)
3. Biles, J.A.: Genjam in perspective: A tentative taxonomy for genetic algorithm music and art systems. Leonardo 36(1), 43–45 (2003)
4. Butz, M.V., Sigaud, O., Gérard, P.: Internal models and anticipations in adaptive learning systems. In: Butz, M.V., Sigaud, O., Gérard, P. (eds.) Anticipatory Behavior in Adaptive Learning Systems. LNCS (LNAI), vol. 2684, pp. 86–109. Springer, Heidelberg (2003)
5. Conklin, D.: Music generation from statistical models. In: Proceedings of Symposium on AI and Creativity in the Arts and Sciences, pp. 30–35 (2003)
6. Conklin, D., Witten, I.: Multiple viewpoint systems for music prediction. In Journal of New. Music Research 24, 51–73 (1995)
7. Davidsson, P.: A framework for preventive state anticipation. In: Butz, M.V., Sigaud, O., Gérard, P. (eds.) Anticipatory Behavior in Adaptive Learning Systems. LNCS (LNAI), vol. 2684, pp. 151–166. Springer, Heidelberg (2003)
8. Dubnov, S.: Spectral anticipations. Computer Music Journal (2006)
9. Dubnov, S., Assayag, G., Lartillot, O., Bejerano, G.: Using machine-learning methods for musical style modeling. IEEE Computer Society 36(10), 73–80 (2003)
10. Dubnov, S., El-Yaniv, R., Assayag, G.: Universal classification applied to musical sequences. In: Proc. of ICMC, pp. 322–340, Michigan (1998)
11. Edelman, G.: Neural Darwinism: The Theory of Neuronal Group Selection. Basic Books (1987)
12. Feder, M., Merhav, N., Gutman, M.: Universal prediction of individual sequences. IEEE Trans. Inform. Theory 38(4), 1258–1270 (July 1992)
13. Franklin, J. A.: Predicting reinforcement of pitch sequences via lstm and td. In: Proc. of International Computer Music Conference, Miami, Florida (2004)

14. Hiller, L.A., Isaacson, L.M.: Experimental Music: Composition with an Electronic Computer. McGraw-Hill Book Company, New York (1959)
15. Huron, D.: Sweet Anticipation: Music and the Psychology of Expectation. MIT Press, Cambridge (2006)
16. Lefebvre, A., Lecroq, T.: Computing repeated factors with a factor oracle. In: Proc. of the Australasian Workshop On Combinatorial Algorithms
17. Martin, A., Seroussi, G., Weinberger, J.: Linear time universal coding and time reversal of tree sources via fsm closure. Information Theory, IEEE Transactions on 50(7), 1442–1468 (July 2004)
18. Meyer, L.B.: Emotion and Meaning in Music. Univ. of Chicago Press, Chicago (1956)
19. Moore, A., Atkeson, C.: Prioritized sweeping: Reinforcement learning with less data and less real time. Machine Learning 13, 103–130 (1993)
20. Pachet, F.: The continuator: Musical interaction with style. In: Proc. of International Computer Music Conference, Gotheborg, Sweden (September 2002)
21. Pearce, M., Conklin, D., Wiggins, G.: Methods for combining statistical models of music. In: Wiil, U.K. (ed.) CMMR 2004. LNCS, vol. 3310, pp. 295–312. Springer, Heidelberg (2005)
22. Ron, D., Singer, Y., Tishby, N.: The power of amnesia: Learning probabilistic automata with variable memory length. Machine Learning 25(2-3), 117–149 (1996)
23. Rosen, R.: Anticipatory Systems of IFSR International Series on Systems Science and Engineering, vol. 1. Pergamon Press, Oxford (1985)
24. Saffran, J.R., Johnson, E.K., Aslin, R.N., Newport, E.L.: Statistical learning of tonal sequences by human infants and adults. cognition. Cognition 70, 27–52 (1999)
25. Saul, L.K., Jordan, M.I.: Mixed memory markov models: Decomposing complex stochastic processes as mixtures of simpler ones. Machine Learning 37(1), 75–87 (1999)
26. Snyder, B.: Music and Memory: An Introduction. MIT Press, New York (2000)
27. Sutton, R.S., Barto, A.G.: Reinforcement Learning: An Introduction. MIT Press, Cambridge (1998)
28. Uchibe, E., Doya, K.: Competitive-cooperative-concurrent reinforcement learning with importance sampling. In: Proc. of International Conference on Simulation of Adaptive Behavior: From Animals and Animats, pp. 287–296 (2004)
29. Xenakis, I.: Formalized Music. University of Indiana Press (1971)
30. Ziv, J., Lempel, A.: Compression of individual sequences via variable-rate coding. IEEE Transactions on Information Theory 24(5), 530–536 (1978)

An Anticipatory Trust Model for Open Distributed Systems

Mario Gómez, Javier Carbó, and Clara Benac-Earle

Group of Applied Artificial Inteligence
Carlos III University of Madrid
Av. Universidad 30, 28911, Leganés (Spain)
mgomez@csd.abdn.ac.uk, {jcarbo,cbenac}@inf.uc3m.es

Abstract. Competitive distributed systems pose a challenge to trust modeling due to the dynamic nature of these systems (e.g. electronic auctions) and the unreliability of self-interested agents. We propose a trust model which does not assume a concrete cognitive model for other agents that an agent may interact with, but uses the discrepancy between the information provided by other agents and its own experience in order to anticipate their actions. By anticipating the behavior of other agents, an agent is able to adapt more effectively to changes in the environment for its own benefit.

1 Introduction and Motivation

Although there are different definitions [10], we can state that trust is an abstract property applied to others that helps to reduce the complexity of decisions. Trust is a universal concept that plays a very important role in social organizations as a mechanism of social control. Therefore, modeling trust in open distributed systems such as agent systems becomes a critical issue since their offline and large-scale nature weaken the social control of direct interactions. For this reason, the agent research community is very interested in this issue.

Often, there are objective and universal criteria to evaluate the quality of interactions (products/services provided by them). In this case, trust can be inferred from certificates issued by third parties that verify such objective criteria. Unfortunately, when a set of universal objective evaluation criteria is not available, this subjective and local trust will not be easily asserted. There are several application domains where interpersonal communications are the main source of trust due to the subjective nature of the evaluation criteria (books, films, web pages, leisure activities, consulting services, technical assistance, etc.).

Although there are several ways to infer trust, numerous studies have shown that in real life one of the most effective channels to avoid deceptions is through reputation-based mechanisms [2]. Usually, the group of people with a good reputation (collaborators, colleagues and friends) that cooperates with a particular person to improve the quality of decisions forms an informal social network [23]. In this context, trust and reputation are strongly linked.

M.V. Butz et al. (Eds.): ABiALS 2006, LNAI 4520, pp. 307–324, 2007.

All in all, many existing approaches to trust modeling have paid little attention to a crucial feature of autonomous agents: their capacity to be pro-active rather than just reactive, i.e., their ability to deal with the future by mental representations or specific forms of learning. For guiding and orienting a future action, a representation of the future, and more precisely, a representation of future effects and of intermediate results of the action, is needed [15]. Anticipatory behavior is an interdisciplinary topic attracting attention from computer scientists, psychologists, philosophers, neuroscientists, and biologists [4]. Anticipation can be seen as mechanism for devising hypotheses that make predictions about future events, conducting experiments to corroborate them and subsequently using the knowledge gained to perform useful behaviors. Anticipatory principles are interesting in the context of trust and reputation modeling because they define a continuing process of discovery and refinement that would allow an agent to adapt more quickly to dynamic environments.

Moreover, in existing frameworks the discrepancy between information and direct experience is used as a source of dishonesty: if agent i says that service s has a quality of service q and agent j has experienced a quality of service r, then $q - r$ is assumed to represent a degree of dishonesty, a source of distrust. Herein, we propose a trust model which does not assume a concrete cognitive model for other agents an agent may interact with, but uses the discrepancy between the information provided by other agents and its own experience in order to anticipate their actions: $q - r$ is not used as a source of dishonesty and distrust, instead, it is used to estimate the quality of service to be obtained in the future. The result is an anticipatory trust modeling framework that allows agents to adapt swiftly to changes in the environment for its own benefit.

In this paper we present the components of an anticipatory model to handle computational trust in dynamic distributed environments. The paper is organized as follows: First, an overview over current trust models is provided. Section 3 describes the different components of trust and presents a synthetic definition of trust as an aggregation of its components, Section 4 describes some experiments to test our model using the ART Testbed, Section 5 discusses the anticipatory mechanisms implied or supported by our approach, and finally, Section 6 sums up the contributions of our work.

2   State of the Art

Several trust models have been proposed; two of the most cited reputation models are SPORAS and HISTOS [25]. SPORAS is inspired in the foundations of the chess players evaluation system called ELOS. The key idea of this model is that trusted agents with very high reputation experience much smaller changes in reputation than agents with low reputation. SPORAS computes the reliability of agents' reputation using the standard deviation of such measure.

HISTOS is designed to complement SPORAS by including a way to deal with witness information (personal recommendations). HISTOS includes witness information as a source of reputation through a recursive computation of weighted means of ratings. It computes reputation of agent i for agent j from the knowledge of all the chain of reputation beliefs corresponding to each possible path connecting i and j. In addition, HISTOS plans to limit the length of paths that are taken into account. To make a fair comparison with other proposals, that limit should be valued as 1, since most of the other views consider that agents communicate only their own beliefs, but not the beliefs of other sources that contributed to their own beliefs of reputation. Based on these principles, the reputation value of a given agent at iteration i, R_i, is obtained recursively from the previous one R_{i-1} and from the subjective evaluation of the direct experience DE_i:

$$R_i = R_{i-1} + \frac{1}{\theta} \cdot \Phi(R_{i-1}) \cdot (DE_i - R_{i-1})$$

Let θ be the effective number of ratings taken into account in an evaluation ($\theta > 1$). The bigger the number of considered ratings, the smaller the change in reputation. Furthermore, Φ stands for a damping function that slows down the changes for very reputable users:

$$\Phi(R_{i-1}) = 1 - \frac{1}{1 + e^{\frac{-(R_{i-1} - Max)}{\sigma}}},$$

where dominion D is the maximum possible reputation value and σ is chosen in a way that the resulting Φ would remain above 0.9 when reputations values were below $\frac{3}{4}$ of D.

Another well known reputation model is due to Singh and Yu. This trust model [24] uses Dempster-Shafer theory of evidence to aggregate recommendations from different witnesses. The main characteristic of this model is the relative importance of fails over success. It assumes that deceptions cause stronger impressions than satisfactions. It then applies different gradients to the curves of gaining/losing reputation in order to lose reputation easily, while it is hard to acquire it. The authors of this trust model define different equations to calculate reputation according to the sign (positive/negative) of the received direct experience (satisfaction/deception) and the sign of the previous reputation corresponding to the given agent.

Instead of Dempster-Shafer theory, Sen's reputation model [22] uses learning to cope with recommendations from different witnesses. Unfortunately learning requires a high number of interactions and a relatively high number of witnesses to avoid colluding agents benefiting from reciprocative agents.

Another remarkable reference in the field is REGRET [21]. The REGRET model takes into account three ways of computing indirect reputation depending on the information source: system, neighborhood and witness reputations. Note that witness reputation is the one that corresponds to the concept of reputation that we are considering. REGRET includes a measure of the social credibility of the agent and a measure of the credibility of the information in the computation of witness reputation. The former is computed from the social relations shared between agents. It is computed in a similar way to neighborhood reputation, using third party references about the recommender directly in the computation of how its recommendations are taken into account. In addition, the latter measure of credibility (information credibility) is computed from the difference between the recommendation and what the agent experienced by itself. The similarity is computed by matching this difference with a triangle fuzzy set centered on 0 (the value 0 stands for no difference at all). The information credibility is considered relevant and taken into account in the experiments of this present comparison. Both decisions are also, to a certain degree, supported by the authors of RE-GRET, who also assume that the accuracy of previous pieces of information (witness) are much more reliable than the credibility based on social relations (neighborhood), and they reduce the use of neighborhood reputation to those situations were there is not enough information on witness reputation. The complete mathematical expression of both measures can be found in [20]. But the key idea of REGRET is that it also considers the role that social relationships may play. It provides a degree of reliability for the reputation values, and it adapts them through the inclusion of a temporal dependent function in computations. The time dependent function ρ gives higher relevance to direct experiences produced at times closer to current time. The reputation held by any part at a iteration i is computed from a weighted mean of the corresponding last θ direct experiences. The general equation is of the form:

$$R_\iota = \sum_{j=i-\theta}^{j=i} \rho(i,j) \cdot W_j,$$

where $\rho(i,j)$ is a normalized value calculated from the next weight function:

$$\rho(i,j) = \frac{f(j,i)}{\sum_{k=i-\theta}^{k=i} f(k,i)},$$

where $i \geq j$. Both represent the time or number of iterations of a direct experience. For instance, a simple example of a time dependent function f is:

$$f(j,i) = \frac{j}{i}$$

REGRET also computes reliability with the standard deviation of reputation values, computed from:

$$STD - DVT_i = 1 - \sum_{j=i-\theta}^{j=i} \rho(i,j) \cdot |W_j - R_i|$$

REGRET, however, defines reliability as a convex combination of this deviation with a measure, $0 < NI < 1$, whether the number of impressions, i, obtained is enough or not. REGRET establishes an intimacy level of interactions, itm, to represent a minimum threshold of experiences to obtain close relationships. More interactions will not increase reliability. The next function models the level of intimacy with a given agent:

$$if(i \in [0, itm]) \rightarrow NI = \sin(\frac{\pi}{2 \cdot itm} \cdot i), Otherwise \rightarrow NI = 1$$

FIRE [14] is a trust and reputation model that integrates four types of information sources: interaction trust, role-based trust, witness reputation and certified reputation. Interaction trust is built from the direct experience of an agent, in particular, the direct trust component of REGRET is exploited in this model. Role-based trust is based on relationships between the agents, which is mostly domain-specific. Witness information is built from reports of witnesses about an agent's behavior. Certified reputation is a novel type of reputation introduced by the authors, which is built from third-party references provided by the agent itself. Certified reputation plays a similar role to what we call advertisements, since in both cases an agent i that has just joined the environment can make some assessment of the trustworthiness of another agent j, based on the certified reputation or advertisements provided by the agent j itself. The main limitation of the FIRE model in [14] is that all agents are assumed to be honest in exchanging information.

Another approach when agents are acting in uncertain environments, is to apply adaptive filters such as Alpha Beta, Kalman and IMM [6,7]. They have been recognized as a reasoning paradigm for time-variable facts within the Artificial Intelligence community [19]. Making time-dependent predictions in noisy environments is not an easy task. They apply a temporal statistical model to the noisy observations perceived through a linear recursive algorithm that estimates a future state variable. Particularly, when they are applied to reputation modeling, the state variable would be the reputation, while observations would be the results from direct experiences.

From the artificial intelligence point of view, reputation models embedded in agents should involve a cognitive approach[17]: enriching the internal model for making cooperative and competitive decisions rather than enriching the exchanged reputation information.

In contrast to the socio-cognitive models, computational models involve a numerical decision making, made up of utility functions, probabilities, and evaluations of past interactions. The combination of both computational models intends to reproduce the reasoning mechanisms behind human decision-making. In this paper we present a trust modeling framework that combines both views, since it assumes the cognitive stance, but uses a numerical approach.

Other researchers have proposed a socio-cognitive view of trust [16] [9], [3]. Schillo's model [16] distinguishes between two types of motivations for trust: honesty and altruism. A more enriched model is from Castelfranchi and Falcone [9] who claim that some other beliefs in addition to reputation are essential to compute the amount of trust of a particular agent: its competence (ability to act as we wish), willingness (intention to cooperate), persistence (consistency along time) and motivation (our contribution to its goals). For the authors, these beliefs should be taken into consideration in determining how much trust is set on this agent. Brainov and Sandholm [3] highlight the relevance of modeling opponent's trust, because, if this outside trust was not taken into account, this would lead to an inefficient trade between agents involved. Thus both agents would be interested in showing the trustworthiness of the counterpart to allocate efficiently resources.

Another example of a socio-cognitive approach from Carbo et al. [8], called a fuzzy reputation agent system (AFRAS), supports the fuzzy nature of the reputation concept itself. It uses fuzzy logic to represent reputation since this concept is built up with vague evaluations (they depend on personal and subjective criteria), uncertain recommendations (malicious agents, different points of view), and incomplete information (untraceability of every agent in open systems). Furthermore, reliability of fuzzy reputation is implicit in the shape of the corresponding fuzzy set. Additionally it also includes other beliefs that intend to represent an emotive characterization of agents including shyness, egoism, and susceptibility. It also includes a global belief and a global adaptation value of agent interaction, referred to as *remembrance*. This attribute determines the relevance given to the last direct interaction when updating trustworthiness. It represents the general confidence of the agent on its own beliefs. The more success is achieved in predicting the behavior of a particular agent, the more relevance is applied to the already asserted beliefs over future experiences with any agent (not only that particular agent).

3   The Anticipatory Trust Model

Typically, a trust model considers two main sources of information to estimate trust: direct experience, sometimes referred to as direct trust or interaction trust, and recommendations, often called witness-information or "word of mouth". In our model we keep this distinction between direct experience and recommendations, but in addition, we distinguish between the recommendations about third party agents and the recommendations provided by an agent about itself, what we call *advertisements*. All in all, our model builds trust upon three components, namely: Direct Trust (DT), Advertisements-based Trust (AT), and Recommendations-based Trust (RT).

In order to adapt more quickly to the dynamic and uncertain nature of an open environment, an agent can anticipate or have expectations (not necessarily rational) about the possible consequences of its actions, therefore, we distinguish between the historic components of trust, based on past information only, and the anticipatory components.

In our model, only the Advertisements-based Trust and the Recommendations-based trust are anticipatory, while trust by direct experience is purely an historic belief. To simplify the dynamics of a multi-agent system, we use a discrete time model made up of time steps. A time step represents the minimal time period an agent requires to make decisions, act, and perceive the result of its actions. We use t to denote a particular time step in the past, T for the current time step, $T+1$ for the next time step, and ΣT for an aggregation of historic beliefs until time step T.

To handle uncertainty and ignorance, we use two dimensions to represent the confidence on a belief, namely: *intimacy*, and *predictability*. Intimacy is a measure of confidence based on the number of data (or interactions) used to calculate a belief, while predictability is a measure of confidence based on the dispersion or variability of data. In our model, all the components of trust have attached a measure of confidence made up of intimacy and predictability. In addition, we propose the use of t-norms for combining intimacy and predictability into a single confidence value, and t-conorms for calculating the confidence coming from several sources of information.

Direct Trust($DT_{ij}^{\Sigma T}$): assesses the Quality of Service i provided from agent j until time step T inclusive.

$$DT_{ij}^{\Sigma T} = \frac{\sum_{t=0}^{T} \varphi(t,T) pDT_{ij}^{t}}{\sum_{t=0}^{T} \varphi(t,T)} \qquad (1)$$

where $pDT_{ij}^{t} : \mathbb{R} \rightarrow [0,1]$ is the partial Direct Trust obtained for agent j and service i in time step t, and $\varphi(T,t) : \mathbb{N} \rightarrow [0,1]$ is a forgetting function used to weight each partial belief according to its age (number of time steps since a belief was obtained, T-t).

Direct Trust Confidence($DTC_{ij}^{\Sigma T}$): assesses the reliance of Direct Trust as an estimator of the Quality of Service i provided by agent j.

$$DTC_{ij}^{\Sigma T} = ITM_{ij}^{DT} \otimes (1 - \upsilon_{dt}(pDT_{ij}^{t})) \qquad (2)$$

where $ITM_{ij}^{DT} \in [0,1]$ is the intimacy level for DT ([21]), a growing function in [0,1] over the number of pDTs used to compute DT, $\upsilon_{dt} \in [0,1]$ is a measure of the variability of pDT_{ij}^{t}, and $\otimes$ is a $T\text{-}norm$ operator.

Advertisements-based Trust(AT_{ij}^{T+1}): assesses the Quality of Service i expected from agent j in the next time step $(T+1)$, based on advertisements.

$$AT_{ij}^{T+1} = \left\{ \begin{array}{cc} 1 & pAT_{ij}^{T+1} + \Delta AT_{ij}^{\Sigma T} \geq 1 \\ 0 & pAT_{ij}^{T+1} + \Delta AT_{ij}^{\Sigma T} \leq 0 \\ pAT_{ij}^{T+1} + \Delta AT_{ij}^{\Sigma T} & 0 < pAT_{ij}^{T+1} + \Delta AT_{ij}^{\Sigma T} < 1 \end{array} \right\} \qquad (3)$$

where $pAT_{ij}^{T+1} : \mathbb{R} \rightarrow [0,1]$ is the most recent advertisement from agent j about service i, and $\Delta AT_{ij}^{\Sigma T}$ (AT-Discrepancy) is the discrepancy between advertisements and experiences obtained in the past (until time step T inclusive).

AT-Discrepancy $\Delta AT_{ij}^{\Sigma T}$: measures the discrepancy between the past advertisements made by agent j about service i and the experiences obtained when that service was requested.

$$\Delta AT_{ij}^{\Sigma T} = \frac{\sum_{t=0}^{T} \varphi(t,T)(pDT_{ij}^{t} - pAT_{ij}^{t})}{\sum_{t=0}^{T} \varphi(t,T)} \qquad (4)$$

where $pAT_{ij}^{t} : \mathbb{R} \rightarrow [0,1]$ is the Partial Advertisements-based Trust for agent j, service i and time step t, and $\varphi(t,T)$ is a time forgetting function.

Note that $\Delta AT_{ij}^{\Sigma T} \in [-1, 1]$, since $pDT_{ij}^t, pAT_{ij}^t, \varphi(t, T) \in [0, 1]$ by definition. Positive values of $\Delta AT_{ij}^{\Sigma T}$ mean that the experiences obtained from agent j and service i were better than advertised, negative values have the opposite meaning, and a zero value means that the experiences matched perfectly with the advertisements.

AT Confidence(ATC_{ij}^{T+1}): assesses the degree of reliance of the Advertisements-based Trust as an estimation of the Quality of Service i to be obtained from agent j in the next time step.

$$ATC_{ij}^{T+1} = ITM_{ij}^{AT} \otimes (1 - \upsilon_{at}(\Delta AT_{ij}^t)) \tag{5}$$

where ITM_{ij}^{AT} is the intimacy level for AT, $\Delta AT_{ij}^t = pDT_{ij}^t - pAT_{ij}^t$ is the partial discrepancy observed between AT and DT in time step t, $\upsilon_{at} \in [0, 1]$ is a measure of the variability of ΔAT, and $\otimes$ is a T-norm operator.

As we have done for Direct Trust and Advertisements-based Trust, we define both partial and historic Recommendations-based Trust (RT). However, RT must handle the fact that there are potentially many providers of information (recommenders) about any other agent. As a result, we have to distinguish between the trust component due to the recommendations provided by a single agent and the trust component due to the recommendations provided by several agents; herein the latter is referred to as combined recommendation.

Recommendations-based Trust(RT_{ijk}^{T+1}): assesses the Quality of Service i expected from agent j in the next time step $(T + 1)$, based on the recommendations from agent k.

$$RT_{ijk}^{T+1} = \begin{cases} 1 & pRT_{ijk}^{T+1} + \Delta RT_{ijk}^{\Sigma T} \geq 1 \\ 0 & pRT_{ijk}^{T+1} + \Delta RT_{ijk}^{\Sigma T} \leq 0 \\ pRT_{ijk}^{T+1} + \Delta RT_{ijk}^{\Sigma T} & 0 < pRT_{ijk}^{T+1} + \Delta RT_{ijk}^{T+1} < 1 \end{cases} \tag{6}$$

where $pRT_{ijk}^t : \mathbb{R} \to [0, 1]$ is the partial Recommendations-based Trust for agent j and service i obtained from agent k, and $\Delta RT_{ijk}^{\Sigma T}$ (RT-Discrepancy) is the discrepancy between past recommendations and experiences about agent i and service j.

RT-Discrepancy($\Delta RT_{ijk}^{\Sigma T}$): measures the discrepancy between the past recommendations by agent k about agent j and service i, and the experiences obtained using that service, until time step T inclusive.

$$\Delta RT_{ijk}^{\Sigma T} = \frac{\sum_{t=0}^T \varphi(t, T)(pDT_{ij}^t - pRT_{ijk}^t)}{\sum_{t=0}^T \varphi(t, T)} \tag{7}$$

where $pRT_{ij}^t : \mathbb{R} \to [0, 1]$ is the partial Recommendations-based Trust for agent j, service i and time step t, and $\varphi(t, T)$ is a time forgetting function.

Note that $\Delta RT_{ijk}^{\Sigma T} \in [-1, 1]$, since $pDT_{ij}^t, pRT_{ij}^t, \varphi(t, T) \in [0, 1]$ by definition. Positive values of $\Delta RT_{ij}^{\Sigma T}$ mean that the experiences obtained from agent j and service i were better than recommended, negative values have the opposite meaning, and a zero value means that the experiences matched perfectly the recommendations.

Combined Recommendations-based Trust(cRT_{ij}^{T+1})**:** assesses the Quality of Service i expected from agent j in the next time step, based on both historic information and the most recent recommendations about that service.

$$cRT_{ij}^{T+1} = \frac{\sum_{k=1}^{N_k} (RT_{ijk}^{T+1} \times RTC_{ij}^{T+1})}{\sum_{k=1}^{N_k} RTC_{ijk}^{T+1}} \tag{8}$$

where RT_{ijk}^{T+1} is the Recommendations-based Trust about agent j and service i obtained from agent k's recommendations, and RTC_{ijk}^{T+1} is the confidence in that belief as an estimation of the Quality of Service i to be obtained from agent j in $T + 1$.

The Combined Recommendations-based Trust aggregates the recommendations obtained from several agents. Similarly, the confidence in cRT is defined as an aggregation of the confidences in every recommendation.

RT Confidence(RTC_{ij}^{T+1})**:** assesses the degree of reliance of the Recommendations-based Trust (RT_{ijk}^{T+1}) obtained from agent k, as an estimation of the Quality of Service i to be obtained from agent j in the next time step.

$$RTC_{ijk}^{T+1} = ITM_{ij}^{RT} \otimes (1 - v_{rt}(\Delta RT_{ijk}^t)) \tag{9}$$

where ITM_{ij}^{RT} is the intimacy level for RT, $\Delta RT_{ijk}^t = pDT_{ij}^t - pRT_{ijk}^t$ is the partial discrepancy observed between DT and RT in time step t, $v_{rt} \in [0, 1]$ is a measure of the variability of ΔRT_{ijk}^t, and $\otimes$ is a T-norm operator.

Combined RT Confidence$(cRTC_{ij}^{T+1})$**:** assesses the degree of reliance of the Combined Recommendations-based Trust as an estimation of the Quality of Service i to be obtained from agent j in the next time step.

$$cRTC_{ijk}^{T+1} = \bigoplus^k (RTC_{ijk}^{T+1}) \tag{10}$$

where $\bigoplus^k$ denotes the aggregation of the confidence associated with each recommender $(RTC_{ijk}^{\Sigma T})$ using a *T-conorm* operator.

Up to now we have defined the components of trust according to the source of information. Now we provide a global measure of trust that integrates the three components into a single belief: the Global Trust.

Global Trust(GT_{ij}^{T+1})**:** assesses the Quality of Service i expected from agent j during the next time step, using all the sources of information.

$$GT_{ij}^{T+1} = \frac{DT_{ij}^{\Sigma T} \times DTC_{ij}^{\Sigma T} + AT_{ij}^{T+1} \times ATC_{ij}^{T+1} + cRT_{ij}^{T+1} \times cRTC_{ij}^{T+1}}{DTC_{ij}^{\Sigma T} + ATC_{ij}^{T+1} + cRTC_{ij}^{T+1}}$$

(11)

where $DT_{ij}^{\Sigma T}$ is the Direct Trust for service i and agent j; AT_{ij}^{T+1} is the Anticipatory Advertisements-based Trust; cRT_{ij}^{T+1} is the Combined Recommendations-based Trust, and $DTC_{ij}^{\Sigma T}$, ATC_{ij}^{T+1}, RTC_{ij}^{T+1} are the confidences associated to DT, AT and cRT respectively.

Global Trust Confidence(GTC_{ij}^{T+1})**:** assesses the reliance of the Global Trust GT_{ij} as an estimation of the Quality of Service i to be obtained in the next time step.

$$GTC_{ij}^{T+1} = DTC_{ij}^{\Sigma T} \oplus ATC_{ij}^{T+1} \oplus cRTC_{ij}^{T+1}$$

(12)

where $\oplus$ is a *T-conorm* operator.

Remark that Global Trust and Global Trust Confidence can be used either independently or combined into a single value (eg. $GT \times GTC$), depending on the specific application domain.

4   Experiments

We have chosen the ART Testbed [12] to test our test model. The ART Testbed is a simulator of the *art appraisals* domain whose goal is twofold: to serve as a competition forum in which researchers can compare their technologies against objective metrics, and as an experimental tool, with flexible parameters, allowing researchers to perform customizable, easily-repeatable experiments. In the art appraisal domain, agents act as painting appraisers with varying levels of expertise in different artistic eras (e.g. classical, impressionist, postmodern). Clients request appraisals for paintings from different eras. Appraisers can use both their own opinions and opinions purchased from other agents, so as to make more accurate appraisals. Appraisers estimate the accuracy of the opinions they send by the cost they choose to invest in generating an opinion, but they may lie about the estimated accuracy of their opinions. Appraisers receive more clients, and thus more profit, for producing more accurate appraisals. Appraisers may also purchase reputation information from other agents. The decisions about which opinion providers and reputation providers to trust strongly impact the accuracy of their final appraisals. In competition mode, the winning agent is selected as the appraiser with the highest bank account balance, which depends on the ability of an agent to (1) estimate the value of its paintings most accurately and (2) purchase more valuable information.

It is easy to map our trust model to the ART Testbed domain because it uses continuous variables and includes both advertisements (named certainties) and recommendations (named reputations). Purchased opinions about the value of a painting are the source of experience and are used to calculate DT, reputations are mapped to recommendations, and finally, advertisements are mapped to certainties, which are values provided by an agent about the accuracy of its opinions. Finally, the concept of a weight in the ART refers to a global measure attached to an agent to represent their opinion's accuracy. In our experiments, we use Global Trust × Global Trust Confidence to obtain those weights.

In order to evaluate the gains and drawbacks of using an anticipatory trust model in dynamic and uncertain environments, we have compared three models to handle trust and reputation: *anticipatory, non anticipatory without honesty,* and *non anticipatory with honesty.* The anticipatory model implements the trust model described in this paper, while the non anticipatory models use only historic information to calculate the global trust; the model with honesty uses the discrepancy between information and experience to calculate the confidence on trust, while the model without honesty simply ignores such discrepancy.

We compare the three trust models introduced above along four variables, namely: *number of appraisals, average error, bank balance,* and *stability.* The number of appraisals (NA) measures the total number of appraisals obtained during an entire simulation, the average error (AE) is the mean of those appraisal's error, the bank balance (BB) is the difference between the revenues and the expenses, and the stability (ST) is the number of time steps in which the average error for the last 5 time steps changes less than a given criterion (|average error increment| < 0.01).

Since in the ART Testbed agents can use their own opinions to appraise a painting, and they know themselves very well, self opinions tend to neutralize the influence of the opinions purchased from other agents. In order to remark the differences between the three trust models being compared, we have enforced all agents to use solely the opinions purchased from other agents, and not their own opinions.

We have conducted three groups of experiments: (a) experiments with *dynamic prices* following a market-like evolving process, (b) experiments varying the degree of *deception* (dishonesty), and (c) experiments varying the prices randomly (a parameter called noise establishes the maximum price variation per time step). The same experimental situation is used as the baseline for the three groups of experiments: a static scenario where agents always provide the best opinions they are able to obtain and are completely sincere. Each experiment varying a parameter is repeated twice. A single experiment involves 9 agents competing during 60 time steps, with the same proportion of agents (3) using each trust model. Each time step, there are 270 paintings belonging to 10 artistic eras to be distributed among appraiser agents according to their relative average error in the previous time step.

Figure 1 summarizes the results of the experiments when evolving the prices dynamically, according to a market-like model consisting of alternating inflation/deflation periods. There are three experimental situations, from left to

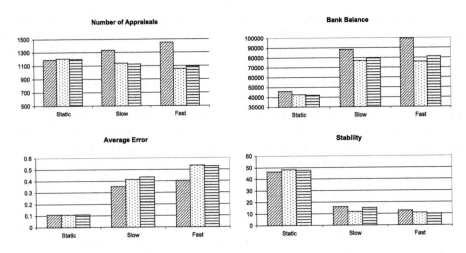

Fig. 1. Experiments with dynamic prices

right: static prices, *slow* price dynamics, and *fast* price dynamics. The bars with diagonal lines represent the average score for the agents using the anticipatory model (ant), dotted bars for the non anticipatory model with honesty (na), and horizontal lines for the non anticipatory model without honesty(nh). Results show that all agents perform similarly in the case of a completely static environment with fixed prices and no deception. However, there are clear differences when considering dynamic prices, and these changes are stronger the more quickly prices change. In particular, the agents using the anticipatory trust model obtain the most accurate estimations of other agents(lower AE), achieve the most clients (higher NA), obtain the best economic results (higher BB), and remain stabilized for the longest amount of time, among the three models compared. The observed differences were significantly different[1]. Although the differences between the two non anticipatory models seem very small as to be generalized at first glance, in some cases the differences have been statically significant; in particular, the difference in the bank balance for the third scenario (fast price dynamics) has reached significance.

Figure 2 sums up the results of our experiments varying the prices randomly according to a specific amount of noise: *no noise* (static prices), *low noise*, and *high noise*. These results are completely consistent with the first group of experiments: the agents with the anticipatory model perform better than the agents using the non anticipatory models in all the variables analyzed. The differences between the anticipatory and the non anticipatory models are statistically significant, but they are not significant between the two non anticipatory models.

---

[1] For this as well as the other results commented, we have tested the statistical significance of the difference between the means by applying *student's bilateral t test* for two samples and a significance criterion $\alpha = 0.05$. When the difference is not statistically significant we apply also *unilateral t tests*.

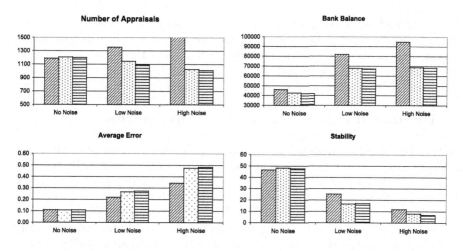

Fig. 2. Experiments with noise (random price changes)

Figure 3 shows the results of our experiments introducing a certain degree of deception concerning both the advertisements (certainties) and the recommendations about the accuracy of agent opinions. We consider three experimental situations: *no deception*, *low deception*, and *high deception*. In this case, both the anticipatory and the non anticipatory models perform very similarly concerning the average error and the number of appraisals achieved (there are no statistically significant differences), but the agents using anticipation achieve better economic results and remain stabilized for longer periods of time (the differences are statistically significant between the anticipatory and both non anticipatory models). The differences between the two non anticipatory models are not relevant (not statistically significant). Anticipatory agents earn more

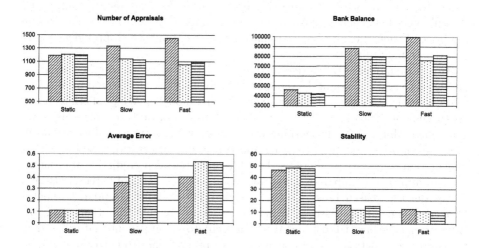

Fig. 3. Experiments with deception

money even when obtaining fewer paintings to appraise, as has been the case for the low deception scenario. This result may seem contradictory at first glance because appraisals are the main source of revenues, but there is an explanation: the anticipatory trust model induces a more efficient behavior, in other words, anticipatory agents purchase less opinions to obtain appraisals of similar quality.

5 Discussion of the Anticipatory Mechanisms

The anticipatory principles are interesting in the context of trust and reputation modeling because they define a continuing process of discovery and refinement that would allow an agent to adapt more quickly to dynamic environments, and we have conducted some experiments that support that claim.

In [5] a preliminary classification of anticipatory mechanisms is proposed with four categories: *implicit anticipatory* mechanisms, in which no actual predictions are made but the behavioral structure is constructed in an anticipatory fashion; *payoff anticipatory* mechanisms, in which the influence of future predictions on behavior is restricted to payoff predictions; *sensory anticipatory* mechanisms, in which future predictions influence sensory (pre-)processing; and *state anticipatory* mechanisms in which predictions about future states directly influence current behavioral decision making. Nevertheless, we cannot easily classify our model framework into one of the former categories; the point is that the kind of anticipatory mechanisms involved in our trust modeling framework are not apparent because anticipatory behavior is more concerned with the architectures and algorithms involved in the decision making process, while our paper is more focused on the cognitive, belief modeling aspects supporting decision making. A more detailed discussion of the kind of anticipatory mechanisms supported by our framework follows:

Since our framework approaches trust as an assessment of the Quality of Service (QoS) expected from an agent in the future, our model is very well suited to support *payoff anticipatory* mechanisms. For example, by comparing the difference between the QoS expected from different providers and considering the price of each one, an agent can make decisions about which providers to use in order to maximize its benefits.

Any trust model is predictive in that the future behavior of other agents is predicted to make the right decisions about which services to use and from whom to request those services. However, existing trust models make predictions based on the use of past observations, while our framework adds a more subjective component that results in learning the difference between the information obtained from other agents in the past and the actual observations, and uses that information to update predictions. The difference between predictive approaches and our approach is apparent when comparing our definition of Direct Trust and Advertisements-based Trust. On the one hand, Direct Trust is a predictive belief altogether, since it is based only on past observations. On the other hand, the Advertisements-based Trust includes an anticipatory component that is based on the comparison between advertisements and observations. To better understand the difference, consider the following example:

Suppose that there are two agents, named P1 and P2, that provide the same service. The QoS provided by P1 is better than the QoS provided by P2. For example, the average QoS for P1 is 0,7 and for P2 is 0,6. Now, suppose that client C has been interacting with P1 and P2 until having a reliable measure (from its subjective point of view) of their QoS, and consequently, has decided to request the given service only from P1. Now, imagine that the QoS provided by P2 has incremented from 0,6 to 0,8. Probably, P2 has changed its own advertisements to reflect the change. If C uses only past observations to assess P2's QoS, it would not react immediately to the new adverts, because it has not interacted with P2 for some time and thus, it has no evidence to update its beliefs about P2. However, if C has learned the pattern of discrepancies between P2's QoS and P2's adverts in the past, then C would be able to infer, from the last adverts by P2, that P2's QoS has improved and it is actually better than P1, so the next time C will request the given service from P2 instead of P1, anticipating the change. That means that C has predicted the change in P2's behavior without having experience it, and has used such a prediction to adopt radically new decisions.

The possibility of making radically new decisions as a consequence of anticipating the future, rather than merely predicting it by observation, is a crucial feature of anticipatory systems. For example, according to [18] an anticipatory system is "...a system containing a predictive model of itself and/or of its environment, which allows it to change state at an instant in accord with the model's predictions pertaining to a later instant." Summing up, anticipation allows an agent to detect changes in the environment before actually experiencing them, thus it protects an agent from suffering bad experiences, as in *preventive state anticipation* [11], that is a kind of *state based anticipatory* mechanism, and enables a swifter adaptation to dynamic environment. All in all, the anticipatory mechanisms proposed herein have some drawbacks. In particular, the use of advertisements and recommendations expose an agent to a higher risk when making radically new decisions, because of the higher uncertainty associated with such information, compared with the uncertainty associated with direct experience and observation.

6   Conclusions

In this paper we have introduced a computational model to handle trust in dynamic and uncertain domains such as electronic market places and distributed information systems. This model extends Rahman [1] notion of *semantic closeness* between experience and information to deal with continuous domains: first of all, we use continuous variables instead of discrete variables; second, our model combines the reputation information and the trust based on direct experiences (Rahman model uses only reputation); and third, we distinguish between two types of information: advertisements and recommendations.

Several frameworks to handle trust in agent societies rely upon the notion of honesty when considering the discrepancy between information and experience. Usually, the discrepancy observed between direct experience and information concerning that experience (witness information) is interpreted as a consequence of the information provider's intentional behavior and, consequently, it is used to estimate the credibility (confidence) of that provider (more discrepancy implying less credibility). In other words, the discrepancy between information and experience is interpreted in terms of the honesty of the information provider.

Sabater [20] argues that although Rahmans approach is useful in some situations, it has some limitations because it is unable to differentiate between lying agents and agents that have a different view of the world. However, there are some reasons to adjust beliefs using the discrepancy between experience and information: on the one hand, in many domains, and specially in real applications, it is actually impossible to know whether an agent is lying or just thinking differently; on the other hand, it it is often more important to estimate the utility expected from an agent than figuring out whether an agent is lying or not.

In our model, the observed discrepancy between information and experience is used to adapt more quickly to changes in the environment by anticipating changes in the world before experiencing them. This approach does not identify discrepancy with bad behavior as such, instead our model uses that information to anticipate the future. However, in order to fully benefit from this approach, discrepancies between information and experience must be relatively consistent over time. That is to say, discrepancies between information and experience must follow a regular pattern so as to be useful.

Our proposal for trust modeling is strongly backed in the following epistemic assumption: the real beliefs of other agents are not knowledgeable, because different agents may use a different framework to represent and reason about the same things. That assumption, which may seem negative at a first glance, is turned into a positive by using the discrepancy between information and experience as valuable knowledge to anticipate the future behavior of an agent. A secondary assumption is that the discrepancy between information and behavior is somewhat consistent over time. If this assumption is violated, our model would perform very similarly to the models being compared in our experiments. In any case, our model is protected against the absence of information-behavior discrepancy patterns by the use of dispersion measures in the computation of confidence values associated with every trust belief.

As a consequence of both assumptions, we think that our framework is appropriate for open systems in general, because heterogeneous agents from different developers would probably use different cognitive models. However, it is especially well suited for competitive environments, such as agent-based e-Commerce applications, since this domain fulfills both assumptions. On the one hand, an e-Commerce application is typically open to heterogeneous agents built by different developers, probably having different views of the same aspects of the world. As a consequence, an agent has well founded reasons to not identify an information-experience discrepancy with a source of distrust. On the other hand, economic domains are expected to show consistent patterns; for instance, we know that advertisements about oneself tend to be positive, while advertisements about the competitors tend to be negative.

We have used controlled experimental conditions to demonstrate the feasibility and utility of the anticipatory mechanism in a market-like simulation environment, the art appraisals domain. On the one hand, we have showed the utility of the anticipatory approach to adapt to changing environments, including both inflationary and deflationary dynamics. On the other hand, we have demonstrated the robustness of the model to deal with deception, both positive (over-valuating), to make an agent believe one is better than he actually is, and negative (under-valuating) deception, to make an agent believe a third agent is worse than himself.

In conclusion, as an extension of our work we have developed a framework based on motivational attitudes to drive the behavior of trust-modeling agents. In particular, we use concepts such as necessity, satisfaction, and curiosity to implement exploratory behavior, which is a key component to achieve adaptation in open and dynamic environment. A discussion of these issues is, unfortunately, out of the scope of this paper, and will be described in forthcoming publications [13].

Acknowledgments

Funded by projects CICYT TSI2005-07344, CICYT TEC2005-07 186 and CAM MADRINET S-0505/TIC/0255

References

1. Abdul-Rahman,S.H.A.: Supporting trust in virtual communities. In: 33th *IEEE International Conference on Systems Sciences*, 2000.
2. Ba, S., Whinston, A., Zhang, H.: Building trust in online auction markets through an economic incentive mechanism. Decision Support System 35, 273–286 (2003)
3. Braynov, S., Sandholm, T.: Trust revelation in multiagent interaction (2002)
4. Butz, M.V., Sigaud, O., Gérard, P. (eds.): Anticipatory Behavior in Adaptive Learning Systems. LNCS (LNAI), vol. 2684. Springer, Heidelberg (2003)
5. Butz, M.V., Sigaud, O., Gérard, P. (eds.): Anticipatory Behavior in Adaptive Learning Systems. LNCS (LNAI), vol. 2684, pp. 86–109. Springer, Heidelberg (2003)
6. Carbo, J., Garcia, J., Molina, J.: Subjective trust inferred by kalman filtering vs. In: Wang, S., Tanaka, K., Zhou, S., Ling, T.-W., Guan, J., Yang, D.-q., Grandi, F., Mangina, E.E., Song, I.-Y., Mayr, H.C. (eds.) Conceptual Modeling for Advanced Application Domains. LNCS, vol. 3289, pp. 496–505. Springer, Heidelberg (2004)
7. Carbo, J., Garcia, J., Molina, J.: Convergence of agent reputation with alpha-beta filtering vs. a fuzzy system. In: International Conference on Intelligent Agents, Web Technologies and Internet Commerce, Wien, Austria (November 2005)
8. Carbo, J., Molina, J., Davila, J.: Trust management through fuzzy reputation. International Journal of Cooperative Information Systems 12(1), 135–155 (March 2003)
9. Castellfranchi, C., Falcone, R.: Principles of trust for multiagent systems: Cognitive anatomy, social importance and quantification. In: Third International Conference on Multi-Agent Systems, pp. 72–79 (1998)

10. Conte, R., Paolucci, M.: Reputation in Artificial Societies. Kluwer Academic Publishers, Dordrecht (2002)
11. Davidsson, P.: A framework for preventive state anticipation. In: Butz, M.V., Sigaud, O., Gérard, P. (eds.) Anticipatory Behavior in Adaptive Learning Systems. LNCS (LNAI), vol. 2684, pp. 151–166. Springer, Heidelberg (2003)
12. Fullam, K., Klos, T., Muller, G., Sabater, J., Schlosser, A., Topol, Z., Barber, K.S., Rosenschein, J., Vercouter, L., Voss, M.: A specification of the agent reputation and trust (art) testbed: Experimentation and competition for trust in agent societies. In: The Fourth International Joint Conference on Autonomous Agents and Multiagent Systems (AAMAS-2005), pp. 512–518 (2005)
13. Gomez, M., Carbo, J., Benac, C.: Trust dynamics, motivational attitudes and epistemic actions: a curiosity-driven exploratory behavior. In: Tenth International Workshop on Trust in Agent Societies (to appear, 2007)
14. Huynh, T.D., Jennings, N.R., Shadbolt, N.R.: An integrated trust and reputation model for open multi-agent systems. Autonomous Agents and Multi-Agent Systems 13(2), 119–154 (2006)
15. Lorini, E., Castelfrnachi, C.: The role of epistemic actions in expectations. In: Second Workshop of Anticipatory Behavior in Adaptive Learning Systems (ABIALS 2004) (2004)
16. Schillo, P.F.M., Rovatsos, M.: Using trust for detecting deceitful agents in artificial societies. Applied Artificial Intelligence, Special Issue on Trust, Deception and Fraud in Agent Societies 14(8), 825–849 (2000)
17. Marsh, S.: Trust in distributed artificial intelligence. In: Castelfranchi, C., Werner, E. (eds.) MAAMAW 1992. LNCS, vol. 830, pp. 94–112. Springer, Heidelberg (1994)
18. Rosen, R.: Anticipatory Systems: Philosophical, Mathematical and Methodological Foundations. Pergamon Press, New York (1985)
19. Russell, S., Norvig, P.: Artificial intelligence: a modern approach. Prentice Hall Pearson Education International (2003)
20. Sabater, J.: Trust and Reputation for agent societies. Consejo Superior de Investigaciones Cientificas, Bellaterra, Spain (2003)
21. Sabater, J., Sierra, C.: Regret: a reputation model for gregarious societies. In: Fourth Workshop on Deception, Fraud and Trust in Agent Societies, pp. 61–69, Montreal, Canada (2001)
22. Sen, S., Biswas, A., Debnath, S.: Believing others: pros and cons. In: Proceedings of the 4th International Conference on MulitAgent Systems, pp. 279–285, Boston, MA, July (2000)
23. Wasserman, S., Galaskiewicz, J.: Advances in Social Network Analysis. Sage Publications, Thousand Oaks, U.S (1994)
24. Yu, B., Singh, M.: A social mechanism for reputation management in electronic communities. In: Klusch, M., Kerschberg, L. (eds.) CIA 2000. LNCS (LNAI), vol. 1860, pp. 154–165. Springer, Heidelberg (2000)
25. Zacharia, G., Maes, P.: Trust management through reputation mechanisms. Applied Artificial Intelligence 14, 881–907 (2000)

Anticipatory Alignment Mechanisms for Behavioral Learning in Multi Agent Systems

Gerben G. Meyer and Nick B. Szirbik

Department of Business & ICT, Faculty of Management and Organization, University of Groningen, Landleven 5, P.O. Box 800, 9700 AV Groningen, The Netherlands, +31 50 363 {7194 / 8125}
{g.g.meyer,n.b.szirbik}@rug.nl

Abstract. In this paper we present a conceptualization and a formalization to define agents' behaviors (as exhibited in agent to agent interactions), via an extension of Petri Nets, and show how behaviors of different agents can be aligned. We explain why these agents can be considered anticipatory, and the link between Business Information Systems and anticipatory systems is elaborated. We show that alignment is a state anticipatory mechanism, where predictions about future states directly influence current behavioral decision making. This results in faster and more reliable interaction execution. Also, alignment provides a mechanism for more direct behavioral learning. We investigated three manners of alignment, individual on-the-fly alignment, pre-interaction alignment, and alignment with the intervention of a third party. This paper explains in some detail how alignment on-the-fly is realized using alignment policies. The features of the other two kinds of alignment are discussed, and future directions for research are pointed out.

1   Introduction

In the anticipatory system research community, the agent based computing area is considered a promising one. However, there is yet little interest in applying the anticipatory agent concept in a real setting. Seminal work of Davidsson, Astor and Ekdahl [5], pointed out that active entities can be characterized as agents when their acting can be described by a social theory. We argue in this paper that business organizations are in fact anticipatory systems themselves. Especially when these use an information system (usually called BIS - Business Information System). Our research group is investigating novel agent-based architectures and development frameworks [13]. We recognize the importance of the anticipatory system concept in this context and position our models of organizations in the initial definition of Rosen ([14], page 339):

> "We tentatively defined the concept of an anticipatory system: a system containing a predictive model of itself and/or of its environment, which allows it to change state at an instant in accord with the models prediction to a latter instant."

M.V. Butz et al. (Eds.): ABiALS 2006, LNAI 4520, pp. 325–344, 2007.

In this paper, we investigate how the anticipatory ability of a single agent can be expressed as an interaction belief and we point out how in some cases this belief can be changed. We describe a policy for alignment that can be applied when the interaction beliefs of two or more interacting agents are not matching. We introduce an extension of Petri Nets to capture the interaction beliefs and also a mechanism to choose the appropriate policy that adapts the beliefs from one agent perspective. Furthermore, we discuss the case when the process of alignment before the actual interaction takes place. Also, alignment by a third-party is investigated. From the anticipatory systems perspective, this research can enable predictive agent model execution (agent-based simulation of organizational models) to be more reliable and necessitate less human intervention in terms of alignment.

1.1 Motivation

Business information systems have evolved from a data centric perspective to a process centric perspective. The role of these systems is to support human activity in a business organization. At a basic level they support information storage and retrieval, information flow and information processing. At a higher level they support human decision making. Depending on the time horizon, the decision can be related to operational management (day-to-day activities), tactical planning (week/month projections), strategic decisions (month/year projections), and even policy implementation (very long term).

The move from data centric to process centric systems did not change the centralistic nature of these systems. The way the system is designed and used ascribes to the notion that there exists an external observer that is able to investigate and understand the processes within the organization. These processes can be identified in a semantic sense and modeled in a syntactic sense, that is, models of the processes can be described in a (semi) formal language. These models can be used to implement systems that support the actors who execute the process in the organization.

The actors who are executing the organizations' processes have only local, often conflicting views, especially in dynamic organizations. If the system is to be designed and implemented by allowing local and different models of the participating actors, a distributed, agent-oriented approach is more suitable. Agent-based modeling and agent-software engineering have been very popular in the last decade and have paved new avenues for the development of the business systems of tomorrow. However, due to the lack of a strict definition of an agent and a clear view about what exactly agent software engineering is, as pointed out by Ekdahl [6], many developmental processes tend to be termed agent-oriented, although they really can be just classified as purely reactive systems. Ekdahl also states:

> "More sophisticated anticipatory systems are those which also contain its own model, are able to change model and to maintain several models, which imply that such systems are able to make hypotheses and also that they can comprehend what is good and bad."

One can infer from this statement that true agent systems are only those that have a clear anticipatory ability, both at the level of the individual agents themselves, and also at the whole multi-agent system level. The ability to reason about a plan in an organization is usually realized via humans. If one tries to simulate a planning organization, a typical barrier is the evaluation of the plans. Such simulations tend to become interactive games, where the "players" (i.e. the expert planners) are becoming decision makers that select the "best" plan. Various plan selection mechanisms can be enacted, but these are usually just models of the behavior of the players. In a monolithic, centralistic system for example, this will be implemented as a single utility function that characterizes the whole organization, which makes explicit the criteria against which a prospective plan is checked. In reality, many expert players are co-operating with the system to adjust and decide for the best plan. The overall behavior of the organization (in terms of planning) is just emerging as a combined behavior of the experts and the system that supports them.

This observation leads to the natural conclusion that it is better to enact decision support structures that mimic the distributed nature of this environment. Attempts to model and implement agent-oriented support for planning and other business processes are still in their infancy, but even simple implementations of crude multi-agent architectures show a higher degree of adaptiveness and flexibility.

We envisage the first wave of applying these kind of agents to repetitive and routine business processes, like the ones in sales and purchasing, financial operations, and operational control of logistic and manufacturing systems [11]. Here, the need for anticipatory based alignment is rather low, but it is easy to execute in an automated way, based on typical policies that have been detected over time. One can say that this application area can be seen as robotics, but in this case the robots are not physical entities, but digital entities (sometimes these are called "softbots").

1.2   Our Approach Towards Anticipatory Agents

Our research team is developing agents via simulation-games, where the behavior of the software agents is deduced from expert players. These human experts can describe their intended behavior, in terms of activities and local goals, but also can describe the behavior they expect from the other agents in the game. These behaviors can be simplified and formally described. From a local perspective the *intended behavior* of self and the *expected behavior* of others can be seen as a specific *interaction belief* of that agent. The organizations' processes can be viewed as a set of running interactions. Each interaction is executed by the agents that play the roles that define the interaction and the execution depends on the (local) interaction beliefs. If the agents have beliefs that are consistent with each other, a coherent execution of the interaction will take place. In an environment where human agents are playing the roles, slight (or even severe) misalignment of these behaviors can be solved by the capacity of the humans to adapt to misunderstandings and information mismatch.

Agents (as humans) develop over time a large base of interaction beliefs, which allow them to cope with a wide range of interaction situations. This is why the organizational processes can be carried out in most contexts and exceptional situations. When using an agent-oriented approach, in order to solve the exceptions that occur but have no resolution beliefs implemented in the software agents, an "escape/intervention" [13] mechanism can be used. Each time an agent cannot find a local solution for a mismatch during an interaction, it can defer control to a higher authority (higher level agent, typically a human). Therefore, such a system will never block, supporting the humans up to the levels it has been programmed to do, but leaving the humans to intervene when the situation is too complex for them to solve.

Interaction beliefs are local anticipatory models. These describe future possible states in a specific interaction from a local perspective of an agent. In an organization, an agent can play various roles by using its "experience" (interaction beliefs that have proved successful in the past), but can also build new ones, depending on the context. Continuous enactment of interaction leads to whole process enactment. In a software multi-agent system, if the captured behaviors are not matching in a given context, the agents will revert to humans. Of course, this can decrease the performance of the system - in terms of support and/or automation - to unacceptable levels. Software agents should also be able to overcome their mismatching behaviors in an anticipatory way. There are multiple ways to tackle behavior mismatches:

- Each agent is individually trying to align its behavior on-the-fly, having only local information.
- The agents try to align their behavior before the interaction starts, by sending and comparing each other's intended behavior.
- There is a third-party agent interfering with the interaction:
  • The third-party agent can be a superior agent that can align and impose a common interaction behavior that is sound, by having full access to the interaction beliefs of the agents. This can happen before the interaction starts.
  • The third-party agent acts as a mediator between the agents.

A typical interaction where such behavioral mismatches can occur, is the buyer-seller interaction, which will be elaborated on in the later parts of the paper. The most encountered behavioral mismatch in the buyer-seller interaction is due to the fact that each party wants to have its output criteria fulfilled first. Basically, that means that the buyer wants to be in possession of the product before he pays, and the seller wants to receive the payment before delivery. In this example, a typical third-party agent interfering with the interaction is a bank, who takes the risk of paying the seller first, and invoicing the buyer after he received the product. Later in the paper, we show how these behavioral mismatches can be solved using various alignment mechanisms.

1.3   Taxonomy and Benefits

In the on-the-fly mechanism, the anticipatory system is the individual agent who tries to align its behavior, based on the limited information it has about the interaction execution. According to the taxonomy of Butz et al. [4], this mechanism is a state anticipatory mechanism, as predictions about future states directly influence current behavioral decision making. In this case, a predictive model must be available to the agent, or it must be learned by the agent. In our approach, such a model is formalized using Behavior Nets, as will be explained in the next section. The Behavior Net captures the planned behavior of an agent for the interaction it intends. The on-the-fly alignment mechanism, as proposed in this paper, will result in faster and more direct model and behavioral learning, as the agent is able to learn new behaviors during the interaction. Furthermore, it will result in improved social skills, as the agent is able to alter its behavior during the interaction, in order to ensure a successful interaction, even when the original behaviors of the interacting agents are not matching.

When aligning behavior before the actual interaction (or pre-interaction alignment), future states do not directly influence current behavioral decision making, instead future states expected by the agents are used to discuss their course of actions, in order to align their beliefs about the interaction. For this reason, it can still be called state anticipation, as the explicit predictions about the future (formalized as Behavior Nets) influence the discussion process, and thus the future behavioral decision making, which is defined in the Behavior Nets of the agents. This form of anticipatory behavior will be beneficial for social behavior within the overall system, as predictive knowledge of other agents is exploited.

We considered that the third choice with a superior agent is "less anticipatory", in the sense that only if viewed from a larger perspective (the system is formed by the participating agents, plus a third-party agent), it becomes a system that investigates a potential scenario for the future. The mediator on the other hand exploits the predictive knowledge of the interacting agents, for aligning them through its own behavior. For this reason, this approach will also be beneficial for social behavior.

1.4   Paper Outline

In the next section, we introduce a representation of the behavior in terms of Behavior Nets. In section 3, we describe a mechanism for individual alignment based on "alignment policies". Section 4 describes the pre-interaction alignment, and section 5 the alignment with a third-party agent interfering. We end with a discussion and conclusions.

2   Behavior Nets

Petri Nets are a class of modeling tools, which originate from the work of Petri [12]. Petri Nets have a well defined mathematical foundation, but also an easily understandable graphical notation [15]. Because of the graphical notation, Petri Nets are powerful design tools, which can be used for communication between the people who are engaged in the design process. On the other hand, because of the mathematical foundation, mathematical models of the behavior of the system can be set up. The mathematical formalism also allows *validation* of the Petri Net by various analysis techniques.

The classical Petri Net is a bipartite graph, with two kind of nodes, *places* and *transitions*, and directed connections between these nodes called *arcs*. A connection between two nodes of the same type is not allowed. A transition is *enabled,* if every input place contains at least one token. An enabled transition may fire, which will change the current *marking* of the Petri Net into a new marking. Firing a transition will *consume* one token from each of its input places, and *produce* one token in each of its output places.

2.1   Definition of Behavior Nets

In the following, the formal definition of Behavior Nets is given, which is a Petri Net extension, based on Workflow Nets [1], Self-Adaptive Recovery Nets [8] and Colored Petri Nets [9]. An example of such a Behavior Net can be seen in figure 2 (a).

Definition 1. *Definition of Behavior Nets*

A Behavior Net is a tuple $BN = (\Sigma, P, Pm, T, Fi, Fo, i, o, L, D, G, B)$ where:

- Σ is a set of data types, also called color sets
- P is a finite set of places
- Pm is a finite set of message places
- T is a finite set of transitions (such that $P \cap Pm = P \cap T = Pm \cap T = \emptyset$)
- $Fi \subseteq ((P \cup Pm) \times T)$ is a finite set of directed incoming arcs, and
- $Fo \subseteq (T \times (P \cup Pm))$ is a finite set of directed outgoing arcs, such that:

$$\forall p \in Pm : \bullet p = \emptyset \oplus p\bullet = \emptyset$$

- i is the input place of the behavior with $\bullet i = \emptyset$ and $i \in P$
- o is the output place of the behavior with $o\bullet = \emptyset$ and $o \in P$
- $L : (P \cup Pm \cup T) \to A$ is the labeling function where A is a set of labels
- $D : Pm \to \Sigma$ denotes which data type the message place may contain
- G is a guard function which is defined from Fi into expressions which must evaluate to a boolean value
- B is a binding function defined from T into a set of bindings b, which binds values (or colors) to the variables of the tokens

The set of types Σ defines the data types tokens can be, and which can be used in guard and binding functions. A data type can be arbitrarily complex, it can for example be a string, an integer, a list of integers, or combinations of variable types.

The places P and Pm and the transitions T are the nodes of the Behavior Net. All three of these sets should be finite. The extension of classical Petri Nets is the addition of the set Pm which are nodes for sending and receiving messages during an interaction. Such a message place is either a place for receiving or for sending messages, it cannot be both.

Fi and Fo are the sets of directed arcs, connecting the nodes with each other. An arc can only be from a place to a transition, or from a transition to a place. By requiring the sets of arcs to be finite, technical problems are avoided, such as the possibility of having a infinite number of arcs between two nodes.

Executing a behavior is part of an interaction process, the behavior is created when the interaction starts, and deleted when the interaction is completed. For this reason, the Behavior Net also has to have one input and one output node, because the Behavior Net initially has one token in the input place when the interaction starts, and can be deleted when there is a token in the output place.

With function L, a label can be assigned to every node. This has no mathematical of formal purpose, but makes the Behavior Net more easily understandable in the graphical representation.

Function D denotes which message place may contain what data type. This is useful for determining which message place an incoming message has to be placed on. Because the two (or more) behaviors in an interaction are distributively executed, message places of both behaviors cannot be connected directly with each other, as the behaviors do not have to be aligned.

Function G is the guard function, which expresses what the content of a token has to be, to let the transition consume the token from the place. Function G is only defined for Fi, because it makes no sense to put constraints on outgoing edges of transitions. In other words, this function defines the preconditions of the transitions.

Transitions can change the content of a token. Binding function B defines per transition, what the content of the tokens produced by the transition will be. Bindings are often written as, for example, $(T1, < x = p, i = 2 >)$, which means that transition $T1$ will bind value p to x and value 2 to i. The values assigned to the variables of the token (which data type must be in Σ) can be constants, but can also be values of the incoming token, or values from the knowledge- or belief-base of the agent.

2.2   Operations

In Behavior Nets, there is a set of *primitive* operations for modifying the net structure, such as adding and deleting places, transitions, arcs and tokens. Besides the primitive operations, there is a set of more *advanced* operations, which also preserve local soundness. By preserving local soundness we mean that after applying the operation, an execution of the behavior will still terminate properly, if the behavior is also terminated properly before the operation. The message places Pm are not taken into account when determining local soundness.

Local soundness refers to a sound behavior, to make the distinction with a sound interaction, which will be referred to as global soundness. More information about soundness can be found in [2]. The used set of advanced operations are:

- division and aggregation*, which divides one transition into two sequential transitions, and vice versa,
- parallelization and sequentialization*, which puts two sequential transitions in parallel, and vice versa,
- specialization and generalization, which divides one transition into two mutual exclusive specializations, and vice versa,
- iteration and noIteration, which replaces a transition with an iteration over a transition, and vice versa,
- receiveMessage and notReceiveMessage, which adds or deletes an incoming message place,
- sendMessage and notSendMessage, which adds or deletes an outgoing message place.

For some of the operations, marked with *, is it not always clear how they can be applied on-the-fly, because of the dynamic change problem [3]. For example, sequentialization, as mentioned above, cannot be applied for every token marking, as it is not always clear on which places the tokens from the old behavior should be placed, when migrating to the new behavior. For modeling the migrations the approach of Ellis et al. [7] is used. By modeling a behavior change as a Petri net, it can be exactly defined how to migrate the tokens from the old behavior to the new behavior. Note that advanced operations can also be described using the primitive operations. For the receiveMessage, notReceiveMessage, sendMessage and notSendMessage, nothing needs to be migrated, as there is no change in the places, except for the message place, which initially do not contain a token. In figure 1 can be seen how the migration for the operation parallelization can be modeled.

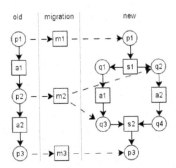

Fig. 1. Migration of old to new behavior

3   Individually Aligning Behaviors On-the-Fly

Before two agents start an interaction, they will both individually choose a behavior they are going to execute, based on what they are expecting out of the interaction. Such behavior is an explicit representation (a Behavior Net) of the course of actions (transitions) and future states (places) the agent is expecting. An interaction however will not terminate, if the planned behaviors of the agents interacting are not matching. When the behaviors are not matching, certain expected future states are unreachable. Therefore, the agent has to alter its predictions about the future. To do so, agents must be able to change their behavior on-the-fly, i.e. during the interaction. For this purpose, alignment policies are used by the agents.

3.1   Alignment Policies

An alignment policy is an ordered set of primitive or advanced operations. In our approach, an agent has a set of policies in his knowledge-base from which it can choose when an interaction for example has deadlocked, i.e. when there is no progression anymore in the execution of the behavior. How an agent will choose an alignment policy (or if it will choose one at all) depends on different factors. The factors used are: problem information, beliefs about the agent being interacted with, and the willingness to change its own behavior.

Problem information. Most of the time, a problem will occur, when the agent is not receiving the message it is expecting. It can also be the case that the agent did not receive a message at all. If it did receive a message, the type of the received message and other information about the problem can be used as attributes for selecting the proper alignment policy.

Beliefs about the agent being interacted with. Beliefs about the other agents can be of great importance when choosing an alignment policy. For example, when the agent completely trusts the other agents, it might be willing to make more change in its behavior than when it distrusts the other agents.

Willingness to change behavior. When an agent has very advanced and fine-tuned behaviors, it is not smart to radically change the behaviors because of one exceptional interaction. On the other hand, when the behavior of the agent is still very primitive, changing it a lot could be a good thing to do. Hence when an agent gets "older", and the behaviors are based on more experience, the willingness to change its behavior will decrease. This approach can be compared with the way humans learn, or with the decrease of the learning rate over time when training a neural network.

Currently, we are experimenting with agents who use a Neural Network to choose an alignment policy [10]. As this approach seems to be promising, other approaches, like Genetic Algorithms, could also be used. When an agent does not know how to handle a certain problem, it can go into escape mode, to learn new ways to overcome its lack of experience. With a neural network, you exactly know the certainty of the agent choosing a specific alignment policy. In this way, it is easy to know when to trigger escape mode. More information about the concept of escape mode can be found in [13].

3.2 Example

This section gives an example of how these alignment policies could work. In this example, as shown in figure 2, the buyer and the seller already agreed on the product the buyer wants to purchase, but, as seen in the figure, they have different ideas about the order of the delivery and the payment. For the sake of the example, we assume that the behavior of the buyer is very advanced, and thus has no willingness to change its behavior. On the other side, the seller's behavior is still primitive and unexperienced, hence we are looking at the problem of how the seller can align its behavior with the buyer, assuming that the seller has trust in the buyer.

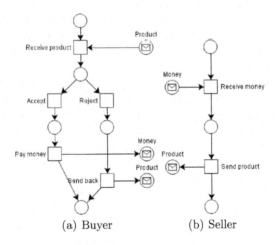

(a) Buyer   (b) Seller

Fig. 2. Behaviors of buyer and seller

When the interaction starts, it immediately deadlocks; the buyer is waiting for the product, and the seller is waiting for the money. A way to overcome this problem would be for the seller to send the product and wait for the money in parallel. Therefore, by using an alignment policy based on the operation `parallelization` the behavior of the seller changes to the behavior as seen in figure 3 (a), and the interaction can continue. However, if the buyer rejects the product, and sends it back, the seller still doesn't have the appropriate behavior to handle this, because the seller is waiting for the money. In case the seller receives the product back, but when it is expecting the money, the seller could use an alignment policy based on the operation `specialization` to overcome this problem, which divides the receive money transition into two separate transitions, *receive money* and *receive product*. The resulting behavior can be seen in figure 3 (b). The behaviors of the buyer (figure 2 (a)) and the behavior of the seller (figure 3 (b)) are now aligned, and thus matching.

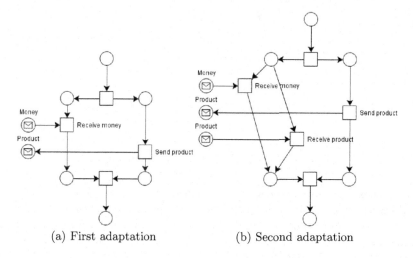

(a) First adaptation   (b) Second adaptation

Fig. 3. Adapted behavior of seller

4   Pre-interaction Alignment

As said before, the agents participating in an interaction will choose a priori a behavior they intend to execute. These behaviors consist of their own intended actions and future states. Conflicting behaviors within an interaction can be solved on-the-fly, as seen in the previous section, but another approach would be to align the behaviors before the actual interaction. In this way, the expected future states of the agents can be aligned beforehand. Such a pre-interaction alignment is a very difficult process. Although it is natural between humans, it is very difficult to formalize and implement a solution that offers pre-interaction alignment for software agents. In this section, we are discussing merely the complexity and problems poised by pre-interaction alignment. An example based on an operational business process illustrates the problem.

4.1   Pre-interaction Alignment with Intended Behaviors Only

There are two ways in which pre-interaction alignment can be achieved. In the first case, we assume that the agents posses only the description of their own intended behavior. Before an interaction starts, the agents that are committed to execute the interaction can be forced by the designer of the interaction-based system to reveal their intended interaction to each other, like in figure 4.

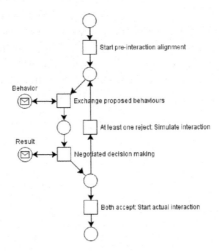

Fig. 4. Behavior for pre-interaction alignment

Each agent can automatically infer whether the interaction will succeed or fail. If both agents realize that their behaviors are not matching and they have internal interaction simulation abilities, each can run simulations of the future interaction and come out with behaviors that matches each other. These anticipatory matching behaviors can be exchanged again, and if the agents agree via negotiation about using one, they can start the actual interaction.

In an interaction with only two agents, there are four possible outcome scenarios of the agents' simulations:

1. Both agents propose new behaviors for each other.
2. One agent proposes a new behavior, and the other accepts it.
3. One agent proposes a new behavior, and the other rejects it.
4. No agent is able to propose a new behavior.

After scenario 1, a following round of negotiation and, if necessary, exchanging new behaviors is needed. In order to reduce the complexity of the pre-interaction negotiation about how to align the behaviors, some rules for selecting a matching behavior can be enacted. For example, only scenario 1 may be allowed, and the selection is based on an "external" assessment, i.e. the simplicity of the result (in terms of the number of places, activities and arcs). Or, alternatively, scenario 1 is applied in successive steps, and via an internal assessment, for example, the results are graded 1 to 10 by each agent, leading to the selection of the one with the highest cumulative grade. This implies of course that the agents should be able to assess independently, without human intervention, the quality of a behavior they have to exhibit.

4.2   Using Interaction Beliefs

Pre-interaction alignment can be improved if the agents also have beliefs about the expected behavior of the other one(s). We are currently extending the language to model behavior and behavioral belief by adding notation for expected behavior. Because this represents the expected course of action and future states of the agent being interacted with, exchange of intended and expected behaviors could more easily reveal potential mismatches in the expected future states of different agents. In this setting, one agent represents for itself the intended behavior for a given interaction, by using a normal Behavior Net, and also Behavior Nets that represent what this agent is expecting from the other agents involved in the interaction.

For example, consider the situation in a small enterprise where the sales manager has to expedite an order for a particular customer. He intends to send a request for re-planning the order to the planner and also a request to the shop-floor scheduler, who is in charge of the actual early re-scheduling of the tasks necessary to execute the order faster. In figure 5, the sales manager intends to send the requests, waits for an answer from the scheduler and depending on the answer (re-scheduling possible or not) continues to handle the negotiation with the customer. He expects that the scheduler will analyze the request, and if feasible, will arrange the re-scheduling and give a positive answer. He also expects that the planner will update the delivery date for the order in its own system.

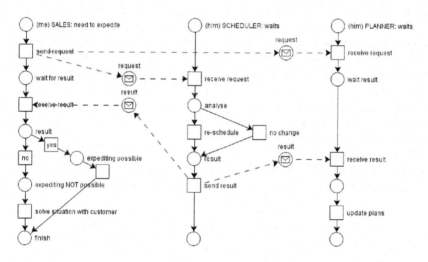

Fig. 5. Interaction Belief of sales manager

The scheduler (as in figure 6), is not cooperative in this interaction. When he receives the request, he is just making a justification for refusal and sends this back to the manager. He is not eager to make himself a new schedule, due to various reasons (e.g. lack of time). His beliefs about the sales manager reflect his unhelpful behavior, because he beliefs that the sales manager made a mistake that resulted in the order expedition and it is his role to solve the problem with the customers. He is not aware for example that expedition of orders can increase the price the customers pay. By behaving in this way, the sales manager

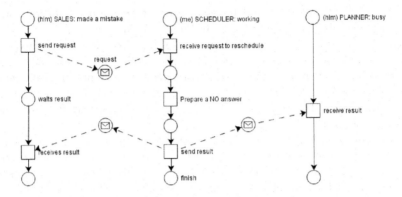

Fig. 6. Interaction Belief of scheduler

will always have a negative response. We call the graphs depicted in figure 5 and figure 6 Interaction Beliefs (IBs), comprising the intended behavior of an agent and also the expected behavior of the others.

On the other hand, the scheduler is aware that the planner has an automatic scheduler used to check the feasibility of the automatically generated plans (capacity check). In figure 7, we depict two matching IBs, in an interaction between the scheduler and the planner. When the scheduler has a problem at the shop-floor level, and needs a new schedule, he asks the planner to provide him with one generated by the schedule-based capacity checker. The scheduler can use this as a blue print for a new schedule and will send the updated schedule to the planner, who will update the plans. However, the planner will not generate a schedule at the expedition request of the sales manager, but will only change the delivery date for the order.

If these three agents would exchange their IBs depicted above, the sales manager will immediately realize that the scheduler will always give him a negative response. Also, he will realize that there is a powerful scheduling capability with

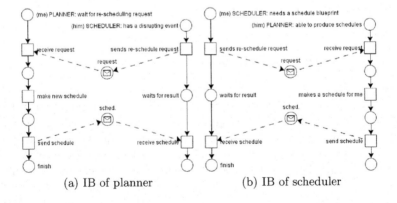

(a) IB of planner   (b) IB of scheduler

Fig. 7. Matching IBs between the scheduler and planner

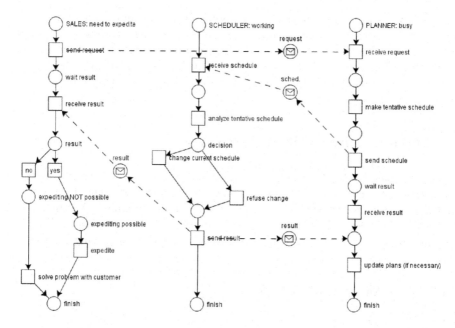

Fig. 8. Human solution

the planner, and the shop-floor scheduler tends to use this capability instead of making its own schedules. The planner will realize also that expedited orders are never realized in the shop-floor, and its own scheduling ability is used only when the scheduler asks for it.

We envisage an alignment via negotiation and re-design process of the IBs, leading to matching that can use all this existent information in the expected behaviors. Currently, such a situation is solved by humans, who can design a solution like in figure 8. Here, the request for expediting is sent directly to the planner, who generates a new prospective schedule for the shop-floor, which is adapted and if feasible is sent back to the planner, who has to update his plans, together with a "green-light" for the sales manager.

It is also possible that the graph in figure 8 is the IB of a new planner who arrived in an organization where the sales manager and the scheduler behave and believe as shown in figures 5 and 6. We can consider that in this situation we have a "knowledgeable" planner agent, and two other "stupid" agents, who are just wasting each other's time. The planner can show how the interaction should be carried out, and the other two agents can "learn" from the IB, which one has better experience. Thus, the expected behavior in IBs (at least from ones of knowledgeable agents), can be used to spread experience to agents who are beginners.

The process of alignment becomes more complex if there are more than two agents involved in the interaction. Also, if the interaction is long, aligning becomes combinatorially complex. For this reason, it is advisable to attempt to align behaviors which are short and are applied in bipolar interactions. Business processes are in fact complex interactions. In order to achieve business process coherence, a method to decompose into simple interaction, as described in [16], is necessary. Furthermore, the use of expected behaviors via IB interchange will increase the complexity of the alignment process compared with the interchange of intended behaviors only.

5   Alignment by a Third-Party Agent

In the previous two kinds of alignment, the interacting agents have to solve the conflict of non-matching behaviors themselves. However, it could be the case that the interacting agents did not manage making an agreement. A logical next step would be to let a third 'neutral' agent intervene in the interaction. This can be done in two ways, the neutral agent being a mediator, or being a superior agent.

5.1   Third Agent as a Mediator

When there is a problem in the execution of a behavior, the agent can choose an alignment policy to solve this conflict, as described in section 3. Such an approach only works if one of the agents is willing to make "sacrifices" to its behavior (e.g. if the seller in the previous section is willing to send the product first, instead of first receiving the money). In such a case, an agent can also ask a mediator, by choosing a special policy. This can also be a policy learned from a superior agent, by the use of escape mode. For this purpose, it has to know an agent who is able to play the mediator-role for this particular interaction.

In the best case scenario, both agents can use their initial non-matching behaviors when interacting with the mediator. For this reason, when one agent calls for a mediator, there is no reason for the other agent in the same interaction to refuse this. But there is still no guarantee that the interaction will be performed successfully. The use of a mediator could require both agents to change their behavior, for example when the mediator wants to have a fee [17]. Individual alignment can be used to alter the behaviors of the initial interacting agents, to make them aligned with the mediator. This "sacrifice", however, can be refused by the agents.

An example of how an interaction between two agents with conflicting behaviors can be solved with a mediator is shown in figure 9. The same behaviors of the buyer and the seller as in the example of section 3 are used. However, for alignment in this case, these agents do not have to change their behavior, because they are only interacting with the mediator. The behavior of the buyer is matching with the behavior of the mediator, and the same is true for the seller. The mediator in this example (which can be a bank) first pays the product, passes the product through to the buyer, and earns the money from the buyer. This is the best-case scenario for this interaction, as the mediator does not want a fee. This is not likely in real life, but this is only an illustrative example.

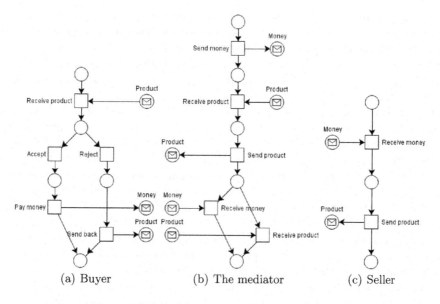

(a) Buyer   (b) The mediator   (c) Seller

Fig. 9. Example of alignment with a mediator

5.2 Third Agent as a Superior Agent

When the interacting agents both have the same superior agent, this agent can be asked to solve the conflict. Again, this can be done by the use of a special policy, as with asking for a mediator. A superior agent can also be asked when pre-interaction alignment has failed. However, in this case, the third agent will not play a role in the interaction. The superior agent will compare the behaviors of both agents, and will "enforce" aligned behaviors to both agents. The superior agent can do this in the same way with simulation as described for the pre-interaction alignment in section 4, but without the negotiation, as the superior agent imposes the agents to use a specific behavior. This can be a way of escape mode itself, as the superior agent does not necessarily need to be a software agent. When the interacting software agents fail, a human can align the behaviors of the agents manually.

6   Discussion and Future Work

As far as we know, this is the first attempt to apply this kind of discrete mathematics to anticipatory agents. Although there are some approaches that apply Petri Nets to model agent interaction, these are mainly concerned with a centralistic view. However, we are taking a distributed approach. This approach has the potential to appeal to two research communities: the one oriented towards Business Information Systems development (who apply Petri Nets for modeling workflows), and also to the growing anticipatory agent community. Some researchers have pointed out that the models used for BIS analysis and design are in fact executable models of the organization they support. Apparently, the inclusion of a executable model of the organization in the organization itself (seen as a system), makes the whole an anticipatory system. Obviously, organizations that use a BIS increase their anticipatory ability. Unfortunately, there is no evidence that the current development of BISs is done with explicit anticipatory ability in mind.

Our strong belief is that agent-oriented BIS that support the business processes of the organization (in terms of interaction support), due to the anticipatory ability of the individual agents, lead to an emergent behavior of the whole system that has an anticipatory nature. Of course, such a statement has to be proven empirically and theoretically. An intuition is that simulation - currently intended for development purposes - can have an important role in the anticipatory architecture of an agent-enabled BIS in an organization. If the executable agent-based model of the organization can perform simulations itself that start from the present state as perceived in the organization, this model can predict future states. The results of these predictive simulations can be used to influence the current state of the organization via an effector sub-system.

Currently, the idea about development simulations is that these are in fact games, where expert players interact with the simulated agents, via the escape/intervention mechanism. An escape is triggered when an agent cannot perform a certain act, and an intervention is when the human supervisor decides that the course of action is not as desired. After the agents are fully developed and are deployed in the organization, the predictive simulations that they could perform should be as automatic as possible (otherwise human intervention would make this anticipatory mechanism inefficient). This observation proves the need for better automatic alignment mechanisms very relevant.

Our future research will be directed towards a number of issues. Mechanisms for triggering the escape mode should be investigated, but also what the human will do after the escape is activated, i.e. how to train the agents. Pre-interaction alignment also has to be further investigated. Mechanisms for simulation resulting in new behaviors, and negotiating what behavior to use, are in a preliminary stage. Furthermore, alignment with the intervention of a third-party agent needs further research as well.

7   Conclusion

As we have shown, it is possible to describe a policy for alignment that can be applied when the interaction beliefs of two or more interacting agents are not matching. We introduced an extension of Petri Nets to capture the interaction beliefs and also a mechanism to choose the appropriate policy that adapts the beliefs from one agent perspective. From the anticipatory systems perspective, this research can enable predictive agent model execution (agent-based simulation of organizational models) to be more reliable and necessitate less human intervention in terms of alignment.

As proposed in this paper, the on-the-fly alignment mechanism results in faster and more direct model and behavioral learning, as the agent is able to learn new behaviors during the interaction. Furthermore, it resulted in improved social skills, as the agent is able to alter its behavior during the interaction, in order to ensure a successful interaction, even when the original behaviors of the interacting agents were not matching.

As we have shown, pre-interaction alignment and alignment with the help of a mediator are beneficial for social behavior, as these approaches exploit predictive knowledge of other agents.

We consider a superior agent aligning the interacting agents forcing them into new "less anticipatory" behavior. For this reason, the other mentioned approaches are favored above this one. This approach can however be used as a last resort, especially because the superior agent can be a human.

Finally, we believe that interdisciplinary work between the BIS research and anticipatory agent research can yield lots of "cross-fertilization" and raise the awareness that BIS enabled organizations are in fact anticipatory systems and also provide test beds for novel anticipatory agent ideas. Today, there are many system architectures that support activities in business processes, like workflow systems and enterprise integrated systems. Most of these architectures are monolithic and centralistic, having a low degree of flexibility. An architecture based on agents that allow change via different mechanisms for behavioral alignment offers a more natural approach to process support.

We consider that business processes that are inherently interactional, like in sale/purchasing, but also planning and scheduling can benefit from these alignment mechanisms. There is a clear uptake in implementing adaptive sale/purchasing systems via web-services, but there is still low emphasis on higher-level issues related to these systems, like conceptual architectures, interaction representations and mechanisms. In contrast with technology-driven approaches, our approach is doing exactly that, taking a top-down, formalism based, research-oriented approach.

References

1. van der Aalst, W.M.P.: Verification of Workflow Nets. In: Azéma, P., Balbo, G. (eds.) ICATPN 1997. LNCS, vol. 1248, pp. 407–426. Springer, Heidelberg (1997)
2. van der Aalst, W.M.P.: Interorganizational Workflows: An approach based on Message Sequence Charts and Petri Nets. Systems analysis, modelling, simulation 37(3), 335–381 (1999)
3. van der Aalst, W.M.P., Jablonski, S.: Dealing with workflow change: identification of issues and solutions. Computer systems science and engineering 5, 267–276 (2000)
4. Butz, M.V., Sigaud, O., Gerard, P.: Internal Modals and Anticipations in Adaptive Learning Systems. In: Butz, M.V., Sigaud, O., Gérard, P. (eds.) Anticipatory Behavior in Adaptive Learning Systems. LNCS (LNAI), vol. 2684, pp. 86–109. Springer, Heidelberg (2003)
5. Davidsson, P., Astor, E., Ekdahl, B.: A Framework for Autonomous Agents Based on the Concept of Anticipatory Systems. In: Proc. of Cybernetics and Systems 1994, pp. 1427–1431. World Scientific, Singapore (1994)

6. Ekdahl, B.: Agents as Anticipatory Systems. In: Proc. of SCI'00 and ISAS'00, pp. 133-137 (2000)
7. Ellis, C.A., Keddara, K.: A Workflow Change is a Workflow. In: van der Aalst, W.M.P., Desel, J., Oberweis, A. (eds.) Business Process Management. LNCS, vol. 1806, pp. 201-217. Springer, Heidelberg (2000)
8. Hamadi, R., Benatallah, B.: Recovery Nets: Towards Self-Adaptive Workflow Systems. In: Zhou, X., Su, S., Papazoglou, M.M.P., Orlowska, M.E., Jeffery, K.G. (eds.) WISE 2004. LNCS, vol. 3306, pp. 439-453. Springer, Heidelberg (2004)
9. Jensen, K.: An Introduction to the Theoretical Aspects of Coloured Petri Nets. In: de Bakker, J.W., de Roever, W.-P., Rozenberg, G. (eds.) A Decade of Concurrency. LNCS, vol. 803, pp. 230-272. Springer, Heidelberg (1993)
10. Meyer, G.G.: Behavior Alignment as a Mechanism for Interaction Belief Matching. University of Groningen (2006)
11. Meyer, G.G.: Intelligent Products: an Application of Agent-Based Ubiquitous Computing. Accepted for: ABUC'07 (2007)
12. Petri, C.A.: Kommunikation mit Automaten. PhD thesis. Institut fur instrumentelle Mathematik, Bonn (1962)
13. Roest, G.B., Szirbik, N.B.: Intervention and Escape Mode. In: Padgham, L., Zambonelli, F. (eds.) AOSE VII / AOSE 2006. LNCS, vol. 4405, pp. 109-120. Springer, Heidelberg (2007)
14. Rosen, R.: Anticipatory Systems. Pergamon, New York (1985)
15. Salimifard, K., Wright, M.: Petri net-based modelling of workflow systems: An overview. European journal of operational research 134(3), 664-678 (2001)
16. Stuit, M., Szirbik, N.B.: Modelling and Executing Complex and Dynamic Business Processes by Reification of Agent Interactions. O'Grady, M.(ed.) Accepted for Lecture Notes in Artificial Intelligence (2007)
17. Stuit, M., Szirbik, N.B., de Snoo, C.: Interaction Beliefs: a Way to Understand Emergent Organisational Behaviour. Accepted for: ICEIS'07, Funchal, Madeira, Portugal (2007)

Backward vs. Forward-Oriented Decision Making in the Iterated Prisoner's Dilemma: A Comparison Between Two Connectionist Models

Emilian Lalev and Maurice Grinberg

Central and East European Center for Cognitive Science,
New Bulgarian University, 21 Montevideo Street, 1618 Sofia, Bulgaria
elalev@cogs.nbu.bg, mgrinberg@nbu.bg

Abstract. We compare the performance of two connectionist models developed to account for some specific aspects of the decision making process in the Iterated Prisoner's Dilemma Game. Both models are based on common recurrent network architecture. The first of them uses a backward-oriented reinforcement learning algorithm for learning to play the game while the second one makes its move decisions based on generated predictions about future games, moves and payoffs. Both models involve prediction of the opponent move and of the expected payoff and have an in-built autoassociator in their architecture aimed at more efficient payoff matrix representation. The results of the simulations show that the model with explicit anticipation about game outcomes could reproduce the experimentally observed dependency of the cooperation rate on the so-called cooperation index thus showing the importance of anticipation in modeling the actual decision making process in human participants. The role of the models' building blocks and mechanisms is investigated and discussed. Comparisons with experiments with human participants are presented.

Keywords: anticipation, cooperation, decision-making, recurrent artificial neural network, reinforcement learning.

1 Introduction

In formal game theory players are described as perfectly rational and possessing perfect information about the game including not only their possible moves and payoffs but also those of their opponents. On the other hand, the bounded rationality view on cognition states that people are almost never perfectly rational (see e.g. [1]). Moreover, they try to minimize the cognitive effort while making decisions. Finally, the results of experiments involving games demonstrate that people rarely play as prescribed by the normative game theory. We have started a series investigations on the cognitive processes involved in decision making in Iterated Prisoner's Dilemma Game (IPDG) from a cognitive science point of view [2-5] using different approaches involving psychological experiments, eye-tracking experiments, and modeling and simulations.

In [3], a simple model based on expected subjective utility theory was put forward. The model used extensively backward reinforcement learning mechanisms and based

M.V. Butz et al. (Eds.): ABiALS 2006, LNAI 4520, pp. 345–364, 2007.

on that made predictions about the move probability of the opponent. Additionally in order to explain some specific characteristics of the decision making process, explicit accounting of the current game was added. The latter allowed for the description of the well known dependency of the cooperation rate and the structure of the payoff matrix expressed by the so called Cooperation Index (CI) (see [6]). This property is not available in typical reinforcement learning based models used to model playing of IPDG and in which the probability for cooperation is based only on past games (see [7] and [8]).

Taking into account the results obtained by Hristova and Grinberg [3], here we propose a connectionist architecture based on a recurrent network which accounts for the payoff structure of the PD game, the past moves and payoffs, and predicts the next moves of the player and his/her opponent, and the expected payoff from the next game. A related attempt, using recurrent neural networks, to model the complexity of IPDG has been made by Taiji and Ikegami [10] but in their model only the moves of the players are used in the recurrent network and only a single payoff matrix is played, so the question of the influence of the different ratios among the payoffs in different game matrices (i.e. dependency on game CI) could not be considered. Two further variants based on the general architecture were explored. The first involved training of the next-move output node using a backward looking reinforcement model (see [9] for details), further referred to as Model B. In the second, the training of the move node was based on evaluation of the future payoffs and thus essentially using anticipation (further referred to as Model A). The analysis and comparisons of the simulation results of the two models with recent experimental results and the discussion of the importance of the mechanisms involved are the main concerns of this paper.

1.1 The Prisoner's Dilemma Game

The Prisoner's dilemma is a two-person game. The payoff table for this game is presented in Table 1. The players simultaneously choose their move – C (cooperate) or D (defect), without knowing their opponent's choice.

Table 1. Payoff table for the PD game. In each cell the comma separated payoffs are the Player I's and Player II's payoffs, respectively.

		Player II	
		C	D
Player I	C	R, R	S, T
	D	T, S	P, P

In Table 1, R is the payoff if both players cooperate (play C), P is the payoff if both players 'defect' (play D), T is the payoff if one defects and the other cooperates, S is the payoff if one cooperates and the other defects.

The payoffs satisfy the inequalities T > R > P > S and 2R > T + S. This structure of the payoff matrix of that game offers a dilemma to the players: there is no obvious best move. The dominant D move (T > R and P > S) would lead to lower payoffs if adopted by all the players (payoff P) although this is the choice prescribed by standard game theory. Cooperation seems to be the best strategy in the long run (R > P) but at the risk of one of the opponents to start to defect and the other to receive the lowest payoff S. This quite complicated situation is at the heart of the dilemma in this game and is the reason for the on-going interest in this game over the past 50 years and continuing today.

Rapoport and Chammah [6] proposed the quantity CI = (R–P)/(T–S), called cooperation index, as a predictor of the probability of C choices, monotonously increasing with CI. In Table 2, two examples of PD games with different CI, 0.1 and 0.9, respectively, are presented.

Table 2. Examples of PD game matrices with different CI – 0.1 and 0.9, respectively. The first payoff in each cell is the payoff of the 'row' player and the second of the 'column' player.

CI=0.1		Player II	
		C	D
Player I	C	56, 56	0, 60
	D	60, 0	50, 50

CI=0.9		Player II	
		C	D
Player I	C	56, 56	0, 60
	D	60, 0	2, 2

1.2   Social Interactions and Modeling of the IPDG

A common assumption is that people build mental models of themselves and other people they interact with (and of the world as a whole) in the long run. Such models include grasping typical aspects of their behavior. This may result in establishing relations of trust or distrust with these people. If more or less pure instances of IPDG are observed in real-life social relations, we keep in mind that, as long as people try to take advantage of these relations, they create the most likely images of the other 'players' and of the environment. These images (or models) are guided by past experience (the history of the relation). The actions assigned to interactions comply with predictions about the others' actions. In other words, past experience and predictions for events based on these experiences are factors which cannot be neglected in understanding human social interactions (e.g., see [12]) and in particular, the IPDG.

From a cognitive modeling point of view the challenge is to understand the decision making mechanisms that would lead to the results observed in the experiments with human participants taking account of all characteristics (like the dependency on CI for instance). We are convinced that the models needed must have a minimal level of complexity and account for playing based on the payoff matrix of the game (e.g. to be sensitive to CI) and on the opponent moves and game outcomes. In the same time human players rely on past experience and predictions of future events. The models presented here are aimed at complying with these requirements.

2 Models – Architectures and Functioning

2.1 Basic Architecture

The core architecture (underlying both presented models) is an Elman recurrent neural network [11] (see Figure 1). In [10], a recurrent network has also been used to model the behaviour of PD game players. However, the network we used has a much more complicated structure including the network input/output structure, the game payoff matrices, the players' moves, and the received payoffs (related to the specific game outcome). The network consists of eight input, thirty hidden-layer, and six output nodes (see Figure 1). The activation functions of the hidden layer and of the output layer are tan-sigmoid and log-sigmoid functions, respectively. Because of the logistic output activation function, a part of the network's outputs could be interpreted as probabilities.

2.1.1 Inputs and Outputs
All the inputs of the network were rescaled within the range [0, 1]. As can be seen in Figure 1, the values of the payoffs from the current game matrix (excluding the payoff S which was always 0), as well as the past game payoff received, the player's and opponent's moves in the previous game were presented at the input nodes at each cycle.

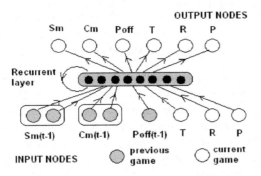

Fig. 1. Schematic view of the recurrent neural network and its inputs and outputs/targets; notation: S_m and C_m are respectively the simulated subject and computer opponent (probability for) moves; Poff(t) is the player's received payoff at time t

The past moves were recoded as [0,1] – for C and [1,0] – for D moves, so that activation would always come from any of the two couples of input nodes, no matter what the moves were – C or D.

The values of the T, R, and P payoffs from the current game had to be reproduced as an output by the model thus implementing an in-built autoassociator. There were two reasons to decide to include this component in the network architecture. The first was that this would force the network to establish representations of the games in its hidden layer which is crucial to account for the game payoff structure in the decision making process. The second one was related to the anticipatory decision mechanism of Model A where the output nodes concerning T, R, and P were used as predictions of the next games' payoffs (see Section 2.2.2 for details).

At the output, the player's move ('S$_m$' node) and the computer-opponent's move ('C$_m$' node) nodes were interpreted as the probability for cooperation for the player and the prediction about the probability of cooperation of his/her opponent in the game at hand. The payoff ('Poff') node represented the expected gain from the current game.

2.1.2 Training

PD games with varying CI – from 0.1 to 0.9 – were presented to the neural network (T was always equal to 1, and S was always 0, R and P were distributed in this interval depending on the CI of a particular game). The games were randomized with respect to CI in the same way as in the experiments with human participants (see e.g. [2,3]) in order to allow comparison with the experimental results (see Section 3).

The network was trained using back-propagation on an input consisting of sequences of overlapping five games – the current game and the four previous games. Such sequences are further called micro-epochs.

In the very beginning of the IPDG, the length of micro-epochs was increasing with each next completed game until it reached five games. The very first inputs were as follows: the first game matrix, the player's move and the prediction of the opponent's move generated with probability 0.5. The first received payoff ('Poff') was obtained from the averaging of the payoffs of the games.

The small number of games that the network dealt with at a time implies sensitivity to local changes in the game and to memory constraints we assumed to exist in real game playing. On the other hand, the micro-epochs were long enough so that specific events in the history of IPDG were able to encode in the memory of the recurrent hidden layer.

The values at the six output nodes were used as predictions when the network was trained within the current micro-epoch. The 'T', 'R', and 'P' output nodes were expected to reproduce the corresponding input values in the input payoff matrices.

The output of the 'S$_m$' node was the model-player's probability for cooperation in the current game. The output at the node 'C$_m$' was the prediction for the cooperation probability of the opponent, and the output at the 'Poff' node meant the expected game payoff.

When both player and opponent had made their moves, and the payoff for the model-player was known, the new target micro-epoch was updated and the network was trained with the inputs it was simulated with and the new targets. For all of the output nodes the training signal is supplied by the game (payoffs) and the opponent moves except for the model-player's move probability. The latter has to be supplied either from experimental data with a human player (if the model is used to fit the behaviour of a real player) or by explicitly modeling the evaluation of the game outcome. Here, we will present results along the latter line based on two different choices of such an evaluation.

2.2 Decision Making of the Models

In order to build a realistic model able to make decisions comparable to the ones made by human subjects, we had to make an assumption for an evaluation mechanism for the outcomes of the player's moves. Hereafter, we discuss two such mechanisms, both based on received payoff maximization, which differ in the emphasis on backward or forward evaluation.

2.2.1 Backward-Looking Model (Model B)

Model B integrates the recurrent network presented in Section 2.1 with the Bush-Mosteller (BM) backward-looking reinforcement learning model in the form proposed by Macy and Flache [9].

We integrated the Macy and Flache model with our recurrent neural network by using the predicted payoff – 'Poff(t)' (see Figure 1) as the player's aspiration level and used it to estimate the target cooperation probability. The current move of the model was generated with a probability equal to the output at the 'S_m' node (see Figure 1). After the game moves were made by the player and its opponent, and the player's payoff was already known, a target probability was calculated using the Macy's and Flache's model [9]. The 'C_m' target node was trained using the actual opponent's move and the 'Poff' output node using the received payoff. The latter was considered to be a kind of aspiration level based on payoff expectation and was used instead of the aspiration update rule from [9]. As explained before, the 'T', 'R', and 'P' output nodes were trained using the values from the input game matrix as targets.

This combination of a neural network model and a reinforcement model was expected to give a model player sensitive to specific game episodes in IPDG and to the payoffs in the game matrix at hand (which could give rise to a CI dependent strategy). Theoretically, when the model encountered an episode, in which all predictions (except for the move prediction) resembled those from any past episode, it would play with the reinforced cooperation probability from that past episode.

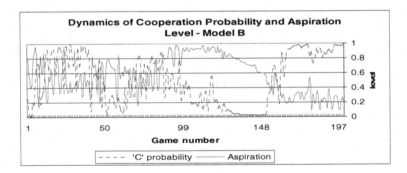

Fig. 2. Dynamics of aspiration (output 'Poff' node) and cooperation probabilities in Model B ('S_m' node)

The dynamics of the decision making process is illustrated in Figure 2, where the fluctuation of the aspiration level together with the player's move probability are shown. It is seen from Figure 2 that as expected low aspiration level leads to high probability of cooperation because the payoffs R are above the aspiration level.

2.2.2 Forward-Looking Model (Model A)

Model A is based on the same neural network architecture as Model B but is aimed at using essentially anticipation mechanisms for deciding about its moves.

Model A uses the predictive properties of the recurrent network in order to "guess" how the game would proceed if its current move were either C or D. An anticipatory module was implemented in the model, so that two sequences of five games predicted by the neural network were produced before making a move. The first sequence began with a C move, and the second one with a D move. Only the first player move was fixed in any sequence. The recurrent network had as first inputs the current game input (together with the other four games from the micro-epoch) including the values of the T, R, and P payoffs, and the players' moves and payoff from the previous game. This is a simpler mechanism than the one used in [10], where all the possible strings of C and D moves are taken into account. Here the first move is chosen and everything else is based on the network predictions.

As the player's move was known in the first fictitious game (C or D), the opponent move was generated with the probability predicted by the network. The payoff for the player from the game was calculated according to the rules of PD game – the autoassociated T, R, P or S based on the moves of both players.

In the second fictitious game the input micro-epoch was updated so that the new T, R, and P values were taken from the output layer of the neural network and considered as prediction about the fictitious game payoffs. The 'Poff(t-1)' node activation got the value of the fictitious payoff from the previous game and the previous moves nodes (the 'S$_m$(t-1)' and 'C$_m$(t-1)' nodes) activations were the fictitious previous game moves. In the next iterations everything was repeated except for that the player move was generated with its predicted probability and was no longer fixed.

So the cycle was completed and the model could predict several future games and related moves and outcomes. The payoffs from both sequences ($Poff_C$ for initial move C and $Poff_D$ for initial move D) were then considered. The obtained payoffs from the five fictitious games for each initial move choice were evaluated using a discount factor as follows:

$$Poff_{C,D} = \sum_{t=1}^{5} Poff_{C,D}(t)\beta^{t-1}, \tag{1}$$

where $Poff_{C,D}(t)$ is the value of the payoff at moment t, for initial move C or D and β is the usual discount parameter that indicates to what extend the remote future game payoffs were important for making decisions at present. If β was 0, only the first fictitious payoff would matter, and if β was 1, all the payoffs would be considered equally important.

A transformation was applied to $Poff_C$ and $Poff_D$ such that the sum of $Poff_C$ and $Poff_D$ became equal to 1 whereas their ratio remained the same.

The probability for cooperation for the current move of the model was then calculated using a soft-max function:

$$P(C) = \frac{e^{Poff_C/k}}{e^{Poff_C/k} + e^{Poff_D/k}}, \tag{2}$$

where $P(C)$ is the calculated cooperation probability and k is a parameter for the sensitivity of the function towards the difference between $Poff_C$ and $Poff_D$. The smaller the value k had, the greater the sensitivity to the difference between the C and D alternative choices became.

3   Game Simulations

3.1   The Computer Opponent

The models were run against a probabilistic Tit-for-two-Tats (Tf2T) computer strategy. Its move depended on the player's two previous moves, thus being adaptive to their temporal cooperativeness without being easily predictable. Depending on the two previous opponent's moves the probability for cooperation was respectively: 0.5 for [C, D] and [D, C], 0.8 for [C, C], and 0.2 for [D, D]. Furthermore, the same computer opponent was used in a series of experiments and such a choice for the simulations here allows for a comparison with the experimental results (see Section 3.2).

They both had the underlying recurrent neural network that provided them with the ability to "recognize" and predict events in the IPDG and, therefore, be able to extract important information such as the strategy of the opponent from the history of the game. Both made their moves probabilistically so that they had the chance to evoke different aspects of their adaptive opponent's strategy, which might have remained invisible otherwise.

3.2   Comparison of Models' and Experimental Results

The results presented in this section are based on 30 IPDG sessions of two-hundred games against the Tf2T computer strategy for each model (B and A). For the comparisons with the experiment, the first 50 games are taken (to match the number of games played by human participants). From the experiment reported in [2], only data from the first part and for the control condition was used in the comparison of Model A, Model B, and participants (see [2] for details). 30 participants played 50 PD games against the computer. The computer used the probabilistic Tf2T strategy described above. This was done to allow the subject to choose his/her own strategy without easily becoming aware of the computer-opponent's strategy. The payoffs were presented as points, which were transformed into real money and paid at the end of the experiment. After each game the subjects got feedback about their and the computer's choice and could always monitor the total number of points they had won and its money equivalent. The subjects received information about the computer's payoff only for the current game and had no information about the computer's total score. This was made to prevent a possible shift of subjects' goal – from trying to maximize the number of points to trying to outperform the computer. In this way, the subjects were stimulated to pay more attention to the payoffs and their relative magnitude and thus indirectly to CI. Games of different CI, ranging from CI = 0.1 to CI = 0.9, were presented both to participants and in simulations with models A and B. Games were presented at random regarding their CI.

In other simulations described in section 3.8., comparisons of participants' and Model A's performance are considered in two more experimental conditions, also taken from [2]: in the first one only games with CI = 0.1 and CI = 0.3 were presented in the IPDG (Low-CI experimental condition). In the second one, games were only with CI = 0.7 and CI = 0.9 (High-CI experimental condition).

The best fit of the experimental results was obtained with the following parameters used for Model A (see equations (1) and (2)): $\beta = 0.7$ and $k = 0.05$.

3.3 Mean Cooperation and Payoffs

In Model B's performance, the payoffs were significantly correlated with the mean level of cooperation in contrast to Model A whose payoffs were not correlated with its cooperation rates against the Tf2T computer player. These results reflect the different nature of the outcome evaluation mechanism in the two models– backward reinforcement learning for Model B and payoff anticipation for Model A.

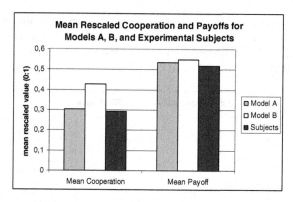

Fig. 3. Comparison of mean cooperation and payoffs between Model A and B, and experimental data from human subjects (taken from [2])

The results for the mean cooperation and payoffs for Model A, Model B, and human participants' experimental data taken from [2] are presented in Figure 3. Regarding mean cooperation, only Model A simulation data and the experimental data were not statistically different (F = 0.121, p = 0.73). The mean cooperation was different for Model B and experiment (F = 5.858, p = 0.019) and for Model B and Model A (F = 6.267, p = 0.015).

For the mean payoff no significant difference was found between the simulations and the experimental data.

3.4 Dependency of Cooperation Rate on CI

First of all, a main effect of CI on cooperation rates was observed in Model A (F = 16.908, p < 0.01) whereas there was no such effect in Model B (F = 0.367, p = 0.83).

In Figure 4, a detailed comparison, concerning the cooperation rate dependency on CI, between the predictions of the two models and the experimental results is shown. It is seen from Figure 4 that Model B gives a completely inadequate description

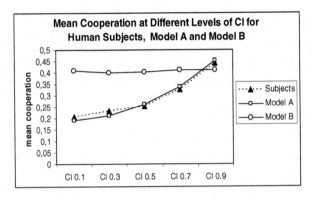

Fig. 4. Model B, Model A, and experimental [2] CI influence on cooperation rates

(no CI dependency) of the experimental results while there were no statistical differences between the mean cooperation of subjects and Model A at all CI levels, and there was no main effect of the type of player (Model A or human) on cooperation (F = 0.386, p = 0.856).

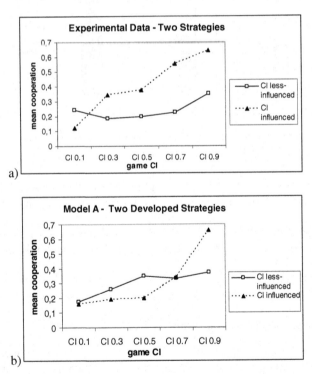

Fig. 5. Comparison of experimental data from: a) human subjects (taken from [2]) and b) Model A simulations, concerning the groups of players with strategies influenced by CI and otherwise

In Figure 5, two groups of players are presented: with strongly and weakly CI influenced strategies for the experiment from [2] and for Model A (see Figure 5, a) and b), respectively). The separation in groups was obtained by cluster analysis based on the monotonous dependency on CI. As seen from Figure 5, there is a qualitative agreement between the model and the experimental data.

As stated earlier our main interest is related to the CI dependency of the cooperation rate in both models. The ability to reproduce such details in the experimental data seems very important to us in order to be able to assess a model's validity. In order to understand the presence and lack of CI dependency in both models we analyzed the hidden layer activations looking for nodes whose activity is correlated or changes with the changes of CI. As discussed earlier in this paper, we included autoassociator nodes in the architecture of the Elman neural network to force the representation of the payoff structure at the hidden layer thus hoping to help the network to account for it (and hopefully for CI). That is why we performed simulations with Model A which essentially used the autoassociator part to make predictions about the payoffs of counterfactual games and with Model B with and without autoassociator nodes. The latter allowed us to see what the responses of these nodes to CI are. The analysis shows that when the autoassociator nodes are present there are hidden nodes whose activity varies with the CI and their number and correlation with CI increases with playing. Such a strong variation of the hidden nodes activations with CI is not observed when the autoassociator nodes are switched off. The conclusion can be made that the inclusion of the autoassociator part is crucial in order to obtain CI dependency in the model. What is the reason for Model B to fail to display CI dependency in its play? One possible explanation is the use of a backward-looking reinforcement mechanism which account mainly the past received payoff and the expected payoff (as aspiration level). Although the network could extract information about the game CI, this information was not useful in determining the playing strategy because it was not needed by the game outcome evaluation mechanism. In the case of Model A, however, the situation is different. The simulation by the model of possible games and moves and outcomes involves the prediction about the payoff structure of the game and thus indirectly of the CI. Thus in the case of Model A the increased sensitivity to CI of the hidden nodes influences the move choice of the model. At closer look however, it turned out that the developed sensitivity of some hidden nodes is only partly responsible for the final dependency of Model A. The largest part is due to the specific anticipatory form of evaluation of the best move involving the payoffs of the game at hand and of anticipated payoffs reflecting the structure of the current game.

3.5   The Role of Anticipation on Cooperation for Model A

As seen from the previous sections, Model A seems much more adequate to account for the experimental findings. Its properties are explored in more detail in this section.

In Figure 6, the CI dependency of cooperation is shown in the case of two different predictions of Model A – that the opponent will cooperate and defect, respectively. It is seen that a CI dependency is present in the case of a predicted D move.

There are several factors working together for the model to behave in such a way: the first one is that the PD games' normalization ($T = 1$, $S = 0$, and R and P distributed in this interval) entailed a strong negative correlation of P with CI ($r = -0.92$, p < 0.01). For example, in the case when the CI was 0.9, P was equal to 0.027, and when CI was 0.1, P was 0.83. Thus, the model, which develops sensitivity to P, develops also sensitivity to CI.

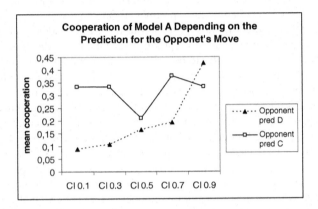

Fig. 6. Cooperation of Model A at different levels of CI according to the rounded prediction for the opponent move (*C* or *D*) for the current game

Another factor is related to the chosen value of the parameter β ($\beta = 0.7$). It ensured that not only the first fictitious payoffs, but also more remote fictitious payoffs would matter for the decision making in the anticipation module of Model A. Thus, when the opponent was predicted to defect by the network, $Poff_C$ (a series with first game outcome *CD* and related payoff S) and $Poff_D$ (a series with first game outcome *DD* and payoff P) would not differ much in their values.

The third factor was found in the soft-max function which would evaluate $Poff_C$ and $Poff_D$. In the case when the opponent was expected to defect in the PD game, it would give rise to relatively high probabilities for cooperation P(C). Thus the cooperation probabilities were higher when P was low (respectively, CI was high) and, vice versa, the likeliness to cooperate decreased when P was high (i.e. CI was low; see Figure 6).

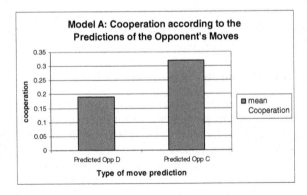

Fig. 7. Mean cooperation of Model A in the case when the predictions for the opponent's move were either *D* or *C*, if 0.5 is taken as a threshold for a *D* or a *C* move

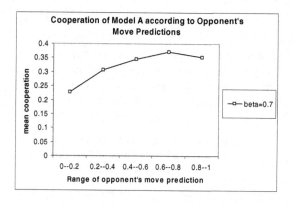

Fig. 8. Cooperation of Model A at five successive ranges of the opponent move predictions

On the other hand, Model A cooperated more as a whole when it predicted that the computer opponent would play a C move (F = 23.12, p < 0.01) (see Figure 7).

Moreover, the cooperation of the model monotonously increased with the expected opponent' probability for cooperation, especially at β = 0.7. To analyze this trend, the predictions for the opponent move probabilities were divided into five ranges in the interval [0, 1] (see Figure 8). As seen from Figure 8, a tendency of increasing cooperation probability with increasing expected cooperation of the opponent can be found.

3.6 Effects Related to Varying β

We discovered that β was of importance for Model A to fit experimental data from human subjects. For example, the overall level of cooperation of the model decreased setting β = 0. The monotonous CI-influence on cooperation was also affected in that case (see Figure 9).

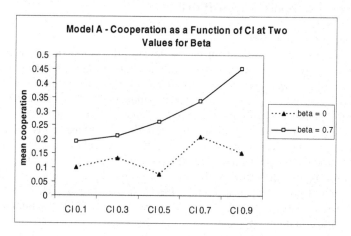

Fig. 9. CI-influence on cooperation of Model A at β = 0 and β = 0.7

Varying β, a qualitative change in the behavior of Model A was found for the relation between the opponent's move predictions and cooperation (see Figure 9). The change was such that when $\beta = 0$, the increase of cooperation with the growth of predicted opponent cooperation probability disappeared. Instead, Model A would cooperate less both when it predicted the opponent would cooperate or defect, and it cooperated more when it was unsure about the other player's moves (see Figure 10).

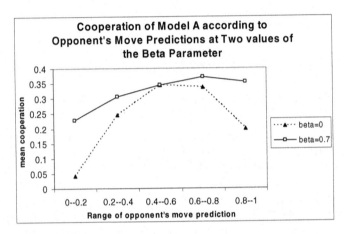

Fig. 10. Influence of predictions about the opponent's moves on cooperation for Model A with $\beta = 0$ and $\beta = 0.7$

As described in section 2.2.2., β is a parameter used to control the depth of the forward-looking of Model A. As seen from Figure 10, it can have a considerable effect on cooperation and its dependency on the game CI.

3.7 Predictions of Model A in the IPDG

3.7.1 Opponent's Move Prediction
The predictions of the network about the moves of a deterministic opponent such as the Tit-For-Tat player became more and more accurate during the IPDG. The beginning of the game served for the model to understand the strategy of its opponent. The overall success in predicting the moves of a deterministic Tit-For-Tat player was 95% and the mean squared error of the respective predicted payoff was 0.04.

This shows that in this easily predictable case Model A consistently predicted two related quantities – prediction about the opponent's move and expected payoff.

Against a stochastic opponent such as the probabilistic Tf2T computer player (see section 3.1. for details) the errors in the opponent's move predictions were larger and the network could predict them in 60% of the cases (see Figure 11). In our opinion, this seems a reasonable estimation of the 'actual' predictability in human playing against the same computer opponent.

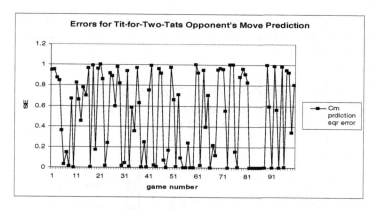

Fig. 11. Network errors in the IPDG related to predictions of the Tf2T stochastic computer opponent, used in the simulations

3.7.2 Predicted Opponent Move and Payoff

In Figure 12 the opponent's move prediction error together with the player's payoff prediction error are given for a fragment of IPDG. Whenever the network didn't manage to predict the move of its opponent, it reflected this in bad predictions of the payoffs it received, and vice versa. This indicates that the network has acquired the rules of PD correctly and associated a given expected game outcome with its corresponding payoff according to the game rules.

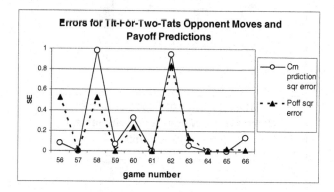

Fig. 12. Plot of squared errors for opponent's move and payoff predictions for a fragment of an IPDG against the probabilistic Tf2T

3.8 Detailed Comparison of Model A with Experimental Data

In the following, we compare the results from simulations with Model A and the results for the three experimental conditions from [2]: the High-CI (CI = 0.7 and CI = 0.9), Low-CI (C = 0.1 and CI = 0.3), and Full-CI conditions (CI covers the range 0.1÷0.9) (see [2] for details). Model A was also tested with settings equivalent to the

High-CI and in the Low-CI conditions (in terms of number and type of games) and it managed to satisfactorily fit the experimental results without any additional parameter fixing as shown bellow.

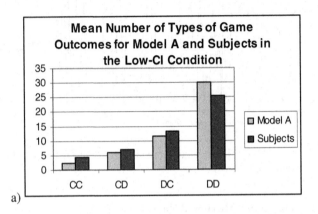

a)

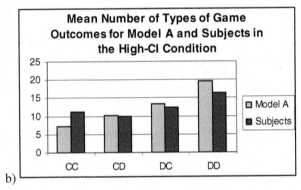

b)

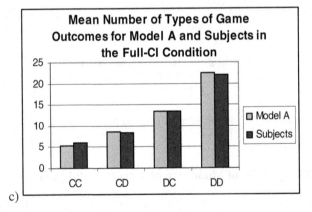

c)

Fig. 13. Comparisons of types of game outcomes for Model A and human subjects in three experimental conditions (see [2]): a) Low-CI condition; b) High-CI condition; c) Full-CI condition

3.8.1 Comparison of Game Outcomes

The numbers of types of game outcomes were compared for Model A and subjects in all experimental conditions. For the Low-CI condition there was a significant difference only for the number of *DD* game outcomes ($F = 5.76$, $p < 0.05$) (see Figure 13a). In the High-CI condition (see Figure 13b) there was also only one difference between the model and subjects for the number of *CC* outcomes ($F = 256.27$, $p < 0.05$). Finally, in the Full-CI condition there were no differences (see Figure 13c).

3.8.2 Comparison of Cooperation and Payoffs

In the experimental study [2] subjects cooperated less in the Low-CI condition than in the High-CI condition ($F = 17.128$, $p < 0.01$), and they cooperated less in the Full-CI condition than in the High-CI condition ($F = 8.299$, $p < 0.01$) (see Figure 14a). These differences were replicated in the simulations with Model A: for the Low-CI and High-CI – $F = 120.46$, $p < 0.01$, and for the Full-CI and High-CI – $F = 18.47$, $p < 0.01$ (see Figure 14b). In Model A the cooperation level was also different between the Low-CI and Full-CI conditions ($F = 41.77$, $p < 0.01$).

As for the received payoffs, both Model A and human participants won more in the Low-CI condition than in the High-CI condition ($F = 85.99$, $p < 0.01$, for Model A and $F = 23.91$, $p < 0.01$, for human participants, respectively; see Figs. 14a, b).

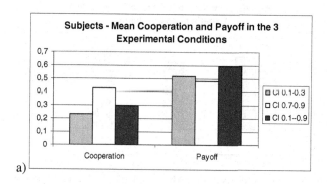

a)

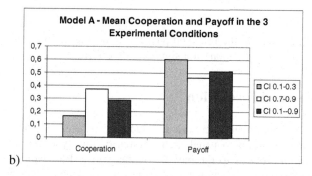

b)

Fig. 14. Comparisons of cooperation and received payoffs within three experimental conditions [2]: a) for human subjects, and b) for simulations with Model A

The mean cooperation between Model A and subjects was different only in the Low-CI condition (F = 4.13, p = 0.047). In the other two conditions there were no significant differences in cooperation as seen from Figure 15.

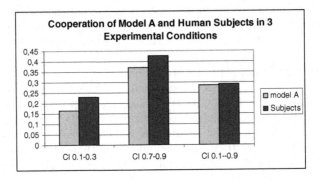

Fig. 15. Comparison of cooperation of Model A and human subjects in the three experimental conditions with respect to CI [2]

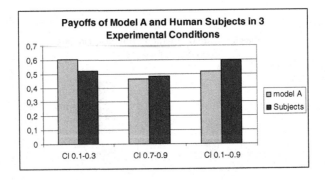

Fig. 16. Comparison of received payoffs for Model A and the subjects in the three experimental conditions

As seen from Figure 16, the received payoffs in the simulations and in the experiments show a similar trend and are not significantly different.

4   Conclusion and Discussion

In this paper, a recurrent neural network architecture was used to simulate IPDG playing. An important addition to usual architectures of this type was the presence of autoassociator nodes related to the payoffs of the games. Based on this architecture two models were explored. They differed in the way the training was performed. The first (Model B) used the reinforcement model of Macy and Flache [9] to evaluate the model player's moves. The second (Model A) used a simple forward-looking mechanism. Although similar with respect to architecture they displayed very different outcomes. The most important difference found between the two models was related to the CI dependency of their moves. The performance of Model A turned out to be very close to human performance (at least with regards to CI dependency of coopera- tion rates) while no such dependency was observed in the moves of Model B. This property was traced down to the appropriate response of the hidden nodes due to ade- quate representation of the game payoff matrix related to the added autoassociator part of the network. However, the availability of the autoassociator part alone was not sufficient to grant CI dependency. It seems that the anticipation mechanism of move evaluation of Model A, based on the generation of counterfactual games, moves, and received payoffs, played a decisive role.

The two models that we developed were connected with our search for a more plausible explanation of the way people play iterated social dilemmas like the Pris- oner's Dilemma game. On the basis of comparison with human subjects' data from the same games, and against the same computer opponent, we came up with the con- clusion that Model A, essentially involving anticipation, accounts better for human performance and can reproduce specific dependencies like the CI dependency. Thus Model A seems to be a more realistic and successful alternative to the schematic model based on subjective utility theory combined with simple reinforcement learning mechanisms proposed in [3]. To our knowledge no other models are able to account for the CI dependency in IPDG exist to date.

Of course, further exploration of the proposed architecture as well as extensive comparisons to experimental results are needed in order to understand and make use of its full potential and clarify completely the role of anticipation in its functioning. Research along these lines is currently in progress and the results will be reported elsewhere.

Acknowledgments. This work was supported by the project MINDRACES, funded by the 6FP of the EC.

References

1. Colman, A.: Cooperation, Psychological Game Theory, and Limitations of Rationality in Social Interaction. Behav. Brain Sci. 26, 139–153 (2003)
2. Hristova, E., Grinberg, M.: Context Effects On Judgment Scales in the Prisoner's Di- lemma Game. In: Proceedings of the 1st European Conference on Cognitive Economics. ECCE1, Gif-sur-Yvette, France (2004)
3. Hristova, E., Grinberg, M.: Investigation of Context Effects in Iterated Prisoner's Di- lemma Game. In: Dey, A.K., Kokinov, B., Leake, D., Turner, R. (eds.) CONTEXT 2005. LNCS (LNAI), vol. 3554, pp. 183–196. Springer, Heidelberg (2005)
4. Hristova, E., Grinberg, M.: Information Acquisition in the Iterated Prisoner's Dilemma Game: An Eye-tracking Study. In: Proceedings of the 27th Annual Conference of the Cognitive Science Society. Erlbaum, Hillsdale, NJ (2005)
5. Grinberg, M., Hristova, E., Popova, M.: Applicability of Eye-Tracking Information Acqui- sition Methods for Studying the Strategy Dynamics in the Iterated Prisoner's Dilemma Game. Position paper in the workshop: What have eye movements told us so far, and what is next. In: CogSci 2006, The 28th Annual Conference of the Cognitive Science Society, Vancouver, July 26-29 (2006)

6. Rapoport, A., Chammah, A.: Prisoner's Dilemma: A Study in Conflict and Cooperation. University of Michigan Press, Ann Arbor (1965)
7. Erev, I., Roth, A.: Simple reinforcement learning models and reciprocation in the prisoner's dilemma game. In: Gigerenzer, G., Selten, R. (eds.) Bounded rationality: the adaptive toolbox, MIT Press, Cambridge, Mass (2001)
8. Camerer, C., Ho, T.-H., Chong, J.: Sophisticated EWA Learning and Strategic Teaching in Repeated Games. J. Econ. Theory 104, 137–188 (2002)
9. Macy, M.W., Flache, A.: Learning Dynamics in Social Dilemmas. PNAS 99(suppl. 3), 7229–7236 (2002)
10. Taiji, M., Ikegami, T.: Dynamics of Internal Models in Game Players. Physica D 134, 253–266 (1999)
11. Elman, J.L.: Finding structure in time. Cognitive Science 14, 179–211 (1990)
12. Leydesdorff, L.L., Dubois, D.: Anticipation in Social Systems. International Journal of Computing Anticipatory Systems 15, 203–216 (2004)

An Experimental Study of Anticipation in Simple Robot Navigation

Birger Johansson and Christian Balkenius

Lunds University Cognitive Science, Kungshuset Lundagård
222 22 Lund, Sweden
{birger.johansson, christian.balkenius}@lucs.lu.se

Abstract. This paper presents an experimental study using two robots. In the experiment, the robots navigated through an area with or without obstacles and had the goal to shift places with each other. Four different approaches (random, reactive, planning, anticipation) were used during the experiment and the times to accomplish the task were compared. The results indicate that the ability to anticipate the behavior of the other robot can be advantageous. However, the results also clearly show that anticipatory and planned behavior are not always better than a purely reactive strategy.

1 Introduction

The ability to anticipate the behaviors of others is something we take more or less for granted and we often do not appreciate the complexity of this ability. When attempting to build robots with anticipatory abilities, it becomes clear that this is far from trivial. Not only does the robot need to control its own movement, it also needs to predict what other robots or possibly humans will do. Moreover, it needs to use the anticipated behaviors of others in a sensible way to change its own behavior.

Consider the following real life situation of the near future. You have sent your personal shopping robot to the "Autonomous supermarket" to get your favorite chocolate cake. To get the cake, located at the other end of shop, your robot cannot chose the straight path toward the cake because of the shelves and other obstacles, including all the other personal shopping robots in the store. Instead some alternative strategy must be used.

One possibility would be to move around randomly in the store until it finds the chocolate cake, but this would probably result in a long period of aimless wandering before it gets to its goal. This type of random behavior is very inefficient and is seldom used in robot navigation, although is rather common in robot exploration.

It is obvious that better methods can be used. Instead of moving at random, the robot may try to move in the direction of the goal. This is a reactive place approach method where the robot reacts to the position of the chocolate cake

M.V. Butz et al. (Eds.): ABiALS 2006, LNAI 4520, pp. 365–378, 2007.

and selects actions accordingly [2]. The problem with this approach is that the robot cannot go straight to the goal because of the shelves and other robots in the store. It needs to apply an obstacle avoidance strategy when there is something in its way. For example, it may turn around and move in some other direction for a short while before turning toward the goal again. This type of reactive navigation has been widely used in robotics where the relation between the stimulus and response is often preprogrammed [8][13]. A number of rules are set up that must be fulfilled for an action to be executed. A problem with such a reactive approach to navigation is that the robot can easily get itself into situations where it becomes trapped.

Although the reactive strategy is more efficient than random movements it would be better to plan a path around the shelves based on knowledge of the layout of the store. This has traditionally been the most common way of dealing with robot navigation. This plan can use grids [21], potential fields [21][1][7], or some symbolic or geometric description of the environment. As long as the map of the shop is correct, the plan will also be correct and can be used to efficiently go to the cake.

Unfortunately, when the personal shopping robot reviews the map after a few seconds of moving according to the planned path, it realizes that the map is no longer accurate. The shelves are still where they are supposed to, but most of the other robots have moved and are not where the map indicates. As with the previous strategy, this makes it necessary to use some obstacle avoidance strategy to avoid colliding with the other robots which may limit the usefulness of the plan.

The solution to this problem is to include the movements of the other robots when the personal shopping robot makes its plan. This is however not trivial as it does not know where the other robots are heading. One reasonable assumption is that they will continue in the direction they have now, although this will only be true for a short while. Better predictions can be made if the robot knows the goals of the other robots. By anticipating the behaviors of others, it will be able to chose a better path and will not have to use the obstacle avoidance strategy as often. The better its ability to anticipate, the less it will need to use its alternative strategies.

Several different types of anticipatory behavior have been used in robotics and AI. First, it is possible to use an anticipatory mechanism to reduce the latency of a control system. For example, Behnke et al. [4] used neural network to reduce the control latency for the FU-fighter team in the soccer RoboCup. The control system had a delay of four frames (132 ms) and with a speed up to 2 m/s, this could result in an error between the actual position of the robot and the tracking of the robots of approximately 20 cm. By feeding a neural network with the position, orientation and motor commands from the last six frames to anticipate the current position, the influence of delay in the system was almost eliminated. A similar method has also been used to predict the location of a moving target for visual tracking [3].

A second type of anticipation concerns anticipation of the environment, for example the movement of other robots. Sharifi et al. [19] describe a system for the simulation league of RoboCup where the future state is used to anticipate which robot will posses the ball next, while Veloso et al. [22] anticipate the state of the whole team. This means that a seemingly passive agent is not passive at all. Instead it actively anticipate opportunities for collaboration.

The anticipation of robot movement can also be based on observation. For example, Stulp et al. [20] model the goal keeper in RoboCup to be able to anticipate its behavior. Ledezma et al. [15] used a similar method to model the behavior of the other players based on their observed input and outputs. Usually, some type of communication between agents are used in anticipation, either a complete knowledge of world or broadcasting of individual plans but there is also work on cooperation without sharing information between agents [18].

Human-robot interaction can also merit from using anticipatory behavior. Sabanovic et al. [17] used a stationary robotic receptionist that provided information to visitors and enhances interaction through story-telling to study human-robot interaction. In this study, the robot receptionist turns toward people passing by and tries to interact with them. To be able to interact in a efficient way, the robot receptionist anticipates the position of people passing by.

The importance of anticipation has also been studied in the domain of computer games [14]. In human activities, Saad [16] pointed out the close connection between driving and anticipation, even stating that "driving is anticipating".

Davidsson [9] used simulations to investigate the benefits of anticipation. Two different types of experiments were conducted. The first investigated competition between agents and in the second experiment, the agents were cooperative. In the experiments, the task of the agents was to pick up targets in a two dimension grid world in a particular order. By using a linearly quasi-anticipatory agent architecture, one agent could realize that it would not reach the target before the other agent and would instead start to move toward the following target. Only one of the robots used anticipatory behavior while the other one used reactive behavior. In the second experiment, the agents cooperate, which leads to a decreased total time for fetching all target objects.

Although simulations can be very valuable in testing different strategies, a simulation must necessarily include a perfect model of the simulated environment. It will thus always be possible to make perfect predictions in a simulation if this is desired. It is well known that this can easily lead to solutions that are not useful when applied to robots that have to operate in the real world [6].

To evaluate the benefit of anticipation in mobile robots in the real world, we tested a number of strategies in three different environments with two robots. We compared a random and a reactive strategy with control methods based on planing with or without anticipation of the behavior of the other robot. The goal of the experiments was to test under what conditions the ability to anticipate would help the robots in a simple task. In addition, we tested three different methods to use the anticipated behavior of the other robot.

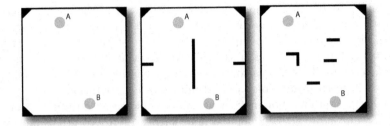

Fig. 1. The three environments with different complexity. A and B: goal locations for the robots.

2   Description of the Experimental System

2.1   Environment

The size of the experimental area was 2×2 m. Bricks marked with white color were placed in the area in two of the experiments as obstacles (Figure 1). Experiment 1 used the empty environment, Experiment 2 used an environment with walls and in Experiment 3, obstacles were placed at random in the environment.

2.2   Robots

The robots used were two modified BoeBots (Parallax Inc., Rocklin, California). These robots are approximately 14 cm long and use a differential steering. No sensors on the robots were used in these experiments. Instead, each robot was marked with two colors that could be detected by a camera mounted 3.5 m above the robot area. This camera transmitted images to a computer that calculated the position and orientation for the two robots four times per second. This computer also performed all the computations for the two robots and controlled the robots via wireless bluetooth communication. In addition, it stored tracking data and collected all statistics for the experiments.

2.3   Control Systems

The control systems of the robots were built using the Ikaros framework[1]. The interface components used included processing of the the video stream from the camera, color tracking to detect the position and orientation of the robots, and wireless communication. In addition, modules where built for reactive robot control, path planning, and anticipation.

Random Control. A random control system was the first tested in the experiments. This system simply transmits random motor commands to the robot until it has reached the goal. The robot is instructed to turn toward a random

---

[1] www.ikaros-project.org

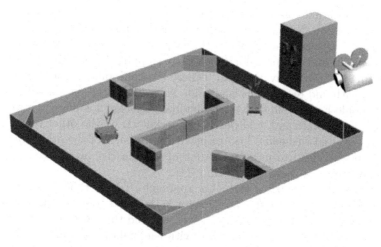

Fig. 2. The computer is using the overhead camera to track the two robots and transmits motor commands via bluetooth

orientation and then travel in this direction until an obstacle activates an obstacle avoidance system, in which case a new random direction is set. This is repeated until both robots have reached their goals.

Reactive Approach. The next control system performs reactive approach where the robot always tries to go directly toward the goal. The desired path is calculated as the straight line between the current location of the robot and the goal location. This strategy will obviously have problems when there are obstacles in the way and to handle this situation a reactive avoidance system was added.

Planning System. The planning system is responsible for path finding within the environment. To accomplish this an A* based navigation algorithm is used [12]. This is a grid-based navigation algorithm with full knowledge of the environment. It finds the shortest path to the goal by testing it in the grid-map. If it is unable to use the shortest path, the second shortest path is tested and so on, until a path has been found. Each robot uses the algorithm to find the best path through the robot area. The grid-map is divided into 32×32 elements with a status of either occupied or free. The planning system takes no account of where the other robots are located and only uses its own position, the desired position and the grid-map to find the path.

Anticipation System. The anticipation system is similar to the planning system but also includes the movements of the other robots. If the other robots were stationary, the A* algorithm could register the other robots as obstacles. When the other robots are moving it becomes necessary to anticipate their position at each time-step in the future. To solve this, each robot has a model of the other robot. This model is built using each robots own planning system, for example, robot A assumes that robot B would use the path that robot A would have used if it were located at the position of robot B and heading for the goal of robot B. Before robot A tries to find its own path, it updates its model of the other robot and uses this to find the path for robot B by stepping forward in the planning and checking if there is any collision. If there is a collision, the robot chooses an alternative path and tests if this is a valid. This is repeated until a valid path is found. It should be stressed that the individual paths are not shared between the robots. Only the start and goal positions are known by the other robot. With noise in the system this could lead to inaccurate models of the other robot and this could in turn lead to more activation of the reactive avoidance system. A similar approach was presented by Guo [11].

An obvious problem arises with this approach. If both robots use the same method to find a valid path, it is possible for both robots to select the alternative path which will result in a collision. A way to avoid this problem is to assign a rank [10] to each robot where the robot with the highest rank always takes the shortest path. For example, let the robot with the longest distance to the goal have the higher rank and let the other robot replan its path around the more highly ranked robot. If the present robot has the lowest rank, we let A* see the other robots as a obstacle but only during that time step. This means that at just that time step there is an object at that position at some time steps later the obstacle has moved and the grid that was occupied in the first time step is free again. In the experiments, we tested three different ways to select the rank of each robot, (1) a fixed rank, (2) the robot closest to its goal would have the highest rank, and (3) the robot with the larger distance to its goal would receive the highest rank. Note that according to the last two strategies, the ranks of the robots may change when the robots move.

Reactive Avoidance. A reactive avoidance system is placed on top of the other navigation systems and is activated if there is an obstacle too close to the robot. We divided the reactive area around the robot into 8 regions (Figure 3). Three in front of the robot, one on each side of the robot and three behind the robot. The robot performs different types of avoidance behaviors depending on in which regions the obstacle was found. If an object is straight ahead, the robot turns on the spot until the obstacle has disappeared from the region and if an object is found to the left of the robot, it steers to the right to obtain a free path. Although the reactive avoidance system mainly helps the robot to reach its goal, it sometimes counteracts the control of the navigation system. For example, when the navigation system instructs the robot to turn right, the reactive avoidance system may detect an obstacle in that area and tell the robot to turn left instead.

2.4 Experimental Procedure

The task for the robots was to switch places with each other. One robot started at position A and the other started at position B (Figure 1). When the first robot had arrived at its goal position, it waited for the other robot to reach

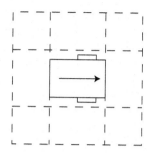

Fig. 3. The robot with the reactive field around it. The reactive fields divide the surrounding into eight regions and different avoidance behaviors are activated depending on the location of the obstacle.

its goal. The goal locations were subsequently switched and same procedure was repeated. With the switching procedure the robots were forced to interact as their paths crossed each other. The presence of interaction is necessary to investigate anticipation in this form.

During this experiment, the time for each position switch was recorded together with the number of times the reactive avoidance system was used. Note that this was a cooperative task where it is the time for both robots to switch places that is recorded.

The worlds were chosen to have multiple possible paths to the goal and the robots had to choose one depending on the strategy they currently used although the chosen path could change during the experiment depending on the position of the robot. This is a form of coordination problem [5].

Six different strategies were tested: (RAND) random behavior, (APPR) reactive approach behavior, (PLAN) planning, (A-fixed) anticipation with fixed rank, (A-short) anticipation with higher rank for the robot closest to the goal, and (A-long) anticipation with lowest rank for the robot closet to the goal.

Each strategy was tested twice before the robots shifted to the next strategy. When all strategies had been tested two times, the procedure was repeated until in total 40 trials with each strategy had been run. In total, there were 240 trials in each experiment.

Ordinary indoor lighting together with daylight from the large laboratory windows are used as light source for the experiment. The fact that no controlled light source was used could influence the tracking performance even if the tracking algorithms are dynamic for change in light intensity. To overcome this, one trial of each approach was run in sequence instead of all trials in one approach one at the time. So if there was an external influence on the tracking system it influenced all the approaches equally.

To ensure that the tracking system was accurate enough, each iteration that had a position or an angle that was impossible for the robot to reach was counted as a tracking error. This statistic was collected to make sure that the tracking system was reliable and not correlated with the performance of the robots.

In all the experiments the time for both robots to reach their goals was stored but also the difference in time between the fastest and the slowest robot. Both times were used to compare the different navigation approaches in each of the different environments. For the statistics, we used the logarithm of the recorded times to minimize the effect of outliers.

3   Results

3.1   Experiment 1

The behaviors of the robots in the different conditions are illustrated in Figure 4. The environment in experiment 1 did not contain any obstacles. In this environment the tracking system had a mean position error of 1.4% and a mean angle error of 4.3%. ANOVA ($F(1,5) = 297.4, \alpha = 0.05$) showed a significant difference between the different strategies and a post hoc Tukey HSD ($df = 234, \alpha = 0.05$) was used to find the effects of the different strategies. As expected, the random behavior was significantly slower than all the other strategies ($p = 0.000020$). The reactive behavior was also significantly faster than the PLAN and the A-short behavior ($p = 0.0027$ and $p = 0.030$ respectively). There was no significant difference between the different anticipatory strategies.

When comparing the time difference between the fastest and slowest robot, ANOVA ($F(1,5) = 73.20, \alpha = 0.05$) was used to find that there was a significant difference between the strategies ($p = 0.000020$) and a post hoc Tukey HSD ($df = 234, \alpha = 0.05$) was used to find the effects of the different strategies (Figure 6). The difference between the times when the two robots reacher their goals was significantly larger for the reactive strategy ($p = 0.000020$).

3.2   Experiment 2

Experiment 2 used an environment with walls. In this environment the tracking error increased due to more obstacles. Shape and shadows from the obstacles make it harder to locate the robot features for exact position and angle tracking. The mean tracking error was 1.7% for the position and 6.8% for the angle of the robots.

ANOVA ($F(1,5) = 303.9, \alpha = 0.05$) showed a significant difference between the strategies and a post hoc Tukey HSD ($df = 234, \alpha = 0.05$) was used to find the effects of the different strategies. Again, the random behavior was significantly slower than all the other strategies ($p = 0.000020$). The reactive approach strategy was significantly slower than the planning and anticipatory strategies ($p = 0.000020$ in all cases, Figure 5 middle). There were no significant difference between the planning and the anticipation approaches.

To analyze the difference in the times when the two robots reached their goals, we used ANOVA ($F(1,5) = 70.50, \alpha = 0.05$) and a post hoc Tukey HSD ($df = 234, \alpha = 0.05$) The time difference for the random approach was significantly larger than for the other approaches ($p = 0.000020$). The reactive approach had

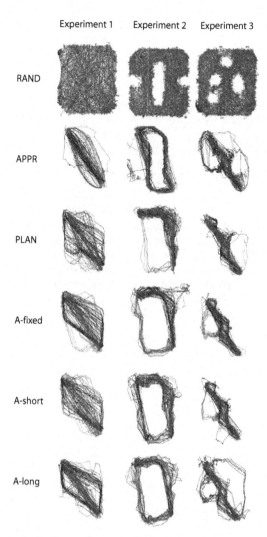

Fig. 4. Illustration of all the movement of the robots. In the experiments with the random behavior, all the available area is covered. It is easy to discern the obstacle location in experiment 2 and 3. Using the reactive approach behavior, less of the area is covered. With this behavior, the required movement has been reduced in comparison to the random approach behavior. Using the planning behavior, the robots will often take the same path which will result in a possible collision and extensive use of the reactive avoidance system. This is most clearly seen in experiment 2 where the robots often both select the top-right path. In the anticipation behaviors, the paths of the robots have more variation because the anticipation causes the robots to use different paths. Note that the robots balance the use of the two path between the two goal locations. In experiment 1, one robot uses the diagonal path while the other moves to the left or right. The same pattern can be seen in experiment 2 and 3, most clearly in A-long in experiment 3.

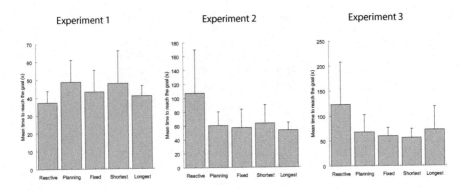

Fig. 5. Left. Time taken to reach the goals for the different strategies in the empty environment in experiment 1. In the empty environment, the reactive approach behavior (APPR) performed best. The error bars show the standard deviation. Middle. Switch time for the different strategies in the environment with walls in experiment 2. The anticipatory strategy where the robot with longest distance to its goal had highest rank (A-long) was most efficient. Right. Switch time for the different strategies in the environment with random obstacles used in experiment 3. Strategy A-short was fastest in this environment.

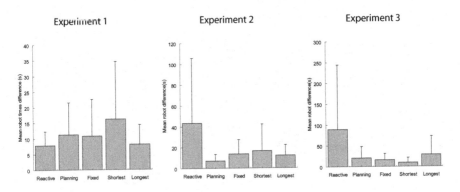

Fig. 6. Mean time difference between the first robot reaching its goal and the second robot reaching its goal

a larger time difference than the planning strategy ($p = 0,000020$), the A-fixed ($p = 0.031$), the A-short ($p = 0.036$), and the A-long strategy ($p = 0.0063$). The planing strategy was significantly faster than the A-fixed ($p = 0.0021$), the A-short ($p = 0.0017$), and the A-fixed strategy ($p = 0.0012$)

3.3   Experiment 3

The environment in the last experiment contained randomly placed obstacles. This environment had a tracking error of 3.5% for the position and 8.4% for the angle. The increase of error, compared to the previous experiments, is due to more shapes and shadows from the randomly placed obstacles.

ANOVA ($F(1,5) = 208.8, \alpha = 0.05$) indicated a significant difference between the strategies and a post hoc Tukey HSD test ($df = 234, \alpha = 0.05$) was used to find the effects of the different strategies. Again, the random strategy was significantly slower than all the other strategies ($p = 0.000020$). The reactive approach strategy was significantly slower than the planing strategy ($p = 0.000021$), the A-fixed ($p = 0.000020$), the A-short ($p = 0.000020$), and the A-long strategy ($p = 0,000045$). We found no significant difference between the different anticipation approaches.

In the time difference between the robots we used ANOVA ($F(1,5) = 30.43$, $\alpha = 0.05$) to find a significant difference between a strategy and post hoc Tukey HSD ($df = 234, \alpha = 0.05$) There was a significant difference between the random approach and the other approaches ($p = 0.000020$). The reactive approach had a larger time difference than PLAN ($p = 0.003870$), A-fixed ($p = 0.009715$), and A-short ($p = 0.000020$). We found no significant difference between the different anticipation approaches.

4   Discussion

We have tested a number of behavioral strategies in robots in three simple environment with varying complexity to investigate the usefulness of anticipatory abilities in a real robot situation. As expected, a reactive approach that always tries to move in the direction of the goal performs well in an empty environment but is much worse when there are obstacles in the way. Also, all strategies were much better than random behavior.

In some cases, some of the anticipatory strategies were more efficient than the planning strategy in the sense that the robots would avoid taking routes where they may collide or interact, but the merit of anticipation clearly depended on how anticipation was used and in what environment. There were no significant differences between the time used for the planning and anticipatory strategies.

With anticipation, the robot will often take a longer path than with only planning and if something goes wrong during the avoidance of the anticipated obstacles, the robot will loose much time. Had the robots been more accurate when they attempted to follow their planned paths, we would expect that anticipation would have been better in most cases as is the case in simulation [9]. It can clearly be seen that under optimal conditions, anticipatory behavior is very efficient, but it is very sensitive to different disturbances.

In experiment 2 and 3, the difference between the time when the first robot reached its goal and the time when the second did was correlated with the complexity of the navigation strategy. The reactive approach had a larger time difference than the planning and anticipatory approaches, except in experiment 3 where there were no significant difference between REACT and A-long. A-long is a strategy that helps the robot that is furthest away from the goal. This would seem to be a very noble approach, where the weakest robot is always allowed to take the shortest path. However, after a while, the weaker robot has advanced closer to the goal and thereby loses the benefit from the other robot. This switching can lead to strange behavior where the robot suddenly stops and turns in another direction because it has lost the rank and the other robot now possesses the area in front of the first robot.

The second experiment indicated that the planning behavior had a larger time difference than the anticipatory behaviors. In experiment 1 we could not find this effect. It is possible that without the obstacles, the effect was not as easy to distinguish even if it was still present.

In the experiments, the two robots had total knowledge of the environment as well as the position and goal of the other robot. They only had to anticipate the movement of the other robot. In such a situation, it may be more advantageous to make a collective plan for both robots. What we are aiming at in the future, however, is the situation where the robots do not have full access to the environment. In this case, the robots must explore the environment to learn about different paths and the position of the other robot. As they will not know the goal of the other robot, it must be inferred from the observed movements. In this case, we will be able to explore different learning methods and different strategies for observing the behavior of the other robot.

In the future, we want to change the experimental set-up in three ways. First, we want to increase the exactness of the control system to allow more precise movement control of the robots. This will probably lead to a greater advantage for the anticipatory strategies. Second, we want to make the environments more complex. In this experiment, we used extremely simple environments to allow all the different behaviors to complete the task. This simplicity reduces the benefit of anticipatory behaviors. If we had instead used a more maze-like environment with lots of long and narrow passages, the anticipatory behaviors would have gained enormously with respect to the other strategies since a collision would be more costly. In the current environments, the cost of interaction between the robots was not high enough to punish behavior when the robots interfered with each other. Third, we will also use a larger number of robots, which is expected to also increase the benefits of anticipation.

In conclusion, we have presented experimental results with two robots in different environments that show that the ability to anticipate the behavior of the other robot has the potential to make the behavior of the robots more efficient. However, this is highly dependent on the complexity of the environment and the accuracy of the control of the robots.

Acknowledgment

We would like to thank Anna Balkenius for helpful comments on the manuscript and Henrik Johansson for the 3D drawing support. This work was supported by the EU project MindRaces, FP6-511931.

References

1. Arkin, R.: Motor schema-based mobile robot navigation. International Journal of Robotics Research 8(4), 92–112 (August 1989)
2. Balkenius, C.: Natural Intelligence in Artificial Creatures. Lund University Cognitive Studies, no. 37 (1995)
3. Balkenius, C., Johansson, B.: Event prediction and object motion estimation in the development of visual attention, In: Berthouze, L., Kaplan, F., Kozima, H., Yano, H., Konczak, J., Metta, G., Nadel, J., Sandini, G., Stojanov, G., Balkenius, C. (eds.) Fifth International Conference on Epigenetic Robotics, pp. 17–22 (2005)
4. Behnke, S., Egorova, A., Gloye, A., Rojas, R., Simon, M.: Predicting away robot control latency. In: Polani, D., Browning, B., Bonarini, A., Yoshida, K. (eds.) RoboCup 2003. LNCS (LNAI), vol. 3020, pp. 712–719. Springer, Heidelberg (2004)
5. Boutilier, C.: Planning, learning and coordination in multiagent decision processes. In: Sixth Conference on Theoretical Aspects of Rationality and Knowledge, pp. 195–210 (1996)
6. Brooks, R.A.: Cambrian intelligence: the early history of the new ai. MIT Press, Cambridge (1999)
7. Brooks, R.: A robust layered control system for a mobile robot. IEEE Journal of Robotics and Automation, RA 2(1), 14–23 (1986)
8. Brooks, R.A.: Intelligence Without Reason. In: Proceedings of the 12th International Joint Conference on Artificial Intelligence (IJCAI-91), Myopoulos, J., Reiter, R.: (eds.) Sydney, Australia, pp. 569–595 Morgan Kaufmann publishers Inc.: San Mateo, CA, USA (1991)
9. Davidsson, P.: Learning by linear anticipation in multi-agent systems. Distributed Artificial Intelligence Meets Machine Learning 1221, 62–72 (1996)
10. Erdmann, M., Lozano-Perez, T.: On multiple moving objects. Algorithmica 2, 477–521 (April 1987)
11. Guo, Y., Parker, L.: A distributed and optimal motion planning approach for multiple mobile robots. In: Proceedings IEEE International Conference on Robotics and Automation (2002)
12. Hart, N.J.R.B., Nilsson, P.E.: A formal basis for the heuristic determination of minimum cost paths. IEEE Transactions on Systems Science and Cybernetics SSC4 2, 100–107 (1968)
13. Korein, J.U., Ish-Shalom, J.: Robotics. IBM Systems Journal 1, 96–106 (1987)
14. Laird, J.E.: It knows what you're going to do: Adding anticipation to a quakebot. In: Proceedings of the Fifth International Conference on Autonomous Agents.Canada, pp. 385–392 ACM Press,New York (May 2001)
15. Ledezma, S.A.B.D.A., Aler, R.: Predicting opponent actions by observation, pp. RobuCup 2004, pp. 286–296 (2004)
16. Saad, F.: In-depth analysis of interactions between drivers and the road environment: contribution of on-board observation and subsequent verbal report, In: Proceedings of the 4th Workshop of international cooperation on theories and concepts in traffic safety, University of Lund (1992)
17. Sabanovic, M.M.S.R.S.: Robots in the wild: Observing human-robot social interaction outside the lab. In: International Workshop on Advanced Motion Control, Istanbul, Turkey (March 2006)
18. Sen, S., Sekaran, M., Hale, J.: Learning to coordinate without sharing information. In: Proceedings of the Twelfth National Conference on Artificial Intelligence, 1997, pp. 509–514 (Reprinted from Proceedings of the NationalConference on Artificial Intelligence (1994)

19. Sharifi, H. A. A., Mousavian, M.: Predicting the future state of the robocup simulation environment: heuristic and neural networks approaches,Systems, Man and Cybernetics, vol. 1, pp. 32–27 (2003)
20. Stulp, M.B.M., Isik, F.: Implicit coordination in robotic teams using learned prediction models. In: IEEE International Conference on Robotics and Automation (2006)
21. Thrun, S., Gutmann, J.-S., Fox, D., Burgard, W., Kuipers, B.: Integrating topological and metric maps for mobile robot navigation: A statistical approach. In: AAAI/IAAI, pp. 989–995 (1998)
22. Veloso, M., Stone, P., Bowling, M.: Anticipation as a key for collaboration in a team of agents: A case study in robotic soccer. In: Schenker, P.S., McKee, G.T. (eds.) Proceedings of SPIE Sensor Fusion and Decentralized Control in Robotic Systems II, Bellingham, vol. 3839, pp. 134–143 (September 1999)

Author Index

Lecture Notes in Artificial Intelligence (LNAI)

Vol. 4434: G. Lakemeyer, E. Sklar, D.G. Sorrenti, T. Takahashi (Eds.), RoboCup 2006: Robot Soccer World Cup X. XIII, 566 pages. 2007.

Vol. 4429: R. Lu, J.H. Siekmann, C. Ullrich (Eds.), Cognitive Systems. X, 161 pages. 2007.

Vol. 4428: S. Edelkamp, A. Lomuscio (Eds.), Model Checking and Artificial Intelligence. IX, 185 pages. 2007.

Vol. 4426: Z.-H. Zhou, H. Li, Q. Yang (Eds.), Advances in Knowledge Discovery and Data Mining. XXV, 1161 pages. 2007.

Vol. 4411: R.H. Bordini, M. Dastani, J. Dix, A.E.F. Seghrouchni (Eds.), Programming Multi-Agent Systems. XIV, 249 pages. 2007.

Vol. 4410: A. Branco (Ed.), Anaphora: Analysis, Algorithms and Applications. X, 191 pages. 2007.

Vol. 4399: T. Kovacs, X. Llorà, K. Takadama, P.L. Lanzi, W. Stolzmann, S.W. Wilson (Eds.), Learning Classifier Systems. XII, 345 pages. 2007.

Vol. 4390: S.O. Kuznetsov, S. Schmidt (Eds.), Formal Concept Analysis. X, 329 pages. 2007.

Vol. 4389: D. Weyns, H.V.D. Parunak, F. Michel (Eds.), Environments for Multi-Agent Systems III. X, 273 pages. 2007.

Vol. 4384: T. Washio, K. Satoh, H. Takeda, A. Inokuchi (Eds.), New Frontiers in Artificial Intelligence. IX, 401 pages. 2007.

Vol. 4371: K. Inoue, K. Satoh, F. Toni (Eds.), Computational Logic in Multi-Agent Systems. X, 315 pages. 2007.

Vol. 4369: M. Umeda, A. Wolf, O. Bartenstein, U. Geske, D. Seipel, O. Takata (Eds.), Declarative Programming for Knowledge Management. X, 229 pages. 2006.

Vol. 4343: C. Müller (Ed.), Speaker Classification. X, 355 pages. 2007.

Vol. 4342: H. de Swart, E. Orlowska, G. Schmidt, M. Roubens (Eds.), Theory and Applications of Relational Structures as Knowledge Instruments II. X, 373 pages. 2006.

Vol. 4335: S.A. Brueckner, S. Hassas, M. Jelasity, D. Yamins (Eds.), Engineering Self-Organising Systems. XII, 212 pages. 2007.

Vol. 4334: B. Beckert, R. Hähnle, P.H. Schmitt (Eds.), Verification of Object-Oriented Software. XXIX, 658 pages. 2007.

Vol. 4333: U. Reimer, D. Karagiannis (Eds.), Practical Aspects of Knowledge Management. XII, 338 pages. 2006.

Vol. 4327: M. Baldoni, U. Endriss (Eds.), Declarative Agent Languages and Technologies IV. VIII, 257 pages. 2006.

Vol. 4314: C. Freksa, M. Kohlhase, K. Schill (Eds.), KI 2006: Advances in Artificial Intelligence. XII, 458 pages. 2007.

Vol. 4304: A. Sattar, B.-h. Kang (Eds.), AI 2006: Advances in Artificial Intelligence. XXVII, 1303 pages. 2006.

Vol. 4303: A. Hoffmann, B.-h. Kang, D. Richards, S. Tsumoto (Eds.), Advances in Knowledge Acquisition and Management. XI, 259 pages. 2006.

Vol. 4293: A. Gelbukh, C.A. Reyes-Garcia (Eds.), MICAI 2006: Advances in Artificial Intelligence. XXVIII, 1232 pages. 2006.

Vol. 4289: M. Ackermann, B. Berendt, M. Grobelnik, A. Hotho, D. Mladenič, G. Semeraro, M. Spiliopoulou, G. Stumme, V. Svátek, M. van Someren (Eds.), Semantics, Web and Mining. X, 197 pages. 2006.

Vol. 4285: Y. Matsumoto, R.W. Sproat, K.-F. Wong, M. Zhang (Eds.), Computer Processing of Oriental Languages. XVII, 544 pages. 2006.

Vol. 4274: Q. Huo, B. Ma, E.-S. Chng, H. Li (Eds.), Chinese Spoken Language Processing. XXIV, 805 pages. 2006.

Vol. 4265: L. Todorovski, N. Lavrač, K.P. Jantke (Eds.), Discovery Science. XIV, 384 pages. 2006.

Vol. 4264: J.L. Balcázar, P.M. Long, F. Stephan (Eds.), Algorithmic Learning Theory. XIII, 393 pages. 2006.

Vol. 4259: S. Greco, Y. Hata, S. Hirano, M. Inuiguchi, S. Miyamoto, H.S. Nguyen, R. Słowiński (Eds.), Rough Sets and Current Trends in Computing. XXII, 951 pages. 2006.

Vol. 4253: B. Gabrys, R.J. Howlett, L.C. Jain (Eds.), Knowledge-Based Intelligent Information and Engineering Systems, Part III. XXXII, 1301 pages. 2006.

Vol. 4252: B. Gabrys, R.J. Howlett, L.C. Jain (Eds.), Knowledge-Based Intelligent Information and Engineering Systems, Part II. XXXIII, 1335 pages. 2006.

Vol. 4251: B. Gabrys, R.J. Howlett, L.C. Jain (Eds.), Knowledge-Based Intelligent Information and Engineering Systems, Part I. LXVI, 1297 pages. 2006.

Vol. 4248: S. Staab, V. Svátek (Eds.), Managing Knowledge in a World of Networks. XIV, 400 pages. 2006.

Vol. 4246: M. Hermann, A. Voronkov (Eds.), Logic for Programming, Artificial Intelligence, and Reasoning. XIII, 588 pages. 2006.

Vol. 4223: L. Wang, L. Jiao, G. Shi, X. Li, J. Liu (Eds.), Fuzzy Systems and Knowledge Discovery. XXVIII, 1335 pages. 2006.

Vol. 4213: J. Fürnkranz, T. Scheffer, M. Spiliopoulou (Eds.), Knowledge Discovery in Databases: PKDD 2006. XXII, 660 pages. 2006.

Vol. 4212: J. Fürnkranz, T. Scheffer, M. Spiliopoulou (Eds.), Machine Learning: ECML 2006. XXIII, 851 pages. 2006.

Vol. 4211: P. Vogt, Y. Sugita, E. Tuci, C.L. Nehaniv (Eds.), Symbol Grounding and Beyond. VIII, 237 pages. 2006.

Vol. 4203: F. Esposito, Z.W. Raś, D. Malerba, G. Semeraro (Eds.), Foundations of Intelligent Systems. XVIII, 767 pages. 2006.

Vol. 4201: Y. Sakakibara, S. Kobayashi, K. Sato, T. Nishino, E. Tomita (Eds.), Grammatical Inference: Algorithms and Applications. XII, 359 pages. 2006.

Vol. 4200: I.F.C. Smith (Ed.), Intelligent Computing in Engineering and Architecture. XIII, 692 pages. 2006.